Springer Theses

Recognizing Outstanding Ph.D. Research

Aims and Scope

The series "Springer Theses" brings together a selection of the very best Ph.D. theses from around the world and across the physical sciences. Nominated and endorsed by two recognized specialists, each published volume has been selected for its scientific excellence and the high impact of its contents for the pertinent field of research. For greater accessibility to non-specialists, the published versions include an extended introduction, as well as a foreword by the student's supervisor explaining the special relevance of the work for the field. As a whole, the series will provide a valuable resource both for newcomers to the research fields described, and for other scientists seeking detailed background information on special questions. Finally, it provides an accredited documentation of the valuable contributions made by today's younger generation of scientists.

Theses are accepted into the series by invited nomination only and must fulfill all of the following criteria

- They must be written in good English.
- The topic should fall within the confines of Chemistry, Physics, Earth Sciences, Engineering and related interdisciplinary fields such as Materials, Nanoscience, Chemical Engineering, Complex Systems and Biophysics.
- The work reported in the thesis must represent a significant scientific advance.
- If the thesis includes previously published material, permission to reproduce this must be gained from the respective copyright holder.
- They must have been examined and passed during the 12 months prior to nomination.
- Each thesis should include a foreword by the supervisor outlining the significance of its content.
- The theses should have a clearly defined structure including an introduction accessible to scientists not expert in that particular field.

More information about this series at http://www.springer.com/series/8790

Pantelis Pnigouras

Saturation of the f-mode Instability in Neutron Stars

Doctoral Thesis accepted by
the Eberhard-Karls University of Tübingen, Tübingen,
Germany

 Springer

Author
Dr. Pantelis Pnigouras
Mathematical Sciences and STAG Research
  Centre
University of Southampton
Southampton, UK

Supervisor
Prof. Kostas D. Kokkotas
Eberhard-Karls University of Tübingen
Tübingen, Germany

ISSN 2190-5053 ISSN 2190-5061 (electronic)
Springer Theses
ISBN 978-3-030-07474-6 ISBN 978-3-319-98258-8 (eBook)
https://doi.org/10.1007/978-3-319-98258-8

This Springer imprint is published by the registered company Springer Nature Switzerland AG
The registered company address is: Gewerbestrasse 11, 6330 Cham, Switzerland

To my father,
for seeking Ithaca.

To my mother,
for relishing the voyage.

Supervisor's Foreword

The detection of gravitational waves in 2015 signalled a new era for theoretical physics and astrophysics. Thanks to the advanced technology developed, the dark side of the Universe is becoming more transparent to humans, with black holes and neutron stars revealing their violent faces in both the gravitational and the electromagnetic spectrum. It will also not be long until the mystery of the Big Bang will be unveiled via gravitational waves. Last year, for the first time in the history of astronomy, a violent event—the merging of two neutron stars—was observed in all bands of the electromagnetic spectrum, following its discovery via gravitational-wave emission.

There are many fundamental questions which can be tackled with neutron star observations. Two of the most compelling ones relate first of all to the dynamics of neutron stars and second to their efficiency as gravitational-wave sources. By answering these two questions, we expect to reveal the equation of state of matter at supranuclear densities for the first time.

Typical, fast-rotating neutron stars, with masses 1–2 times larger than the solar mass, are expected to be the outcome of the gravitational collapse of a massive star. Furthermore, the merging of two such neutron stars will generically produce, most of the times, a supermassive neutron star. The newly formed object (born from either of the above two formation channels) will be hot and will rotate fast until, possibly, the magnetic field brakes its rotation down to the spins that we observe in the known, old neutron stars. During this early stage, rotational instabilities may take place, deforming the nascent object and emitting copious amounts of gravitational waves at specific frequencies, which will carry information about the physical parameters that describe neutron stars.

One of the oscillation modes that may become unstable due to rotation is the so-called f-mode, which is associated with the fundamental oscillation frequency. The unstable patterns of pulsation of the f-mode grow in time and, when a specific amplitude is reached, they saturate by transferring oscillation energy to other modes via nonlinear mode coupling. The saturation amplitude of the f-mode is directly related to the efficiency with which gravitational waves are emitted. Since 2002 there have been systematic studies about the saturation of another potentially unstable mode, the so-called r-mode. But, until now, there had been no such study for the saturation of the f-mode instability. A plausible reason could be the complicated nature of the problem, compared to similar studies for the r-mode. This thesis is the first systematic and successful attempt in addressing this question. It includes the basic theory of nonlinear mode coupling and develops the methodology for the f-mode. Finally, it arrives at concrete estimations by applying the above to the two promising astrophysical sources of gravitational waves, namely typical and supramassive neutron stars, and examines the possibility of observing the cosmological stochastic background of gravitational waves due to f-mode instabilities throughout the Universe.

The most important application of this work, which is extremely well-timed, is related to the post-merger supramassive neutron star born from the collision, which, apart from its large mass, acquires extremely high spin. This combination favours the onset of rotational instabilities and the neutron star can become unstable for a quite wide range of temperatures and spins. In addition, the instability grows on very short timescales, of the order of a few seconds, during which even the strongest magnetic field will not be able to drain significant amounts of angular momentum. This scenario was studied in detail in Doneva et al. (2015), and these initial results were very promising. The existence of this specific "gravitational-wave afterglow" can be correlated with the observed light curves of short γ-ray bursts, which, in many cases, acquire a plateau lasting hundreds to thousands of seconds and suggesting the survival of the post-merger neutron star remnant for minutes to hours before collapsing to a black hole. During this phase, the star is unstable and the emitted gravitational radiation should be detectable up to a few tens of Mpc with current gravitational-wave detectors, Advanced LIGO and Virgo, and up to a few hundreds of Mpc with the planned next-generation detectors, the Einstein Telescope and the Cosmic Explorer.

I believe that the thesis includes an excellent review on oscillations and instabilities of neutron stars that can be pleasantly read by anyone, while in the appendices one will find the detailed analytic calculations and the extensive formulae derived for the problem.

Concluding, I consider Dr. Pnigouras's thesis an excellent piece of scientific work, written in an elegant, inspiring, and easy-to-read way. The results are sound, timely, and came out of a combination of analytical and computational work.

Tübingen, Germany   Prof. Kostas D. Kokkotas
June 2018

Reference

Doneva, D. D., Kokkotas, K. D. & Pnigouras, P. (2015). Gravitational wave afterglow in binary neutron star mergers. *Physical Review D, 92*, 104040. https://doi.org/10.1103/PhysRevD.92. 124004, arXiv:1510.00673.

Preface

Since their theoretical prediction in 1934 and the serendipitous discovery of the first pulsar in 1967, neutron stars remain among the most challenging objects in the Universe. Thanks to the advancement of theory, experiments, and observations, many aspects of their nature have been deciphered, yet their inner structure is still unknown. Gravitational waves emitted by neutron star oscillations can be used to obtain information about their equation of state, that is, the equation of state of dense nuclear matter. As discovered in the 1970s, certain oscillation modes can be secularly unstable to the emission of gravitational radiation, via the so-called Chandrasekhar-Friedman-Schutz (CFS) mechanism, thus rendering gravitational-wave asteroseismology a promising probe of the neutron star interior, especially after the recent birth of gravitational-wave astronomy.

After its initial growth phase, the instability is expected to saturate, due to nonlinear effects. The saturation amplitude of the unstable mode determines the detectability of the generated gravitational-wave signal, but also affects the evolution of the neutron star through the instability window, namely the region where the instability is active. In this work, we study the saturation of CFS-unstable f-modes (fundamental modes), due to low-order nonlinear mode coupling. Using the quadratic-perturbation approximation, we show that the unstable (parent) mode resonantly couples to pairs of stable (daughter) modes, which drain the parent's energy and make it saturate, via a mechanism called parametric resonance instability. The saturation amplitude of the most unstable f-mode multipoles is calculated throughout their instability windows, for typical and supramassive newborn neutron stars, simply modelled as polytropes in a Newtonian context.

Contrary to previous studies, where the saturation amplitude is treated as a constant, we find that it changes significantly throughout the instability window and, hence, during the neutron star evolution. Using the highest values obtained for the saturation amplitude, a signal from an unstable f-mode may even lie above the sensitivity of current, second-generation, gravitational-wave detectors.

In Chap. 1, we present a brief history of the field and the reasons which motivate such an enterprise, starting from the concept of asteroseismology and how it can be applied in neutron stars, so that the equation of state of dense nuclear matter is determined. Then, we discuss neutron stars as gravitational-wave sources, focusing on the presence of unstable oscillation modes and reviewing their significance both for gravitational-wave asteroseismology and neutron star evolution. In the rest of the chapters, we provide detailed information about the concepts introduced here.

In Chap. 2 we are going to derive the linear perturbation formalism, with the help of which the various classes of modes emerge, like polar (e.g., f-modes) and axial (e.g., r-modes). Chapter 3 is devoted to the f-mode CFS instability, where we will see how the instability works. In Chap. 4, we will obtain the nonlinear perturbation formalism, needed to introduce mode coupling, and discuss the mechanism responsible for the saturation of unstable modes, the so-called parametric resonance instability. The application of the mode coupling analysis to CFS-unstable f-modes, in both typical and supramassive neutron stars governed by polytropic equations of state, is presented in Chap. 5. We should note that, throughout this work, we use Newtonian gravity, with gravitational radiation introduced via post-Newtonian analysis (see Chap. 3). Chapter 6 concludes our study with a summary and some final remarks. At the beginning of each chapter, we review their contents in more detail.

Most of the lengthy derivations of formulae used throughout the chapters are addressed in appendices. In Appendix A, we present the Lane-Emden formalism for polytropic stars, together with Chandrasekhar's extension for rotating configurations. Low-order rotational corrections to the eigenfrequencies and eigenfunctions of polar modes are derived in Appendix B. Explicit formulae for the polar mode growth and damping rates, due to gravitational waves and viscosity, are given in Appendix C. In Appendices D and E, we obtain the equations of motion of nonlinear perturbations and an expression for the polar mode coupling coefficient, respectively. Several important results for a parametrically unstable coupled mode network are derived in Appendix F. Finally, the couplings responsible for the saturation of an unstable f-mode, for one of the models used in Chap. 5, is presented in Appendix G.

This doctoral dissertation is the product of my graduate studies in the Theoretical Astrophysics section of the Institute for Astronomy and Astrophysics in the University of Tübingen, Germany, under the supervision of Prof. Kostas D. Kokkotas. The thesis was submitted on the 11th of July 2016 and the defence took place on the 4th of May 2017. This work resulted in four published papers in refereed journals (Pnigouras and Kokkotas 2015; Doneva et al. 2015; Surace et al. 2016; Pnigouras and Kokkotas 2016) and two papers in conference proceedings (Pnigouras et al. 2016, 2017). Excerpts and figures have been reproduced and reprinted with permission from Pnigouras and Kokkotas (2015), Doneva et al. (2015), and Pnigouras and Kokkotas (2016); copyright (2015, 2016) by the American Physical Society (APS).

The dissertation consists of six main chapters and seven appendices. Sections marked with an asterisk (*) are intended as optional, but some results derived there may still be used in the mandatory sections (albeit always referenced). The same applies to appendices. Following the usual typographic convention, boldface symbols represent vectors.

Computations were performed with an original code, written in Fortran 95 and parallelized with OpenMP, executed in the Theoretical Astrophysics computing centre of the University of Tübingen. Individual runs lasted from a few hours to a few days, depending on the model and the workload. The graphs were produced with Mathematica 10.4.

The title page drawing is an artist's conception of a neutron star, deformed by the hexadecapole f-mode and emitting gravitational waves, courtesy of Christos Tsirvoulis. Finally, the musical score of the Epilogue was created with Noteflight.

Southampton, UK   Dr. Pantelis Pnigouras
June 2018

References

Doneva, D. D., Kokkotas, K. D. & Pnigouras, P. (2015). Gravitational wave afterglow in binary neutron star mergers. *Physical Review D, 92,* 104040. https://doi.org/10.1103/PhysRevD.92. 124004, arXiv:1510.00673.

Pnigouras, P. & Kokkotas, K. D. (2015). Saturation of the f-mode instability in neutron stars. I. Theoretical framework. *Physical Review D, 92,* 084018. https://doi.org/10.1103/PhysRevD. 92.084018, arXiv:1509.01453.

Pnigouras, P. & Kokkotas, K. D. (2016). Saturation of the f-mode instability in neutron stars. II. Applications and results. *Physical Review D, 94,* 024053. https://doi.org/10.1103/PhysRevD. 94.024053, arXiv:1607.03059.

Pnigouras, P., Kokkotas, K. D. & Doneva, D. D. (2017). Saturation of the f-mode instability in neutron stars. In *Proceedings of the 14th Marcel Grossmann Meeting on General Relativity.* Rome, Italy: World Scientific. https://doi.org/10.1142/9789813226609_0552.

Pnigouras, P., Kokkotas, K. D., Doneva, D. D. & Surace, M. (2016). Saturation of the f-mode instability in neutron stars. In *Proceedings of the 28th Texas Symposium on Relativistic Astrophysics.* Geneva, Switzerland. https://indico.cern.ch/event/336103/contributions/786952/ attachments/1205025/1783754/proc_221.pdf.

Surace, M., Kokkotas, K. D. & Pnigouras, P. (2016). The stochastic background of gravitational waves due to the f-mode instability in neutron stars. *Astronomy & Astrophysics, 586,* A86. https://doi.org/10.1051/0004-6361/201527197, arXiv:1512.02502.

Acknowledgements

According to Homer (*Odyssey*, Book II), during the absence of Odysseus from Ithaca, Athena, the goddess of wisdom, disguised herself as Odysseus's friend Mentor, in order to counsel his son Telemachus. I cannot think of a more appropriate simile to describe my relationship with Kostas Kokkotas, who, like a true mentor, never imparted his wisdom in an authoritative way, but conveyed it in the form of advice, always giving me the freedom to question it and learn from my own mistakes. I am grateful not only for his guidance as a teacher, but also for his concern about my well-being and his support.

I would also like to thank all the people who have worked in the Theoretical Astrophysics section of the University of Tübingen during the past few years, and especially Daniela Doneva and Marco Surace, for our fertile collaboration, Kai Schwenzer, for sharing his insight and generously offering his help whenever requested, Marlene Herbrik, for enlightening discussions, Andreas Boden, for tech support, Kostas Glampedakis and Andrea Maselli, for distracting me from the never-ending workload, and Heike Fricke, for explaining the German ways and helping me deal with the ferocious monster that is bureaucracy.

I am also deeply indebted to my teachers Christos Tsagas, from the Aristotle University of Thessaloniki, Maria Missa, and Dimitris Pegios, to whom a part of this work is credited.

For their moral support, I express my gratitude to Athina Valera, Elli Kalaki, and my family. Finally, I am thankful to Christos Tsirvoulis, for the title page drawing, and Marianna Ntinou, for her help on the Epilogue's musical score.

Contents

List of Figures

List of Tables

Chapter 1
Introduction

You will smile and hold my hands tight.
A star will ring on the damp sky.
I may
cry.

Tasos Livaditis, *Simple talk* (1950)
English translation: P. P.

1.1 Prologue: Music of the Spheres

Driven by his vision of a harmonious Cosmos, sculpted by music and mathematics, Pythagoras of Samos (c. 569–475 BC) believed that the celestial bodies are carried around the sky by crystalline spheres, which produce musical sounds with their motion. This idea became key for ancient cosmogony, with Plato (c. 427–347 BC) stating that:

> Above, on each of its circles, is perched a Siren, accompanying its revolution, uttering a single sound, one note; from all eight[1] is produced the accord of a single harmony.[2]

Due to their elegance, the Pythagorean views survived until the 17th century, when Johannes Kepler (1571–1630) struggled to prove their validity, by studying the planetary motions. Unfortunately, the inspiring ideas in his *Mysterium Cosmographicum* (1597) were for naught. The rapid development of astronomy, to which Kepler himself contributed greatly, revealed that the sheet where the "Music of the spheres" was written is blank.

However, the Universe is far from being a quiet place. Modern science has shown that the celestial bodies do produce sounds, albeit different from the ones Pythagoras postulated. Like every object in the Universe, stars possess characteristic oscillation frequencies. Quakes or other cataclysmic events can excite these oscillations and make the star "ring", much like the sound one gets when ringing a bell. Alas, even

---

[1] Five spheres for the planets known at the time (Mercury, Venus, Mars, Jupiter, Saturn), two spheres for the Sun and Moon, plus one for the distant stars.

[2] *The Republic of Plato*, X617b. Translated by Allan Bloom. New York: Basic Books, 2nd ed. (1991).

© Springer Nature Switzerland AG 2018
P. Pnigouras, *Saturation of the f-mode Instability in Neutron Stars*,
Springer Theses, https://doi.org/10.1007/978-3-319-98258-8_1

though the ringing of a star may bring tears to the Greek poet Tasos Livaditis (1922–1988),[3] the vacuum of space makes it inaudible, so we cannot really hear the sounds of the stars. Or can we?

1.2 Asteroseismology

Even though we are not able to directly hear the stars ringing, we may instead *see* them! In fact, the theory of stellar pulsation was developed in order to explain the observations of classical variable stars, such as the Cepheids and RR Lyrae stars.

There are two general types of stellar oscillations: *radial*, where the star expands and contracts while maintaining its spherical shape, and *nonradial*, where the shape of the star deviates from sphericity (Fig. 1.1). Nonradial pulsation theory can be traced back to the work of Lord Kelvin (Thomson 1863), preceding that of radial pulsation, developed by Ritter (1879). An analytic solution for the nonradial oscillations of a nonrotating, homogeneous, compressible stellar model was obtained by Pekeris (1938), whereas Cowling (1941) studied the nonradial oscillations of polytropic stars. Nonradial pulsation did not draw much attention until the work of Ledoux (1951), who proposed nonradial oscillations as an explanation for the double periodicity and the large temporal variations in the broadening of spectral lines observed in β Canis Majoris (a prototype of β-Cephei–type variable stars). For a description of the above studies, the reader is referred to the work of Ledoux and Walraven (1958).

Since then, nonradial stellar pulsation has been suggested as an explanation for several observations. For instance, the solar five-minute oscillations, discovered by Leighton et al. (1962), were interpreted by Ulrich (1970) and Leibacher and Stein (1971) as nonradial oscillation modes, a picture which was later confirmed by Deubner (1975), when he managed to match a series of observations with theoretical estimates. As a result of this discovery, many observations in the past few decades unveiled thousands of modes in the Sun.

But, apart from explaining the observations, there is more we can learn from stellar oscillations. Quoting Eddington (1926, § 1):

> At first sight it would seem that the deep interior of the sun and stars is less accessible to scientific investigation than any other region of the universe. Our telescopes may probe farther and farther into the depths of space; but how can we ever obtain certain knowledge of that which is hidden behind substantial barriers? What appliance can pierce through the outer layers of a star and test the conditions within?

Today, we know that this "which is hidden behind substantial barriers" can be studied with the help of oscillation modes. *Helioseismology* uses solar oscillations to probe the internal structure of the Sun, via the comparison of theoretical models with the observed oscillation spectrum, just like seismic wave data can be used to study the Earth's interior. One of the most famous programs dedicated to this kind of research is

---

[3]Τάσος Λειβαδίτης, Ἁπλῆ κουβέντα (Μακρόνησος, 1950). *Ποίηση*, Τόμος πρώτος, σελ. 105. Αθήνα: Κέδρος, δέκατη έκδοση (2002).

Fig. 1.1 Radial ($l = 0$) and nonradial ($l \neq 0$) stellar oscillation modes, depicted as spherical harmonics Y_l^m

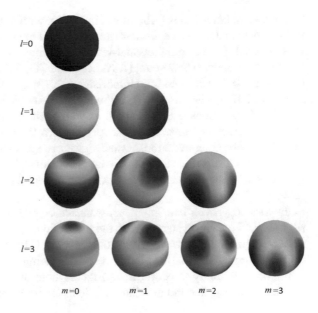

the Global Oscillation Network Group (GONG), a community-based project which has developed a "network of extremely sensitive, and stable solar velocity imagers located around the Earth to obtain nearly continuous observations of the Sun's five-minute oscillations" (http://gong.nso.edu).

In principle, this concept can be extended to other stars too. Various oscillation modes have been found in stars which were previously considered as nonpulsating, such as white dwarfs, Ap stars, and early type-O and B stars. Since distant stars cannot be resolved into a two-dimensional disk image, like the Sun, the identification of modes is very hard. However, much information can be (and has been) obtained by observations of stellar oscillations. After all, *asteroseismology* provides the only observational method for studying the stellar interior. More details on the studies and methods briefly outlined above can be found in the work of Unno et al. (1989) and Aerts et al. (2010).

So far, we have described applications of asteroseismology via electromagnetic signals. Another promising possibility is the utilisation of gravitational waves. Predicted by the general theory of relativity (Einstein 1916, 1918), gravitational waves are spacetime oscillations, due to the change in a body's gravitational field. Since any system with a time-varying quadrupole moment—which measures the system's deviation from axisymmetry—emits gravitational radiation, nonradial oscillations of stars should generate gravitational waves.

Large interferometers, with arm lengths up to 4 km, have been set up around the world, in order to detect minute length changes, associated with gravitational waves impinging on the Earth. On the 14th of September 2015, the Advanced Laser Interferometer Gravitational-wave Observatory (LIGO; Abramovici et al. 1992; Aasi et al. 2015a) in the USA detected gravitational waves directly for the first time, from a pair of coalescing black holes (Abbott et al. 2016b), with a second detection following a few months later, from a similar source (Abbott et al. 2016a), thus signifying the long-awaited onset of gravitational-wave astronomy. More detectors are scheduled to operate in the next few years (Advanced VIRGO, Italy; Caron et al. 1997; Acernese et al. 2015; LIGO-India; Unnikrishnan 2013; KAGRA, Japan; Somiya 2012; Aso et al. 2013), whereas there are already plans about space-borne interferometers (eLISA; Amaro-Seoane et al. 2012), as well as more advanced (third-generation) terrestrial detectors (Einstein Telescope; Punturo et al. 2010; Sathyaprakash et al. 2012). In addition, gravitational-wave signatures are sought in pulse arriving times from known pulsars (Pulsar Timing Array; Hobbs et al. 2010). An extensive review on gravitational-wave sources and detectors can be found in Sathyaprakash and Schutz (2009).

In practice, *gravitational-wave asteroseismology* can only be applied in very compact stars, like neutron stars, where the gravitational fields involved are large (for a review on gravitational waves from compact objects the reader is referred to Kokkotas and Stergioulas 2006). This is a very interesting prospect, because it gives us a unique opportunity to peep through the keyhole of a largely unexplored area of physics, where matter and gravitational fields are found at their extremes.

1.3   Neutron Stars Inside Out

With a mass of order a solar mass and a radius of about 10 km, neutron stars constitute nature's high-energy laboratories. Their theoretical prediction is usually attributed to Baade and Zwicky (1934b, a) who, just two years after the discovery of the neutron by James Chadwick, proposed that highly compact objects, consisting of closely packed neutrons, could be formed after a supernova explosion. However, there is evidence that Lev Landau had already postulated their existence, even before the discovery of the neutron (Yakovlev et al. 2013), albeit not in such an explicit and prescient manner. A few years later, Tolman (1939) and Oppenheimer and Volkoff (1939) derived the general relativistic hydrostatic equilibrium equations and then, modelling a neutron star as a degenerate cold Fermi gas of neutrons, obtained an upper limit for its mass equal to $\approx 0.7 \, M_{\odot}$, where $M_{\odot}$ is the solar mass. Considering more sophisticated equations of state, this upper limit is shifted to higher masses, which is necessary to explain the observations, as we will see below.

The observation of the first neutron star did not come until 1967, when Jocelyn Bell and Antony Hewish detected a radio source emitting pulses with a period of 1.33 s (Hewish et al. 1968). The unexpected source was initially given the nickname LGM-1, which stands for "Little Green Men", humorously suggesting that the signal originated from an extraterrestrial civilisation. Discoveries of more *pulsars* (pulsating stars) followed in the next years, dissolving the mystery and associating the pulses with fast rotation rates and large magnetic fields ($\sim 10^8 - 10^{12}$ G), thus classifying pulsars as neutron stars, since they are the only theoretically predicted objects which can support such properties.

In 1974, Russell Hulse and Joseph Taylor made another important discovery, by detecting the first pulsar in a neutron star binary system (Hulse and Taylor 1975; see also Damour 2015). As it was immediately realised (Wagoner 1975), the Hulse-Taylor binary could serve as an indirect test for the existence of gravitational waves, which should lead to a decrease of the system's orbital period. The prediction of general relativity was indeed confirmed with remarkable accuracy, with Hulse and Taylor being awarded the Nobel prize in 1993. Ever since, a number of pulsar binaries have been discovered and are still being used to test Einstein's theory of gravity (for a review, see Will 2014).

Currently, we know of more than 2500 pulsars and neutron stars have been related to many types of systems. Soft γ-ray repeaters (SGRs) and anomalous X-ray pulsars (AXPs) are thought to be types of *magnetars*, namely neutron stars with extremely high magnetic fields (up to $\sim 10^{15}$ G; Duncan and Thompson 1992). Neutron stars are also believed to be the engines powering low-mass X-ray binaries (LMXBs), where a neutron star accretes matter from a companion star and spins up. This mechanism could be the origin of millisecond pulsars, which, as their name suggests, rotate with periods $\sim 1 - 10$ ms. The fastest-spinning pulsar known today is rotating at 716 Hz and is a member of such a binary system (Hessels et al. 2006).

From the above we see that much information has been gathered so far about neutron stars. However, it is not enough to infer the biggest unknown that plagues neutron star physics: their equation of state. Equations of state are often categorised as soft or stiff, depending on how fast pressure increases as a function of density (the faster, the stiffer). Stiff equations of state are less "compressible" and can support larger masses than soft ones (for instance, see Glendenning 2000, Sects. 5.3.12 and 5.3.13), meaning that neutron star mass measurements could serve as tests for the proposed equations of state (e.g., see Kiziltan et al. 2013). This can be seen in Fig. 1.2, where the mass–radius curve is plotted for various popular equations of state. Typical neutron star masses lie around 1.4 $M_\odot$, but the recent observations of pulsars with masses $1.97 \pm 0.04 \, M_\odot$ (Demorest et al. 2010) and $2.01 \pm 0.04 \, M_\odot$ (Antoniadis et al. 2013) rule out some soft equations of state.

The final word about the equation of state of matter at supranuclear densities has to come from the collaboration of theory [i.e., quantum chromodynamics (QCD), where many complications still need to be tamed], experiments, and observations, with the latter being, at the moment, the most promising source of information (Lattimer and Prakash 2007). Theoretically, mass–radius measurements from a certain number of sources could allow us to construct the equation of state of neutron stars (Lindblom 1992). However, as it can be seen from Fig. 1.2, it would be rather hard to pick the correct equation of state, unless very accurate measurements are provided for the mass and the radius, with the latter being currently much less precisely known than the former. The additional information channel of gravitational radiation can, in principle, contribute towards this direction, via the application of gravitational-wave asteroseismology (Ferrari 2010; Andersson et al. 2011).

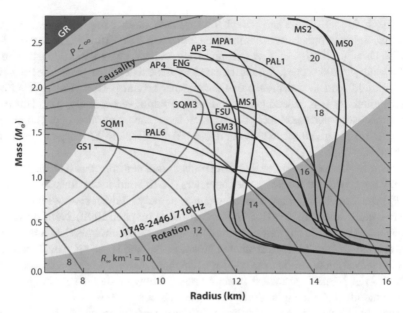

Fig. 1.2 Mass–radius curves for popular hadronic (black lines) and strange quark (green lines) matter equations of state. The maximum allowed mass depends on the stiffness of the equation of state. Also shown are: the general relativity limit (GR; the star collapses to a black hole for any equation of state), the Buchdahl limit ($P < \infty$; the maximum mass for an incompressible star), the causality limit (the speed of sound exceeds the speed of light), and the rotation limit from the fastest known pulsar, spinning at 716 Hz (Hessels et al. 2006). Orange lines correspond to redshifted radius values. Credit: Lattimer (2012). Republished with permission of Annual Reviews, Inc.; permission conveyed through Copyright Clearance Center, Inc.

1.4 Odes from Modes

Gravitational waves from neutron stars may occur in the form of either a *burst* or a *continuous* emission.

The main scenarios associated with burst emission are: (i) Binary neutron star mergers, where a pair of neutron stars, orbiting around each other, emit (continuous) gravitational radiation, leading to the gradual shrinking of their orbit, until they finally coalesce and produce a gravitational-wave burst. The oscillations induced on the massive post-merger neutron star remnant should leave their imprint on the gravitational-wave signal, thus allowing asteroseismological studies to be performed (Stergioulas et al. 2011; Bauswein and Janka 2012; Bauswein et al. 2012, 2014). (ii) Pulsar *glitches*, where a small and sudden increase in the neutron star rotation rate occurs. According to current understanding, glitches can be explained as the redistribution of the star's angular momentum, resulting from a starquake or the spin lag between the superfluid core and the solid crust. This process can induce a time-varying quadrupole moment and, hence, emission of gravitational waves (Warszawski and Melatos 2012). (iii) Giant magnetar flares, which are rare, but very powerful, events (Mazets et al. 1979; Hurley et al. 1999, 2005; Palmer et al. 2005), attributed to the reconfiguration of the star's magnetic field and potentially accompanied by gravitational-wave emission (Zink et al. 2012).

The basic mechanisms related to continuous gravitational-wave emission are: (i) Nonaxisymmetric deformations, often dubbed "mountains", which can be caused, for example, by magnetic forces (Gualtieri et al. 2011; Mastrano et al. 2011) or elastic forces in the crust (Bildsten 1998; Ushomirsky et al. 2000; for a review, see Jones 2002). (ii) Nonradial oscillation modes, driven unstable by the emission of gravitational radiation due to the Chandrasekhar-Friedman-Schutz (CFS) mechanism. The latter is the case we are going to focus on.

The CFS mechanism (Chandrasekhar 1970; Friedman and Schutz 1978a, b) results in a *secular* instability, where the mode amplitude grows exponentially on a time scale which is related to the mode's gravitational-wave emission rate. Among the best gravitational-wave emitters are the f-modes, which are the *fundamental* oscillations of the star and become unstable for relatively large rotation rates (Ipser and Lindblom 1990). On the other hand, viscosity, which mainly depends on the temperature, acts against the instability by damping the modes, giving rise to the *instability window*, namely a region in the temperature–spin plane where the instability is active (Ipser and Lindblom 1991).

Later studies showed that another class of oscillations, the r-modes, become unstable for any rotation rate of the star (Andersson 1998; Friedman and Morsink 1998). The r-modes are horizontal fluid motions driven by rotation (Papaloizou and Pringle 1978) and, due to their much larger instability windows (Lindblom et al. 1998; Owen et al. 1998; Andersson and Kokkotas 2001), they are considered a very promising gravitational-wave source.

Subsequent studies on the r-mode instability naturally raised the question of the maximum amplitude that the oscillation can attain, before it is halted by nonlinear effects. As it was shown, coupling of the unstable r-mode to other modes of the star can work as an energy drain and saturate the instability at low levels (Schenk et al. 2001; Morsink 2002; Arras et al. 2003; Brink et al. 2004b, a, 2005), rendering a possible detection quite hard—in the most optimistic cases, the signal may be detectable with Advanced LIGO from within the Local Galactic Group ($\sim$1 Mpc; Bondarescu et al. 2007, 2009; see also Aasi et al. 2015b).

A similar mechanism ought to operate for the f-mode instability as well. So far, the evolution of unstable f-modes in the nonlinear regime has been studied only via hydrodynamic simulations (Shibata and Karino 2004; Ou et al. 2004; Kastaun et al. 2010; see also Kastaun 2011 for a similar study on the r-mode instability). However, the CFS instability sets in on secular time scales, way beyond the current capabilities of nonlinear simulations. To circumvent this issue, they either artificially increase the gravitational-wave growth rate of the mode, or start with a very high mode amplitude and then track its decay. In both cases, the mode amplitude acquires large values, which may never be actually achieved, because the coupling mechanism is expected to operate at lower amplitudes.

Determining the saturation amplitude of CFS-unstable modes is also important for neutron star evolution. The instability is expected to be relevant in fast rotators, e.g., newborn neutron stars or members of LMXBs. The r-mode instability can affect both types of systems (Levin 1999; Bondarescu et al. 2007, 2009), whereas the f-mode instability is mainly applicable in nascent stars (Passamonti et al. 2013). According to observations of LMXBs, many sources seem to reside well inside the theoretically expected r-mode instability window (e.g., see Gusakov et al. 2014a, Fig. 2 and Table I). On the other hand, evolutionary scenarios suggest that the star spends a limited amount of its life inside the window, if one assumes the saturation amplitudes obtained from the mode coupling studies (Bondarescu et al. 2007). This paradox could possibly be resolved if (a) the instability window has a different shape and/or size than expected, implying additional sources of dissipation or even exotic forms of matter (Andersson and Kokkotas 2001, Sects. 4.3–4.6), or (b) the saturation amplitude is even smaller than the estimates of the mode coupling analysis, which manifests the need for a refined calculation, using more realistic models, or the action of more effective saturation mechanisms. It has been shown that large-amplitude viscous effects should also play a role in the saturation of unstable r- (Alford et al. 2012) and f-modes (Passamonti and Glampedakis 2012), but the predicted amplitudes are higher than those achieved via mode coupling. Another proposed mechanism involves the interaction of superfluid vortices with superconducting flux tubes, saturating the r-mode to lower amplitudes (Haskell et al. 2014), but still not low enough to explain the paradox (this mechanism is relevant only for mature, cold stars, where superfluidity operates, so it does not apply to f-modes; for the interested reader, a review on superfluidity of nuclear matter can be found in Lombardo and Schulze 2001). An alternative scenario explaining the observations suggests the occurrence of resonances between "normal" and "superfluid" modes, which alter the instability region (Gusakov et al. 2014a, b).

Another important issue that has to be addressed by neutron star evolutionary scenarios is their maximum spin rate. If angular momentum is conserved during the core collapse of the neutron star progenitor, then angular velocities close to the break-up limit should be theoretically feasible ($\sim$1 ms; e.g., see Heger et al. 2000). Observations of young pulsars, however, imply initial rotation periods $\sim$10–100 ms, with the fastest known young pulsar rotating with a period of only 16 ms (Marshall et al. 1998). This contradiction suggests the involvement of some mechanism which either spins down the star at an early stage in its life or makes it spin slowly from the outset. The same problem applies to LMXBs, where accretion should spin the neutron star up to frequencies close to break-up, contrary to observations, which indicates that some process prevents further spin-up of the star (Chakrabarty et al. 2003).

As first suggested by Friedman (1983), the onset of a CFS instability could serve as an upper limit for neutron star angular velocities. If the star enters the instability window of an unstable mode, then the emission of gravitational waves results in angular momentum loss, which spins down a newborn star or balances the accretion torque in an LMXB (Lindblom et al. 1998; Andersson et al. 1999a, b, 2000; Bondarescu and Wasserman 2013; Alford and Schwenzer 2014a, 2015). The saturation amplitude of the unstable mode determines the evolutionary route of the neutron star inside the instability window. In LMXBs, the same process could also work with neutron star mountains (Bildsten 1998; Ushomirsky et al. 2000). For more mechanisms which may be responsible for the spin-down of nascent neutron stars, see Ott et al. (2006).

From the discussion above, it becomes apparent that the calculation of the saturation amplitude of CFS-unstable modes is an important ingredient for neutron star physics. Asteroseismology can provide much information about neutron star bulk and internal properties, using either r-modes (Alford and Schwenzer 2014b; Mytidis et al. 2015; Kokkotas and Schwenzer 2016), or f- and, possibly, w-modes (Andersson and Kokkotas 1996, 1998; Kokkotas et al. 2001; Benhar et al. 2004; Gaertig and Kokkotas 2011; Doneva et al. 2013; Doneva and Kokkotas 2015)—the latter, also called gravitational-wave modes, result from the coupling of fluid and gravitational-wave oscillations, and exist only in a general relativistic framework (Kokkotas and Schutz 1986, 1992).

Even though f-modes have much smaller instability windows and larger growth time scales than r-modes, they could still play a significant role in the evolution of newborn neutron stars (Passamonti et al. 2013). In fact, according to recent work (Doneva et al. 2015), unstable f-modes with large instability windows and very short growth time scales can develop in "supramassive" neutron stars, which are supported against gravitational collapse due to their fast rotation (Cook et al. 1992, 1994). A possible gravitational-wave detection from such objects has auspicious implications for neutron star physics, since only a few known equations of state are consistent with their existence (e.g., see Lasky et al. 2014). But, in order to reach inarguable conclusions, the estimation of the f-mode saturation amplitude is an important step.

References

Aasi, J., et al. (2015a). Advanced LIGO. *Classical Quantum Gravity, 32*, 074001. https://doi.org/10.1088/0264-9381/32/7/074001, arXiv:1411.4547.

Aasi, J., et al. (2015b). Searches for continuous gravitational waves from nine young supernova remnants. *The Astrophysical Journal, 813*, 39. https://doi.org/10.1088/0004-637X/813/1/39, arXiv:1412.5942.

Abbott, B. P., et al. (2016a). GW151226: Observation of gravitational waves from a 22-solar-mass binary black hole coalescence. *Physical Review Letters, 116*, 241103. https://doi.org/10.1103/PhysRevLett.116.241103.

Abbott, B. P., et al. (2016b). Observation of gravitational waves from a binary black hole merger. *Physical Review Letters, 116*, 061102. https://doi.org/10.1103/PhysRevLett.116.061102, arXiv:1602.03837.

Abramovici, A., Althouse, W. E., Drever, R. W. P., Gürsel, Y., Kawamura, S., & Raab, F. J., et al. (1992). LIGO: The laser interferometer gravitational-wave observatory. *Science, 256*, 325–333. https://doi.org/10.1126/science.256.5055.325.

Acernese, F., et al. (2015). Advanced Virgo: a second-generation interferometric gravitational wave detector. *Classical Quantum Gravity*, *32*, 024001. https://doi.org/10.1088/0264-9381/32/2/024001, arXiv:1408.3978.

Aerts, C., Christensen-Dalsgaard, J., & Kurtz, D. W. (2010). *Asteroseismology*. Astronomy and astrophysics library. New York: Springer. http://adsabs.harvard.edu/abs/2010aste.book....A.

Alford, M. G., & Schwenzer, K. (2014a). Gravitational wave emission and spin-down of young pulsars. *The Astrophysical Journal*, *781*, 26. https://doi.org/10.1088/0004-637X/781/1/26, arXiv:1210.6091.

Alford, M. G., & Schwenzer, K. (2014b). What the timing of millisecond pulsars can teach us about their interior. *Physical Review Letters*, *113*, 251102. https://doi.org/10.1103/PhysRevLett.113.251102, arXiv:1310.3524.

Alford, M. G., & Schwenzer, K. (2015). Gravitational wave emission from oscillating millisecond pulsars. *Monthly Notices of the Royal Astronomical Society*, *446*, 3631–3641. https://doi.org/10.1093/mnras/stu2361, arXiv:1403.7500.

Alford, M. G., Mahmoodifar, S., & Schwenzer, K. (2012). Viscous damping of r-modes: Large amplitude saturation. *Physical Review D*, *85*, 044051. https://doi.org/10.1103/PhysRevD.85.044051, arXiv:1103.3521.

Amaro-Seoane, P., et al. (2012). Low-frequency gravitational-wave science with eLISA/NGO. *Classical Quantum Gravity*, *29*, 124016. https://doi.org/10.1088/0264-9381/29/12/124016, arXiv:1202.0839.

Andersson, N. (1998). A new class of unstable modes of rotating relativistic stars. *The Astrophysical Journal*, *502*, 708–713. https://doi.org/10.1086/305919, arXiv:gr-qc/9706075.

Andersson, N., & Kokkotas, K. D. (1996). Gravitational waves and pulsating stars: What can we learn from future observations? *Physical Review Letters*, *77*, 4134–4137. https://doi.org/10.1103/PhysRevLett.77.4134, arXiv:gr-qc/9610035.

Andersson, N., & Kokkotas, K. D. (1998). Towards gravitational wave asteroseismology. *Monthly Notices of the Royal Astronomical Society*, *299*, 1059–1068. https://doi.org/10.1046/j.1365-8711.1998.01840.x, arXiv:gr-qc/9711088.

Andersson, N., & Kokkotas, K. D. (2001). The r-mode instability in rotating neutron stars. *International Journal of Modern Physics D*, *10*, 381–441. https://doi.org/10.1142/S0218271801001062, arXiv:gr-qc/0010102.

Andersson, N., Ferrari, V., Jones, D. I., Kokkotas, K. D., Krishnan, B., & Read, J. S., et al. (2011). Gravitational waves from neutron stars: promises and challenges. *General Relativity and Gravitation*, *43*, 409–436. https://doi.org/10.1007/s10714-010-1059-4, arXiv:0912.0384.

Andersson, N., Jones, D. I., Kokkotas, K. D., & Stergioulas, N. (2000). R-mode runaway and rapidly rotating neutron stars. *The Astrophysical Journal*, *534*, L75–L78. https://doi.org/10.1086/312643, arXiv:astro-ph/0002114.

Andersson, N., Kokkotas, K., & Schutz, B. F. (1999a). Gravitational radiation limit on the spin of young neutron stars. *The Astrophysical Journal*, *510*, 846–853. https://doi.org/10.1086/306625, arXiv:astro-ph/9805225.

Andersson, N., Kokkotas, K. D., & Stergioulas, N. (1999b). On the relevance of the r-mode instability for accreting neutron stars and white dwarfs. *The Astrophysical Journal*, *516*, 307–314. https://doi.org/10.1086/307082, arXiv:astro-ph/9806089.

Antoniadis, J., et al. (2013). A massive pulsar in a compact relativistic binary. *Science*, *340*, 448. https://doi.org/10.1126/science.1233232, arXiv:1304.6875.

Arras, P., Flanagan, É. É., Morsink, S. M., Schenk, A. K., Teukolsky, S. A., & Wasserman, I. (2003). Saturation of the r-mode instability. *The Astrophysical Journal*, *591*, 1129–1151. https://doi.org/10.1086/374657, arXiv:astro-ph/0202345.

Aso, Y., Michimura, Y., Somiya, K., Ando, M., Miyakawa, O., & Sekiguchi, T., et al. (2013). Interferometer design of the KAGRA gravitational wave detector. *Physical Review D*, *88*, 043007. https://doi.org/10.1103/PhysRevD.88.043007, arXiv:1306.6747.

Baade, W., & Zwicky, F. (1934a). On super-novae. *Proceedings of the National Academy of Sciences of the United States of America, 20*, 254–259. https://doi.org/10.1073/pnas.20.5.254.

Baade, W., & Zwicky, F. (1934b). Cosmic rays from super-novae. *Proceedings of the National Academy of Sciences of the United States of America, 20*, 259–263. https://doi.org/10.1073/pnas.20.5.259.

Bauswein, A., & Janka, H.-T. (2012). Measuring neutron-star properties via gravitational waves from neutron-star mergers. *Physical Review Letters, 108*, 011101. https://doi.org/10.1103/PhysRevLett.108.011101, arXiv:1106.1616.

Bauswein, A., Janka, H.-T., Hebeler, K., & Schwenk, A. (2012). Equation-of-state dependence of the gravitational-wave signal from the ring-down phase of neutron-star mergers. *Physical Review D, 86*, 063001. https://doi.org/10.1103/PhysRevD.86.063001, arXiv:1204.1888.

Bauswein, A., Stergioulas, N., & Janka, H.-T. (2014). Revealing the high-density equation of state through binary neutron star mergers. *Physical Review D, 90*, 023002. https://doi.org/10.1103/PhysRevD.90.023002, arXiv:1403.5301.

Benhar, O., Ferrari, V., & Gualtieri, L. (2004). Gravitational wave asteroseismology reexamined. *Physical Review D, 70*, 124015. https://doi.org/10.1103/PhysRevD.70.124015, arXiv:astro-ph/0407529.

Bildsten, L. (1998). Gravitational radiation and rotation of accreting neutron stars. *The Astrophysical Journal, 501*, L89–L93. https://doi.org/10.1086/311440, arXiv:astro-ph/9804325.

Bondarescu, R., & Wasserman, I. (2013). Nonlinear Development of the r-mode Instability and the Maximum Rotation Rate of Neutron Stars. *The Astrophysical Journal, 778*, 9. https://doi.org/10.1088/0004-637X/778/1/9, arXiv:1305.2335.

Bondarescu, R., Teukolsky, S. A., & Wasserman, I. (2007). Spin evolution of accreting neutron stars: Nonlinear development of the r-mode instability. *Physical Review D, 76*, 064019. https://doi.org/10.1103/PhysRevD.76.064019, arXiv:0704.0799.

Bondarescu, R., Teukolsky, S. A., & Wasserman, I. (2009). Spinning down newborn neutron stars: Nonlinear development of the r-mode instability. *Physical Review D, 79*, 104003. https://doi.org/10.1103/PhysRevD.79.104003, arXiv:0809.3448.

Brink, J., Teukolsky, S. A., & Wasserman, I. (2004a). Nonlinear coupling network to simulate the development of the r-mode instability in neutron stars. I. Construction. *Physical Review D, 70*, 124017. https://doi.org/10.1103/PhysRevD.70.124017, arXiv:gr-qc/0409048.

Brink, J., Teukolsky, S. A., & Wasserman, I. (2004b). Nonlinear couplings of r-modes: Energy transfer and saturation amplitudes at realistic timescales. *Physical Review D, 70*, 121501. https://doi.org/10.1103/PhysRevD.70.121501, arXiv:gr-qc/0406085.

Brink, J., Teukolsky, S. A., & Wasserman, I. (2005). Nonlinear coupling network to simulate the development of the r mode instability in neutron stars II. Dynamics. *Physical Review D, 71*, 064029. https://doi.org/10.1103/PhysRevD.71.064029, arXiv:gr-qc/0410072.

Caron, B., et al. (1997). The Virgo interferometer. *Classical Quantum Gravity, 14*, 1461–1469. https://doi.org/10.1088/0264-9381/14/6/011.

Chakrabarty, D., Morgan, E. H., Muno, M. P., Galloway, D. K., Wijnands, R., & van der Klis, M., et al. (2003). Nuclear-powered millisecond pulsars and the maximum spin frequency of neutron stars. *Nature, 424*, 42–44. https://doi.org/10.1038/nature01732, arXiv:astro-ph/0307029.

Chandrasekhar, S. (1970). Solutions of two problems in the theory of gravitational radiation. *Physical Review Letters, 24*, 611–615. https://doi.org/10.1103/PhysRevLett.24.611.

Cook, G. B., Shapiro, S. L., & Teukolsky, S. A. (1992). Spin-up of a rapidly rotating star by angular momentum loss: Effects of general relativity. *The Astrophysical Journal, 398*, 203–223. https://doi.org/10.1086/171849.

Cook, G. B., Shapiro, S. L., & Teukolsky, S. A. (1994). Rapidly rotating neutron stars in general relativity: Realistic equations of state. *The Astrophysical Journal, 424*, 823–845. https://doi.org/10.1086/173934.

Cowling, T. G. (1941). The non-radial oscillations of polytropic stars. *Monthly Notices of the Royal Astronomical Society, 101*, 367. https://doi.org/10.1093/mnras/101.8.367.

Damour, T. (2015). 1974: The discovery of the first binary pulsar. *Classical Quantum Gravity, 32*, 124009. https://doi.org/10.1088/0264-9381/32/12/124009, arXiv:1411.3930.

Demorest, P. B., Pennucci, T., Ransom, S. M., Roberts, M. S. E., & Hessels, J. W. T. (2010). A two-solar-mass neutron star measured using Shapiro delay. *Nature, 467*, 1081–1083. https://doi.org/10.1038/nature09466, arXiv:1010.5788.

Deubner, F.-L. (1975). Observations of low wavenumber nonradial eigenmodes of the sun. *Astronomy & Astrophysics, 44*, 371–375. http://adsabs.harvard.edu/abs/1975A%26A....44.371D.

Doneva, D. D. & Kokkotas, K. D. (2015). Asteroseismology of rapidly rotating neutron stars: An alternative approach. *Physical Review D, 92*, 124004. https://doi.org/10.1103/PhysRevD.92.124004, arXiv:1507.06606.

Doneva, D. D., Gaertig, E., Kokkotas, K. D., & Krüger, C. (2013). Gravitational wave asteroseismology of fast rotating neutron stars with realistic equations of state. *Physical Review D, 88*, 044052. https://doi.org/10.1103/PhysRevD.88.044052, arXiv:1305.7197.

Doneva, D. D., Kokkotas, K. D. & Pnigouras, P. (2015). Gravitational wave afterglow in binary neutron star mergers. *Physical Review D, 92*, 104040. https://doi.org/10.1103/PhysRevD.92.104040, arXiv:1510.00673.

Duncan, R. C., & Thompson, C. (1992). Formation of very strongly magnetized neutron stars: Implications for gamma-ray bursts. *The Astrophysical Journal, 392*, L9–L13. https://doi.org/10.1086/186413.

Eddington, A. S. (1926). *The internal constitution of the stars*. Cambridge science classics. Cambridge: Cambridge University Press. https://doi.org/10.1017/CBO9780511600005.

Einstein, A. (1916). Näherungsweise Integration der Feldgleichungen der Gravitation. *Sitzungsberichte der Königlich Preussischen Akademie der Wissenschaften, 1*, 688. http://adsabs.harvard.edu/abs/1916SPAW.......688E.

Einstein, A. (1918). Über Gravitationswellen. *Sitzungsberichte der Königlich Preussischen Akademie der Wissenschaften, 1*, 154. http://adsabs.harvard.edu/abs/1918SPAW.......154E.

Ferrari, V. (2010). Probing the physics of neutron stars with gravitational waves. *Classical Quantum Gravity, 27*, 194006. https://doi.org/10.1088/0264-9381/27/19/194006.

Friedman, J. L., & Morsink, S. M. (1998). Axial instability of rotating relativistic stars. *The Astrophysical Journal, 502*, 714–720. https://doi.org/10.1086/305920, arXiv:gr-qc/9706073.

Friedman, J. L., & Schutz, B. F. (1978a). Lagrangian perturbation theory of nonrelativistic fluids. *The Astrophysical Journal, 221*, 937–957. https://doi.org/10.1086/156098.

Friedman, J. L., & Schutz, B. F. (1978b). Secular instability of rotating Newtonian stars. *The Astrophysical Journal, 222*, 281–296. https://doi.org/10.1086/156143.

Friedman, J. L. (1983). Upper limit on the frequency of pulsars. *Physical Review Letters, 51*, 11–14. https://doi.org/10.1103/PhysRevLett.51.11.

Gaertig, E., & Kokkotas, K. D. (2011). Gravitational wave asteroseismology with fast rotating neutron stars. *Physical Review D, 83*, 064031. https://doi.org/10.1103/PhysRevD.83.064031, arXiv:1005.5228.

Glendenning, N. K. (2000). *Compact stars: nuclear physics, particle physics, and general relativity* (2nd ed.). Astronomy and astrophysics library. New York: Springer. http://adsabs.harvard.edu/abs/2000csnp.conf....G.

Gualtieri, L., Ciolfi, R., & Ferrari, V. (2011). Structure, deformations and gravitational wave emission of magnetars. *Classical Quantum Gravity, 28*, 114014. https://doi.org/10.1088/0264-9381/28/11/114014, arXiv:1011.2778.

Gusakov, M. E., Chugunov, A. I., & Kantor, E. M. (2014a). Explaining observations of rapidly rotating neutron stars in low-mass x-ray binaries. *Physical Review D, 90*, 063001. https://doi.org/10.1103/PhysRevD.90.063001, arXiv:1305.3825.

Gusakov, M. E., Chugunov, A. I., & Kantor, E. M. (2014b). Instability windows and evolution of rapidly rotating neutron stars. *Physical Review Letters*, *112*, 151101. https://doi.org/10.1103/PhysRevLett.112.151101, arXiv:1310.8103.

Haskell, B., Glampedakis, K., & Andersson, N. (2014). A new mechanism for saturating unstable r modes in neutron stars. *Monthly Notices of the Royal Astronomical Society*, *441*, 1662–1668. https://doi.org/10.1093/mnras/stu535. arXiv:1307.0985.

Heger, A., Langer, N., & Woosley, S. E. (2000). Presupernova evolution of rotating massive stars. I. Numerical method and evolution of the internal stellar structure. *The Astrophysical Journal*, *528*, 368–396. https://doi.org/10.1086/308158, arXiv:astro-ph/9904132.

Hessels, J. W. T., Ransom, S. M., Stairs, I. H., Freire, P. C. C., Kaspi, V. M., & Camilo, F. (2006). A radio pulsar spinning at 716 Hz. *Science*, *311*, 1901–1904. https://doi.org/10.1126/science.1123430, arXiv:astro-ph/0601337.

Hewish, A., Bell, S. J., Pilkington, J. D. H., Scott, P. F., & Collins, R. A. (1968). Observation of a rapidly pulsating radio source. *Nature*, *217*, 709–713. https://doi.org/10.1038/217709a0.

Hobbs, G., et al. (2010). The international pulsar timing array project: Using pulsars as a gravitational wave detector. *Classical Quantum Gravity*, *27*, 084013. https://doi.org/10.1088/0264-9381/27/8/084013, arXiv:0911.5206.

Hulse, R. A., & Taylor, J. H. (1975). Discovery of a pulsar in a binary system. *The Astrophysical Journal*, *195*, L51–L53. https://doi.org/10.1086/181708.

Hurley, K., Boggs, S. E., Smith, D. M., Duncan, R. C., Lin, R., & Zoglauer, A., et al. (2005). An exceptionally bright flare from SGR 1806–20 and the origins of short-duration γ-ray bursts. *Nature*, *434*, 1098–1103. https://doi.org/10.1038/nature03519, arXiv:astro-ph/0502329.

Hurley, K., Cline, T., Mazets, E., Barthelmy, S., Butterworth, P., & Marshall, F., et al. (1999). A giant periodic flare from the soft γ-ray repeater SGR1900+14. *Nature*, *397*, 41–43. https://doi.org/10.1038/16199, arXiv:astro-ph/9811443.

Ipser, J. R., & Lindblom, L. (1990). The oscillations of rapidly rotating Newtonian stellar models. *The Astrophysical Journal*, *355*, 226–240. https://doi.org/10.1086/168757.

Ipser, J. R., & Lindblom, L. (1991). The oscillations of rapidly rotating Newtonian stellar models II. Dissipative effects. *The Astrophysical Journal*, *373*, 213–221. https://doi.org/10.1086/170039.

Jones, D. I. (2002). Gravitational waves from rotating strained neutron stars. *Classical Quantum Gravity*, *19*, 1255–1265. https://doi.org/10.1088/0264-9381/19/7/304, arXiv:gr-qc/0111007.

Kastaun, W. (2011). Nonlinear decay of r modes in rapidly rotating neutron stars. *Physical Review D*, *84*, 124036. https://doi.org/10.1103/PhysRevD.84.124036, arXiv:1109.4839.

Kastaun, W., Willburger, B., & Kokkotas, K. D. (2010). Saturation amplitude of the f-mode instability. *Physical Review D*, *82*, 104036. https://doi.org/10.1103/PhysRevD.82.104036, arXiv:1006.3885.

Kiziltan, B., Kottas, A., De Yoreo, M., & Thorsett, S. E. (2013). The neutron star mass distribution. *The Astrophysical Journal*, *778*, 66. https://doi.org/10.1088/0004-637X/778/1/66, arXiv:1011.4291.

Kokkotas, K. D. & Schwenzer, K. (2016). R-mode astronomy. *The European Physical Journal A*, *52*, 38. https://doi.org/10.1140/epja/i2016-16038-9, arXiv:1510.07051.

Kokkotas, K. D., & Stergioulas, N. (2006). Gravitational waves from compact sources. In A. M. Mourão, M. Pimenta, R. Potting & P. M. Sá (Eds.), *Proceedings of the Fifth International Workshop "New Worlds in Astroparticle Physics", Faro, Portugal*. arXiv:gr-qc/0506083.

Kokkotas, K. D., Apostolatos, T. A., & Andersson, N. (2001). The inverse problem for pulsating neutron stars: A 'fingerprint analysis' for the supranuclear equation of state. *Monthly Notices of the Royal Astronomical Society*, *320*, 307–315. https://doi.org/10.1046/j.1365-8711.2001.03945.x, arXiv:gr-qc/9901072.

Kokkotas, K. D., & Schutz, B. F. (1986). Normal modes of a model radiating system. *General Relativity and Gravitation*, *18*, 913–921. https://doi.org/10.1007/BF00773556.

Kokkotas, K. D., & Schutz, B. F. (1992). W-modes: A new family of normal modes of pulsating relativistic Stars. *Monthly Notices of the Royal Astronomical Society, 255*, 119–128. https://doi.org/10.1093/mnras/255.1.119.

Lasky, P. D., Haskell, B., Ravi, V., Howell, E. J., & Coward, D. M. (2014). Nuclear equation of state from observations of short gamma-ray burst remnants. *Physical Review D, 89*, 047302. https://doi.org/10.1103/PhysRevD.89.047302, arXiv:1311.1352.

Lattimer, J. M. (2012). The nuclear equation of state and neutron star masses. *Annual Review of Nuclear and Particle Science, 62*, 485–515. https://doi.org/10.1146/annurev-nucl-102711-095018, arXiv:1305.3510.

Lattimer, J. M., & Prakash, M. (2007). Neutron star observations: Prognosis for equation of state constraints. *Physics Reports, 442*, 109–165. https://doi.org/10.1016/j.physrep.2007.02.003, arXiv:astro-ph/0612440.

Ledoux, P., & Walraven, T. (1958). Variable stars. *Encyclopedia of physics (Handbuch der Physik)* (Vol. 51, pp. 353–604). Berlin: Springer. http://adsabs.harvard.edu/abs/1958HDP....51.353L.

Ledoux, P. (1951). The nonradial oscillations of gaseous stars and the problem of beta Canis Majoris. *The Astrophysical Journal, 114*, 373. https://doi.org/10.1086/145477.

Leibacher, J. W., & Stein, R. F. (1971). A new description of the solar five-minute oscillation. *Astrophysical Letters, 7*, 191–192. http://adsabs.harvard.edu/abs/1971ApL....7.191L.

Leighton, R. B., Noyes, R. W., & Simon, G. W. (1962). Velocity fields in the solar atmosphere I. Preliminary report. *The Astrophysical Journal, 135*, 474. https://doi.org/10.1086/147285.

Levin, Y. (1999). Runaway heating by r-modes of neutron stars in low-mass x-ray binaries. *The Astrophysical Journal, 517*, 328–333. https://doi.org/10.1086/307196, arXiv:astro-ph/9810471.

Lindblom, L. (1992). Determining the nuclear equation of state from neutron-star masses and radii. *The Astrophysical Journal, 398*, 569–573. https://doi.org/10.1086/171882.

Lindblom, L., Owen, B. J., & Morsink, S. M. (1998). Gravitational radiation instability in hot young neutron stars. *Physical Review Letters, 80*, 4843–4846. https://doi.org/10.1103/PhysRevLett.80.4843, arXiv:gr-qc/9803053.

Lombardo, U. & Schulze, H.-J. (2001). Superfluidity in neutron star matter. In D. Blaschke, N. K. Glendenning & A. Sedrakian (Eds.), *Physics of neutron star interiors* (vol. 578). Lecture notes in physics. Berlin: Springer. http://adsabs.harvard.edu/abs/2001LNP...578...30L, arXiv:astro-ph/0012209.

Marshall, F. E., Gotthelf, E. V., Zhang, W., Middleditch, J., & Wang, Q. D. (1998). Discovery of an ultrafast x-ray pulsar in the supernova remnant N157B. *The Astrophysical Journal, 499*, L179–L182. https://doi.org/10.1086/311381, arXiv:astro-ph/9803214.

Mastrano, A., Melatos, A., Reisenegger, A., & Akgün, T. (2011). Gravitational wave emission from a magnetically deformed non-barotropic neutron star. *Monthly Notices of the Royal Astronomical Society, 417*, 2288–2299. https://doi.org/10.1111/j.1365-2966.2011.19410.x, arXiv:1108.0219.

Mazets, E. P., Golenskii, S. V., Ilinskii, V. N., Aptekar, R. L., & Guryan, I. A. (1979). Observations of a flaring X-ray pulsar in Dorado. *Nature, 282*, 587–589. https://doi.org/10.1038/282587a0.

Morsink, S. M. (2002). Nonlinear couplings between r-modes of rotating neutron stars. *The Astrophysical Journal, 571*, 435–446. https://doi.org/10.1086/339858, arXiv:astro-ph/0202051.

Mytidis, A., Coughlin, M. & Whiting, B. (2015). Constraining the r-mode saturation amplitude from a hypothetical detection of r-mode gravitational waves from a newborn neutron star: Sensitivity study. *The Astrophysical Journal, 810*, 27. https://doi.org/10.1088/0004-637X/810/1/27, arXiv:1505.03191.

Oppenheimer, J. R., & Volkoff, G. M. (1939). On massive neutron cores. *Physical Review, 55*, 374–381. https://doi.org/10.1103/PhysRev.55.374.

Ott, C. D., Burrows, A., Thompson, T. A., Livne, E., & Walder, R. (2006). The spin periods and rotational profiles of neutron stars at birth. *The Astrophysical Journal. Supplement Series, 164*, 130–155. https://doi.org/10.1086/500832, arXiv:astro-ph/0508462.

Ou, S., Tohline, J. E., & Lindblom, L. (2004). Nonlinear development of the secular bar-mode instability in rotating neutron stars. *The Astrophysical Journal, 617*, 490–499. https://doi.org/10.1086/425296, arXiv:astro-ph/0406037.

Owen, B. J., Lindblom, L., Cutler, C., Schutz, B. F., Vecchio, A., & Andersson, N. (1998). Gravitational waves from hot young rapidly rotating neutron stars. *Physical Review D*, *58*, 084020. https://doi.org/10.1103/PhysRevD.58.084020, arXiv:gr-qc/9804044.

Palmer, D. M., et al. (2005). A giant γ-ray flare from the magnetar SGR 1806–1820. *Nature*, *434*, 1107–1109. https://doi.org/10.1038/nature03525, arXiv:astro-ph/0503030.

Papaloizou, J. & Pringle, J. E. (1978). Non-radial oscillations of rotating stars and their relevance to the short-period oscillations of cataclysmic variables. *Monthly Notices of the Royal Astronomical Society*, *182*, 423–442. http://adsabs.harvard.edu/abs/1978MNRAS.182..423P.

Passamonti, A., & Glampedakis, K. (2012). Non-linear viscous damping and gravitational wave detectability of the f-mode instability in neutron stars. *Monthly Notices of the Royal Astronomical Society*, *422*, 3327–3338. https://doi.org/10.1111/j.1365-2966.2012.20849.x, arXiv:1112.3931.

Passamonti, A., Gaertig, E., Kokkotas, K. D., & Doneva, D. (2013). Evolution of the f-mode instability in neutron stars and gravitational wave detectability. *Physical Review D*, *87*, 084010. https://doi.org/10.1103/PhysRevD.87.084010, arXiv:1209.5308.

Pekeris, C. L. (1938). Nonradial oscillations of stars. *The Astrophysical Journal*, *88*, 189. https://doi.org/10.1086/143971.

Punturo, M., et al. (2010). The Einstein telescope: A third-generation gravitational wave observatory. *Classical Quantum Gravity*, *27*, 194002. https://doi.org/10.1088/0264-9381/27/19/194002.

Ritter, A. (1879). *Wiedemann's Annalen* (vol. 8, p. 179).

Sathyaprakash, B. S., & Schutz, B. F. (2009). Physics, astrophysics and cosmology with gravitational waves. *Living Reviews in Relativity*, *12*, 2. https://doi.org/10.12942/lrr-2009-2, arXiv:0903.0338.

Sathyaprakash, B., et al. (2012). Scientific objectives of Einstein telescope. *Classical Quantum Gravity*, *29*, 124013. https://doi.org/10.1088/0264-9381/29/12/124013, arXiv:1206.0331.

Schenk, A. K., Arras, P., Flanagan, É. É., Teukolsky, S. A., & Wasserman, I. (2001). Nonlinear mode coupling in rotating stars and the r-mode instability in neutron stars. *Physical Review D*, *65*, 024001. https://doi.org/10.1103/PhysRevD.65.024001, arXiv:gr-qc/0101092.

Shibata, M., & Karino, S. (2004). Numerical evolution of secular bar-mode instability induced by the gravitational radiation reaction in rapidly rotating neutron stars. *Physical Review D*, *70*, 084022. https://doi.org/10.1103/PhysRevD.70.084022, arXiv:astro-ph/0408016.

Somiya, K. (2012). Detector configuration of KAGRA-the Japanese cryogenic gravitational-wave detector. *Classical Quantum Gravity*, *29*, 124007. https://doi.org/10.1088/0264-9381/29/12/124007, arXiv:1111.7185.

Stergioulas, N., Bauswein, A., Zagkouris, K., & Janka, H.-T. (2011). Gravitational waves and non-axisymmetric oscillation modes in mergers of compact object binaries. *Monthly Notices of the Royal Astronomical Society*, *418*, 427–436. https://doi.org/10.1111/j.1365-2966.2011.19493.x, arXiv:1105.0368.

Thomson, W. (1863). Dynamical problems regarding elastic spheroidal shells and spheroids of incompressible liquid. *Philosophical Transactions of the Royal Society of London*, *153*, 583–616. https://doi.org/10.1098/rstl.1863.0028.

Tolman, R. C. (1939). Static solutions of Einstein's field equations for spheres of fluid. *Physical Review*, *55*, 364–373. https://doi.org/10.1103/PhysRev.55.364.

Ulrich, R. K. (1970). The five-minute oscillations on the solar surface. *The Astrophysical Journal*, *162*, 993. https://doi.org/10.1086/150731.

Unnikrishnan, C. S. (2013). IndIGO and Ligo-India scope and plans for gravitational wave research and precision metrology in India. *International Journal of Modern Physics D*, *22*, 1341010. https://doi.org/10.1142/S0218271813410101, arXiv:1510.06059.

Unno, W., Osaki, Y., Ando, H., Saio, H., & Shibahashi, H. (1989). *Nonradial oscillations of stars* (2nd ed.). Tokyo: University of Tokyo Press. http://adsabs.harvard.edu/abs/1989nos.book....U.

Ushomirsky, G., Cutler, C., & Bildsten, L. (2000). Deformations of accreting neutron star crusts and gravitational wave emission. *Monthly Notices of the Royal Astronomical Society*, *319*, 902–932. https://doi.org/10.1046/j.1365-8711.2000.03938.x, arXiv:astro-ph/0001136.

Wagoner, R. V. (1975). Test for the existence of gravitational radiation. *The Astrophysical Journal*, *196*, L63–L65. https://doi.org/10.1086/181745.

Warszawski, L., & Melatos, A. (2012). Gravitational-wave bursts and stochastic background from superfluid vortex avalanches during pulsar glitches. *Monthly Notices of the Royal Astronomical Society, 423*, 2058–2074. https://doi.org/10.1111/j.1365-2966.2012.20977.x, arXiv:1203.4466.

Will, C. M. (2014). The confrontation between general relativity and experiment. *Living Reviews in Relativity, 17*, 4. https://doi.org/10.12942/lrr-2014-4, arXiv:1403.7377.

Yakovlev, D. G., Haensel, P., Baym, G., & Pethick, C. (2013). Lev Landau and the concept of neutron stars. *Physics-Uspekhi, 56*, 289–295. https://doi.org/10.3367/UFNe.0183.201303f.0307, arXiv:1210.0682.

Zink, B., Lasky, P. D., & Kokkotas, K. D. (2012). Are gravitational waves from giant magnetar flares observable? *Physical Review D, 85*, 024030. https://doi.org/10.1103/PhysRevD.85.024030, arXiv:1107.1689.

Chapter 2
The Oscillation Modes: Linear Perturbation Scheme

Starting with the standard equations of hydrodynamics (Sect. 2.1), we are going to derive the linear perturbation formalism (Sect. 2.2), used to obtain the oscillation modes of a star. We will first consider the simple case of a nonrotating star (Sect. 2.3), in which the various classes of modes are defined (Sect. 2.4), namely *polar* and *axial* modes. The former class contains the (fundamental) f-modes, the (pressure) p-modes, and the (buoyancy) g-modes, and the latter comprises the (rotational) r-modes. An additional class of modes, called *hybrid* modes, will also be discussed. For the sake of completeness, we provide the analytic formulae for the eigenfrequencies and eigenfunctions of the modes of a homogeneous star (Sect. 2.5), even though they are not used throughout this study. Finally, we consider the general case of a rotating star (Sect. 2.6), making use of the slow-rotation approximation, i.e., introducing (up to second-order) rotational corrections to the eigenfrequencies.

2.1 The Fluid Equations

Assuming a star which is uniformly rotating with an angular velocity $\mathbf{\Omega}$, then the inviscid fluid equations, in a reference frame rotating with the star, are expressed as

$$\frac{\partial \rho}{\partial t} + \nabla \cdot (\rho \mathbf{v}) = 0, \tag{2.1.1}$$

$$\frac{\partial \mathbf{v}}{\partial t} + (\mathbf{v} \cdot \nabla)\mathbf{v} + 2\mathbf{\Omega} \times \mathbf{v} + \mathbf{\Omega} \times (\mathbf{\Omega} \times \mathbf{r}) = -\frac{\nabla p}{\rho} - \nabla \Phi, \tag{2.1.2}$$

and

$$\nabla^2 \Phi = 4\pi G \rho, \tag{2.1.3}$$

© Springer Nature Switzerland AG 2018
P. Pnigouras, *Saturation of the f-mode Instability in Neutron Stars*,
Springer Theses, https://doi.org/10.1007/978-3-319-98258-8_2

where ρ is the density, p the pressure, Φ the gravitational potential, r the position, v the velocity, and G the gravitational constant. In order to close, the system above has to be supplemented with an equation of state, namely

$$p = p(\rho, \mu). \tag{2.1.4}$$

In neutron stars, μ usually corresponds to entropy or composition. If the star has a finite temperature,[1] it deviates from isentropy, whereas stratification leads to composition gradients, both affecting the pressure distribution throughout the star.

For a star in equilibrium, time derivatives and velocities vanish, so the Euler equation (2.1.2) becomes

$$\frac{\nabla p}{\rho} + \mathbf{\Omega} \times (\mathbf{\Omega} \times r) = -\nabla \Phi. \tag{2.1.5}$$

Alternatively, one can express the centrifugal term as $\mathbf{\Omega} \times (\mathbf{\Omega} \times r) = -\nabla \Phi_{\text{rot}}$, which makes Eq. (2.1.5)

$$\frac{\nabla p}{\rho} = -\nabla (\Phi - \Phi_{\text{rot}}), \tag{2.1.6}$$

where

$$\Phi_{\text{rot}} = \frac{1}{2} |\mathbf{\Omega} \times r|^2. \tag{2.1.7}$$

Thus, $\Phi - \Phi_{\text{rot}}$ is the "effective" gravitational potential, due to the contribution of the centrifugal force in hydrostatic equilibrium. As for the rest of the equations, the continuity equation (2.1.1) vanishes identically, whereas the Poisson equation (2.1.3) and the equation of state (2.1.4) remain unchanged.

2.2 Linear Perturbation Formalism

There are two ways of describing a fluid in hydrodynamics. The evolution of a physical quantity can be defined either at a given position in space (*Eulerian* description), or in a certain fluid element (*Lagrangian* description). Fluid perturbations can be defined using either framework (see, for instance, Lynden-Bell and Ostriker 1967 and Unno et al. 1989, Chap. III).

The Eulerian perturbation, denoted by δ, can be thought of as the perturbation of a physical quantity at a given position and is defined as

$$f(r_0, t) = f_0(r_0) + \delta f(r_0, t), \tag{2.2.1}$$

---

[1]That is, a temperature higher than the Fermi temperature of the system, so that thermal pressure is comparable to degeneracy pressure.

where f_0 denotes the equilibrium value of the physical quantity f, at the given position r_0. On the other hand, the Lagrangian perturbation, denoted by Δ, is the perturbation of this physical quantity in a certain fluid element and can be expressed as

$$f(\boldsymbol{r}, t) = f_0(\boldsymbol{r}_0) + \Delta f(\boldsymbol{r}, t), \qquad (2.2.2)$$

where $\boldsymbol{r}_0$ and $\boldsymbol{r}$ are the initial and final positions of the fluid element respectively. Defining the displacement vector as $\boldsymbol{\xi} = \boldsymbol{r} - \boldsymbol{r}_0$, then the relation between Lagrangian and Eulerian perturbations can be found as

$$\Delta f(\boldsymbol{r}, t) = f(\boldsymbol{r}_0 + \boldsymbol{\xi}, t) - f_0(\boldsymbol{r}_0) = f(\boldsymbol{r}_0, t) + (\boldsymbol{\xi} \cdot \nabla_0) \, f(\boldsymbol{r}, t) - f_0(\boldsymbol{r}_0) + \mathcal{O}\left(\boldsymbol{\xi}^2\right).$$

Thus, to linear order in $\boldsymbol{\xi}$,

$$\Delta f(\boldsymbol{r}, t) = \delta f(\boldsymbol{r}_0, t) + (\boldsymbol{\xi} \cdot \nabla_0) \, f(\boldsymbol{r}, t), \qquad (2.2.3)$$

with $\nabla_0 f(\boldsymbol{r}, t) = \nabla \, f(\boldsymbol{r}, t)|_{r=r_0}$. The subscript 0 for equilibrium quantities will be omitted below.

Imposing "small" perturbations on the equilibrium and retaining only first-order perturbative terms, Eqs. (2.1.1)–(2.1.4) become

$$\frac{\partial \delta \rho}{\partial t} + \nabla \cdot (\rho \delta \boldsymbol{v}) = 0, \qquad (2.2.4)$$

$$\frac{\partial \delta \boldsymbol{v}}{\partial t} + 2\boldsymbol{\Omega} \times \delta \boldsymbol{v} = -\frac{\nabla \delta p}{\rho} + \frac{\nabla p}{\rho^2}\delta\rho - \nabla \delta \Phi, \qquad (2.2.5)$$

$$\nabla^2 \delta \Phi = 4\pi G \delta \rho, \qquad (2.2.6)$$

and

$$\frac{\Delta p}{p} = \Gamma_1 \frac{\Delta \rho}{\rho} + \left(\frac{\partial \ln p}{\partial \ln \mu}\right)_\rho \frac{\Delta \mu}{\mu}, \qquad (2.2.7)$$

where Γ_1 is the *adiabatic exponent* and is defined as

$$\Gamma_1 = \left(\frac{\partial \ln p}{\partial \ln \rho}\right)_\mu. \qquad (2.2.8)$$

We will consider perturbations for which $\Delta \mu \approx 0$. If μ corresponds to entropy, this means that the fluid displacement occurs adiabatically, because the thermal relaxation time scale is much longer than the oscillation period (however, cf. Gualtieri et al. 2004). On the other hand, if μ is associated with composition, then this condition means that the composition of the displaced fluid element is "frozen"; weak interaction processes need much more time than an oscillation period to restore

chemical equilibrium between the displaced fluid element and the surrounding matter (Reisenegger and Goldreich 1992).

By definition, $\Delta v = d\xi/dt = \dot{\xi} + (v \cdot \nabla)\xi$, where the dot indicates partial differentiation with respect to time. But, since we work on the rotating frame, $v = 0$ in
the background. By also using Eq. (2.2.3), this means that $\Delta v = \dot{\xi} = \delta v$. Then, the
perturbed Euler equation (2.2.5) can be written as

$$\ddot{\xi} + \mathcal{B}(\dot{\xi}) + \mathcal{C}(\xi) = 0, \tag{2.2.9}$$

where

$$\mathcal{B}(\xi) = 2\mathbf{\Omega} \times \xi \tag{2.2.10}$$

and

$$\mathcal{C}(\xi) = \frac{\nabla \delta p}{\rho} - \frac{\nabla p}{\rho^2}\delta\rho + \nabla\delta\Phi. \tag{2.2.11}$$

Operator $\mathcal{C}$ can be written in terms of ξ by using Eqs. (2.2.4), (2.2.6), (2.2.7) to
replace the perturbations $\delta\rho$, $\delta\Phi$, and δp, respectively (see, for example, Schenk
et al. 2001, Sect. II B, or Lynden-Bell and Ostriker 1967, Sect. 2.1).

Equation (2.2.9) is the equation of motion for linear perturbations. Seeking harmonic solutions, of the form $\xi(r, t) = \xi(r)e^{i\omega t}$, it becomes

$$-\omega^2\xi + i\omega\mathcal{B}(\xi) + \mathcal{C}(\xi) = 0. \tag{2.2.12}$$

This is the equation which, supplemented with the appropriate boundary conditions,
forms an eigenvalue problem, which needs to be solved in order to obtain the eigenfrequencies ω and eigenfunctions ξ of stellar oscillation modes.

2.3 Nonrotating Stars

In the case of a nonrotating star ($\Omega = 0$), the centrifugal term in Eq. (2.1.5) is zero
and the star is spherically symmetric. Using spherical polar coordinates (r, θ, ϕ),
this means that all equilibrium quantities (i.e., density, pressure, and gravitational
potential) depend only on the radial coordinate r. This, together with the fact that
the Coriolis term vanishes in Eq. (2.2.5), simplifies the calculation greatly.

Based on Unno et al. (1989, Chap. III) and Aerts et al. (2010, Chap. 3), we are
going to derive the eigenvalue equation system and the corresponding boundary
conditions, from which we can get the eigenfrequencies ω and eigenfunctions ξ of
the oscillation modes of a nonrotating star. The fast-track reader may go straight
to Sect. 2.3.7, after taking a look at the most important results of Sects. 2.3.1–
2.3.6, which are: the eigenvalue equation system (2.3.16)–(2.3.18), its boundary
conditions (2.3.24)–(2.3.27), and the expressions for the eigenfunctions of polar and
axial modes, given by Eqs. (2.3.19) and (2.3.53) respectively.

2.3.1 *Separation of Variables

Time Separation

As discussed in the previous section, we can separate the harmonic time dependence of perturbed quantities, i.e., $\delta f(r, \theta, \phi, t) = \delta f(r, \theta, \phi)e^{i\omega t}$. We proceed with splitting the equation of motion (2.2.5) into its radial and horizontal parts, namely

$$-\omega^2 \xi_r + \frac{1}{\rho}\frac{\partial \delta p}{\partial r} + \frac{\partial \delta \Phi}{\partial r} + \frac{\delta \rho}{\rho}\frac{d\Phi}{dr} = 0 \qquad (2.3.1)$$

and

$$-\omega^2 \boldsymbol{\xi}_\perp + \nabla_\perp \left(\frac{\delta p}{\rho} + \delta\Phi \right) = \mathbf{0}, \qquad (2.3.2)$$

where we used Eq. (2.1.5), ξ_r and $\boldsymbol{\xi}_\perp$ are the radial and horizontal components of the displacement vector $\boldsymbol{\xi}$, and the horizontal component of the gradient operator is defined as

$$\nabla_\perp = \frac{1}{r}\left(0, \frac{\partial}{\partial \theta}, \frac{1}{\sin\theta}\frac{\partial}{\partial \phi} \right). \qquad (2.3.3)$$

Using Eq. (2.3.2), the perturbed continuity equation (2.2.4) is written as

$$\frac{\delta \rho}{\rho} + \frac{\xi_r}{\rho}\frac{d\rho}{dr} + \frac{1}{r^2}\frac{\partial}{\partial r}\left(r^2 \xi_r \right) + \frac{1}{\omega^2}\nabla_\perp^2 \left(\frac{\delta p}{\rho} + \delta\Phi \right) = 0, \qquad (2.3.4)$$

where the angular part of the Laplacian operator is given by

$$\nabla_\perp^2 = \frac{1}{r^2 \sin^2\theta}\left[\sin\theta \frac{\partial}{\partial \theta}\left(\sin\theta \frac{\partial}{\partial \theta} \right) + \frac{\partial^2}{\partial \phi^2} \right]. \qquad (2.3.5)$$

In a similar manner, the perturbed Poisson equation (2.2.6) and the perturbed equation of state (2.2.7) become

$$\frac{1}{r^2}\frac{\partial}{\partial r}\left(r^2 \frac{\partial \delta\Phi}{\partial r} \right) + \nabla_\perp^2 \delta\Phi = 4\pi G \delta\rho \qquad (2.3.6)$$

and

$$\frac{\delta\rho}{\rho} = \frac{1}{\Gamma_1}\frac{\delta p}{p} - A\xi_r \qquad (2.3.7)$$

respectively, where we assumed $\Delta\mu \approx 0$ (see Sect. 2.2) and A is the so-called *Schwarzschild discriminant*, given by

$$A = \frac{d\ln\rho}{dr} - \frac{1}{\Gamma_1}\frac{d\ln p}{dr}. \qquad (2.3.8)$$

We can now use Eq. (2.3.7) to eliminate $\delta\rho$ from Eqs. (2.3.4), (2.3.1), and (2.3.6), to obtain

$$\frac{1}{r^2}\frac{\partial}{\partial r}\left(r^2\xi_r\right) + \frac{1}{\Gamma_1}\frac{d\ln p}{dr}\xi_r + \left(\frac{\rho}{\Gamma_1 p} + \frac{\nabla_\perp^2}{\omega^2}\right)\frac{\delta p}{\rho} + \frac{1}{\omega^2}\nabla_\perp^2\delta\Phi = 0, \qquad (2.3.9)$$

$$\frac{1}{\rho}\left(\frac{\partial}{\partial r} + \frac{\rho g}{\Gamma_1 p}\right)\delta p - \left(\omega^2 + gA\right)\xi_r + \frac{\partial\delta\Phi}{\partial r} = 0, \qquad (2.3.10)$$

and

$$\left[\frac{1}{r^2}\frac{\partial}{\partial r}\left(r^2\frac{\partial}{\partial r}\right) + \nabla_\perp^2\right]\delta\Phi - 4\pi G\rho\left(\frac{\delta p}{\Gamma_1 p} - A\xi_r\right) = 0, \qquad (2.3.11)$$

where

$$g = \frac{d\Phi}{dr}, \qquad (2.3.12)$$

which is the local gravitational acceleration.

Angular Separation

Since the coefficients in Eqs. (2.3.9)–(2.3.11) depend only on the radial coordinate r and the only differential operator with respect to the horizontal coordinates θ and ϕ is the angular part of the Laplacian operator $\nabla_\perp^2$, a separation of radial and horizontal variables is possible. The angular dependence of perturbed quantities [say, $f(\theta, \phi)$] has to be an eigenfunction of the angular part of the Laplacian operator, namely

$$r^2\nabla_\perp^2 f(\theta, \phi) = -\Lambda f(\theta, \phi), \qquad (2.3.13)$$

or

$$\frac{1}{\sin\theta}\frac{\partial}{\partial\theta}\left(\sin\theta\frac{\partial f}{\partial\theta}\right) + \frac{1}{\sin^2\theta}\frac{\partial^2 f}{\partial\phi^2} = -\Lambda f,$$

where $-\Lambda$ is the eigenvalue.

The coefficients of this partial differential equation depend only on the colatitude variable θ, so a separation between the angular variables themselves may be pursued. The angular dependence is written as $f(\theta, \phi) = f_1(\theta)f_2(\phi)$ and, substituted to the equation above, gives

$$\frac{1}{f_1(\theta)}\sin\theta\frac{d}{d\theta}\left(\sin\theta\frac{df_1}{d\theta}\right) + \Lambda\sin^2\theta = -\frac{1}{f_2(\phi)}\frac{d^2 f_2}{d\phi^2}.$$

Both sides of this equation have to be equal to a constant (say, α), since they are expressions of different variables. Starting from the right-hand side, we get

$$\frac{d^2 f_2}{d\phi^2} = -\alpha f_2(\phi),$$

which yields the solution

$$f_2(\phi) = e^{\pm\sqrt{-\alpha}\phi}.$$

This solution has to be periodic in ϕ, i.e., satisfy the constraint $f_2(\phi) = f_2(\phi + 2\pi)$, so that it is unique for a certain position in space. This implies that $\alpha = m^2$, where $m \in \mathbb{Z}$. So, the corresponding expression for $f_1(\theta)$ is

$$\frac{d}{dx}\left[(1 - x^2)\frac{df_1}{dx}\right] + \left(\Lambda - \frac{m^2}{1 - x^2}\right) f_1(x) = 0,$$

where $x = \cos\theta$. This is known as the associated Legendre equation and it can be shown that it has a regular solution when $\Lambda = l(l + 1)$, where l is a non-negative integer and $|m| \leq l$. This solution is

$$f_1(\theta) = P_l^m(\cos\theta),$$

where P_l^m are the associated Legendre polynomials (e.g., see Abramowitz and Stegun 1972, Chap. 8).

After normalising the angular dependence function $f(\theta, \phi)$, so that

$$\int_{\theta=0}^{\pi}\int_{\phi=0}^{2\pi} f_l^m(\theta, \phi) f_{l'}^{m'*}(\theta, \phi) \sin\theta d\theta d\phi = \delta_{ll'}\delta_{mm'}, \qquad (2.3.14)$$

where δ_{ab} is the Kronecker delta and the star denotes complex conjugation, we obtain the final expression for $f(\theta, \phi)$, namely

$$f_l^m(\theta, \phi) = (-1)^m\sqrt{\frac{2l + 1}{4\pi}\frac{(l - m)!}{(l + m)!}}P_l^m(\cos\theta)e^{im\phi} \equiv Y_l^m(\theta, \phi), \qquad (2.3.15)$$

which are the well known spherical harmonics $Y_l^m(\theta, \phi)$.

Hence, as expected, perturbations on a spherical star assume the form of spherical harmonics. Replacing ξ_r, δp, and $\delta\Phi$ in Eqs. (2.3.9)–(2.3.11) as $\xi_r(r, \theta, \phi) = \xi_r(r)Y_l^m(\theta, \phi)$ etc. we get

$$\frac{1}{r^2}\frac{d}{dr}(r^2\xi_r) - \frac{\rho g}{\Gamma_1 p}\xi_r + \left[\frac{\rho}{\Gamma_1 p} - \frac{l(l + 1)}{\omega^2 r^2}\right]\frac{\delta p}{\rho} - \frac{l(l + 1)}{\omega^2 r^2}\delta\Phi = 0, \qquad (2.3.16)$$

$$\frac{1}{\rho}\left(\frac{d}{dr} + \frac{\rho g}{\Gamma_1 p}\right)\delta p - (\omega^2 + gA)\xi_r + \frac{d\delta\Phi}{dr} = 0, \qquad (2.3.17)$$

and

$$\frac{1}{r^2}\frac{d}{dr}\left(r^2\frac{d\delta\Phi}{dr}\right) - \frac{l(l + 1)}{r^2}\delta\Phi - 4\pi G\rho\left(\frac{\delta p}{\Gamma_1 p} - A\xi_r\right) = 0. \qquad (2.3.18)$$

We now have a system of three equations for the variables ξ_r, δp, and $\delta\Phi$, which depend only on r. From Eq. (2.3.2) we can derive the expression for the displacement vector $\boldsymbol{\xi}$ as

$$\boldsymbol{\xi}(r, \theta, \phi) = \left[\xi_r(r),\ \xi_h(r)\frac{\partial}{\partial\theta},\ \xi_h(r)\frac{1}{\sin\theta}\frac{\partial}{\partial\phi} \right] Y_l^m(\theta, \phi), \qquad (2.3.19)$$

where

$$\xi_h = \frac{1}{\omega^2 r}\left(\frac{\delta p}{\rho} + \delta\Phi \right). \qquad (2.3.20)$$

2.3.2 *Boundary Conditions

In order to solve Eqs. (2.3.16)–(2.3.18) as an eigenvalue problem, we need some boundary conditions at the centre and the surface of the star.

Near the origin $(r \to 0)$, we have

$$g \to 0,\ \ \rho, p \approx const.,\ \ A \to 0.$$

Equations (2.3.16)–(2.3.18) then become

$$\frac{d}{dr}\left(r^2\xi_r\right) - \frac{l(l+1)}{\omega^2}\left(\frac{\delta p}{\rho} + \delta\Phi \right) \approx 0, \qquad (2.3.21)$$

$$\frac{d}{dr}\left(\frac{\delta p}{\rho} + \delta\Phi \right) - \omega^2\xi_r \approx 0, \qquad (2.3.22)$$

and

$$\frac{d}{dr}\left(r^2\frac{d\delta\Phi}{dr} \right) - l(l+1)\delta\Phi \approx 0. \qquad (2.3.23)$$

The solution of Eq. (2.3.23) is

$$\delta\Phi = c_1 r^l + c_2 r^{-(l+1)},$$

where c_1, c_2 are constants. Since $\delta\Phi$ must be regular at the centre, $c_2 = 0$. This yields the first boundary condition for the centre, which can be alternatively written as

$$\frac{d\delta\Phi}{dr} - \frac{l\delta\Phi}{r} = 0\ \ \text{at}\ \ r \to 0. \qquad (2.3.24)$$

This form is more useful for the dimensionless formulation, presented in Sect. 2.3.3 below. The same practice is applied to the variables ξ_r and $(\delta p/\rho + \delta\Phi)$ in Eqs. (2.3.21) and (2.3.22), and their solutions are

$$\left(\frac{\delta p}{\rho} + \delta\Phi\right) \propto r^l \quad \text{and} \quad \xi_r \propto r^{l-1}.$$

Therefore, the second boundary condition at the centre can be written as

$$\xi_r - \frac{l}{\omega^2 r}\left(\frac{\delta p}{\rho} + \delta\Phi\right) = 0 \quad \text{at} \quad r \to 0. \tag{2.3.25}$$

At the surface ($r = R$; R being the stellar radius) the density and the pressure vanish, so Eq. (2.3.18) is written again like Eq. (2.3.23). The solution which vanishes at infinity is

$$\delta\Phi \propto r^{-(l+1)},$$

so the first boundary condition at the surface is

$$\frac{d\delta\Phi}{dr} + \frac{l+1}{r}\delta\Phi = 0 \quad \text{at} \quad r = R. \tag{2.3.26}$$

The second boundary condition at the surface depends on the treatment of the stellar atmosphere and may, consequently, be quite complicated. Assuming that the star is assigned a definite boundary at $r = R$, then the pressure is zero at the perturbed surface. Thus, the Lagrangian pressure perturbation must vanish at the surface, i.e.,

$$\Delta p = 0 \quad \text{at} \quad r = R. \tag{2.3.27}$$

Equations (2.3.16)–(2.3.18), together with the boundary conditions (2.3.24)–(2.3.27), form a boundary value problem with ω^2 as an eigenvalue. Each solution corresponds to a mode of oscillation, with eigenfrequency ω and eigenfunctions ($\xi_r, \delta p, \delta\Phi$).

2.3.3 *Dimensionless Formulation

A particularly helpful practice, especially when numerical calculations are involved, is the development of a dimensionless formulation for the equations governing the problem. We introduce the following dimensionless variables:

$$y_1 = \frac{\xi_r}{r}, \tag{2.3.28}$$

$$y_2 = \frac{1}{gr}\left(\frac{\delta p}{\rho} + \delta\Phi\right), \tag{2.3.29}$$

$$y_3 = \frac{1}{gr}\delta\Phi, \tag{2.3.30}$$

and

$$y_4 = \frac{1}{g} \frac{d\delta\Phi}{dr}.$$ (2.3.31)

By also defining the dimensionless radius

$$x = r/R,$$ (2.3.32)

we can rewrite Eqs. (2.3.16)–(2.3.18) as four first-order differential equations, namely,

$$x\frac{dy_1}{dx} = (V_g - 3) y_1 + \left[\frac{l(l+1)}{c_1\tilde{\omega}^2} - V_g\right] y_2 + V_g y_3,$$ (2.3.33)

$$x\frac{dy_2}{dx} = (c_1\tilde{\omega}^2 - A^*) y_1 + (A^* - U + 1) y_2 - A^* y_3,$$ (2.3.34)

$$x\frac{dy_3}{dx} = (1 - U) y_3 + y_4,$$ (2.3.35)

and

$$x\frac{dy_4}{dx} = U A^* y_1 + U V_g y_2 + \left[l(l+1) - U V_g\right] y_3 - U y_4,$$ (2.3.36)

where

$$V_g = \frac{V}{\Gamma_1} = -\frac{1}{\Gamma_1}\frac{d \ln p}{d \ln r} = \frac{\rho g r}{\Gamma_1 p},$$ (2.3.37)

$$U = \frac{d \ln M_r}{d \ln r} = \frac{4\pi\rho r^3}{M_r},$$ (2.3.38)

$$c_1 = \left(\frac{r}{R}\right)^3 \frac{M}{M_r},$$ (2.3.39)

$$A^* = -r A,$$ (2.3.40)

and

$$\tilde{\omega}^2 = \frac{\omega^2}{GM/R^3},$$ (2.3.41)

with

$$M_r = \int_0^r 4\pi\rho r^2 dr,$$ (2.3.42)

denoting the mass enclosed inside a radius r, and M being the total mass of the star.

Accordingly, the boundary conditions (2.3.24)–(2.3.27) are rewritten as

$$ly_3 - y_4 = 0, \tag{2.3.43}$$

$$c_1 \tilde{\omega}^2 y_1 - ly_2 = 0, \tag{2.3.44}$$

at the centre ($x \to 0$), and

$$(l+1)y_3 + y_4 = 0, \tag{2.3.45}$$

$$y_1 - y_2 + y_3 = 0, \tag{2.3.46}$$

at the surface ($x \to 1$).

2.3.4 *The Cowling Approximation

A useful way to simplify the calculation of mode eigenfrequencies and eigenfunctions is to neglect the perturbation of the gravitational potential, i.e., $\delta\Phi \approx 0$. This is known as the *Cowling approximation* (Cowling 1941) and is mainly valid for modes with large degrees l and overtones n (see Sect. 2.4 below). The major advantage of this technique is that it reduces the order of the system (2.3.16)–(2.3.18) by two, with a corresponding reduction in the number of boundary conditions. In particular, the last terms of Eqs. (2.3.16) and (2.3.17) are ignored, while Eqs. (2.3.18), (2.3.24), and (2.3.26) are no longer needed.

Applying a transformation of the variables ξ_r and δp to the new variables $\tilde{\xi}$ and $\tilde{\eta}$, defined by

$$\tilde{\xi} = r^2 \xi_r \exp\left(-\int_0^r \frac{\rho g}{\Gamma_1 p} dr\right)$$

and

$$\tilde{\eta} = \frac{\delta p}{\rho} \exp\left(\int_0^r A\,dr\right) = \omega^2 r \xi_h \exp\left(\int_0^r A\,dr\right)$$

respectively, Eqs. (2.3.16) and (2.3.17) result in the simple canonical form

$$\frac{d\tilde{\xi}}{dr} = h(r)\left[\frac{l(l+1)}{\omega^2} - \frac{\rho r^2}{\Gamma_1 p}\right]\tilde{\eta} \tag{2.3.47}$$

and

$$\frac{d\tilde{\eta}}{dr} = \frac{1}{r^2 h(r)}\left(\omega^2 + gA\right)\tilde{\xi}, \tag{2.3.48}$$

where

$$h(r) = \exp\left[-\int_0^r \left(A + \frac{\rho g}{\Gamma_1 p}\right) dr\right].$$

If we further define the speed of sound as

$$c_s = \sqrt{\frac{\Gamma_1 p}{\rho}},$$

then Eqs. (2.3.47) and (2.3.48) can take the form

$$\frac{d\tilde{\xi}}{dr} = h(r)\frac{r^2}{c_s^2}\left(\frac{L_l^2}{\omega^2} - 1\right)\tilde{\eta} \qquad (2.3.49)$$

and

$$\frac{d\tilde{\eta}}{dr} = \frac{1}{r^2 h(r)}\left(\omega^2 - N^2\right)\tilde{\xi}, \qquad (2.3.50)$$

where N is known as the Brunt-Väisälä frequency and L_l is the so-called Lamb frequency, given by

$$N^2 = -gA$$

and

$$L_l^2 = \frac{l(l+1)c_s^2}{r^2}$$

respectively. Also known as buoyancy frequency, N is the angular frequency at which a vertically displaced fluid parcel will oscillate, around its equilibrium position, within a stratified environment.

2.3.5 *Mode Trapping

By performing a local treatment of Eqs. (2.3.49) and (2.3.50), their coefficients can be treated as constants. Then, joining the two equations, we get

$$\frac{d^2\tilde{\xi}}{dr^2} + \frac{1}{\omega^2 c_s^2}\left(\omega^2 - L_l^2\right)\left(\omega^2 - N^2\right)\tilde{\xi} = 0.$$

This is a Helmholtz equation,[2] so

---

[2]The Helmholtz equation describes the spatial part of the wave equation and its general form is $\nabla^2 u(r) + k^2 u(r) = 0$, where k represents the wavenumber.

$$k_r^2 = \frac{1}{\omega^2 c_s^2} \left(\omega^2 - L_l^2\right)\left(\omega^2 - N^2\right) \tag{2.3.51}$$

and

$$\tilde{\xi}(r), \ \tilde{\eta}(r) \propto e^{ik_r r},$$

where k_r represents the radial wavenumber. Thus, Eq. (2.3.51) is a dispersion relation. The following cases might occur:

$$\left.\begin{array}{c} \omega^2 > L_l^2, N^2 \\ \omega^2 < L_l^2, N^2 \end{array}\right\} \Rightarrow k_r^2 > 0,$$

or

$$\left.\begin{array}{c} L_l^2 < \omega^2 < N^2 \\ N^2 < \omega^2 < L_l^2 \end{array}\right\} \Rightarrow k_r^2 < 0.$$

If $k_r^2 > 0$, the radial wavenumber is real, so waves can propagate in the radial direction or, equivalently, the variables display an oscillatory behaviour. On the other hand, if $k_r^2 < 0$, the radial wavenumber is imaginary and the variables are exponentially damped. These modes are often called *evanescent*.

Since the angular dependence of the oscillation variables is of the spherical harmonic type, they all satisfy Eq. (2.3.13). This is a Helmholtz equation as well, with the help of which we can introduce the horizontal wavenumber

$$k_h^2 = \frac{l(l+1)}{r^2} = \frac{L_l^2}{c_s^2}. \tag{2.3.52}$$

Points where $k_r = 0$ indicate that there is no motion in the radial direction for a certain mode, which only travels in the horizontal direction with a wavenumber k_h, given by Eq. (2.3.52). These points are known as turning points and imply that the various modes are actually trapped in specific stellar regions, namely regions where either $\omega^2 > L_l^2, N^2$ or $\omega^2 < L_l^2, N^2$. Pressure and buoyancy modes, which will be introduced later in Sect. 2.4, are confined in the high- and low-frequency region respectively. In general, the turning points can be determined by the equation $k_r = 0$, which yields $\omega = L_l$ or $\omega = N$. From these two equations we acquire the so-called propagation diagram of the star (for instance, see Fig. 1 in Osaki 1975 and Fig. 15.2 in Unno et al. 1989).

Mode trapping is visualised in a more intuitive way in Fig. 2.1, where a ray representation is used for pressure and buoyancy waves. As the pressure waves propagate into the star, the deeper parts of the wavefronts experience a higher sound speed and, therefore, travel faster. Hence, the direction of propagation is bent, in the same way as light rays are refracted when experiencing a higher speed of light in a medium.

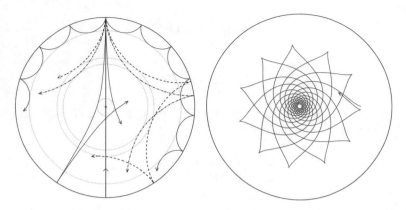

Fig. 2.1 Propagation of pressure (*left*) and buoyancy (*right*) waves, in a Sun-like star. Here, each ray represents a wave and is always vertical to the corresponding wavefront. The pressure (sound) rays are bent by the increase in sound speed with depth, until they reach the inner turning point (indicated by the dotted circles). At the surface, the waves are reflected due to the rapid decrease in density. On the other hand, the buoyancy waves are confined around the centre of the star. The pressure waves correspond to modes with a frequency of 3 mHz and degrees (in order of increasing penetration depth) $l = 75, 25, 20$, and 2. The line passing through the centre illustrates the radial mode. The buoyancy wave corresponds to a mode with frequency $190 \, \mu$Hz and degree $l = 5$. Credit: Cunha et al. (2007). Reprinted by permission from Springer Customer Service Centre GmbH: Springer Nature, © 2007

2.3.6 *A Trivial Solution

In the equations derived so far in this section we implicitly assumed that the eigenvalues are nonzero. However, there is also a trivial solution, for which $\omega = \xi_r = \delta\rho = \delta p = \delta\Phi = 0$, but $\boldsymbol{\xi}_\perp \neq \mathbf{0}$.

In order to derive Eq. (2.3.4), we used Eq. (2.3.2) to calculate the divergence of $\boldsymbol{\xi}_\perp$, assuming of course that δp, $\delta\Phi \neq 0$. If we take the trivial solution above, we get

$$\nabla_\perp \cdot \boldsymbol{\xi}_\perp = 0,$$

or

$$\boldsymbol{\xi}_\perp = \nabla_\perp \times (U \boldsymbol{e}_r),$$

where $\boldsymbol{e}_r$ is the radial unit vector and U an arbitrary scalar function. Expanding U into spherical harmonics, we get

$$U = \sum_l \sum_{m=-l}^{l} U_l^m(r) Y_l^m(\theta, \phi).$$

Then, replacing into the equation for $\boldsymbol{\xi}_\perp$, we obtain, for a single "mode",

$$\xi(r, \theta, \phi) = \frac{U_l^m(r)}{r} \left(0, \ \frac{1}{\sin\theta} \frac{\partial}{\partial\phi}, \ -\frac{\partial}{\partial\theta} \right) Y_l^m(\theta, \phi). \tag{2.3.53}$$

This represents a steady eddy motion of the fluid, confined on spherical surfaces. In a rotating star, these modes become oscillatory, driven by the Coriolis force, as we will discuss in Sect. 2.4.

2.3.7 Mode Orthogonality, Decomposition, and Energy

For a nonrotating star, the eigenvalue equation (2.2.12) becomes

$$\mathcal{C}(\xi) = \omega^2 \xi. \tag{2.3.54}$$

Operator $\mathcal{C}$ is Hermitian[3] (Lynden-Bell and Ostriker 1967), which means that the solutions to Eq. (2.3.54) are orthogonal, i.e.,

$$\langle \xi_\alpha, \xi_\beta \rangle \equiv \int \xi_\alpha^* \cdot \xi_\beta \rho \mathrm{d}^3 r = I_\alpha \delta_{\alpha\beta}, \tag{2.3.55}$$

where the indices correspond to different solutions and the parameter I_α can be thought of as the "moment of inertia" of the perturbation. Replacing the eigenfunction of polar modes (2.3.19) in Eq. (2.3.55) and using the orthogonality relation of spherical harmonics (2.3.14), we obtain

$$
\begin{aligned}
I_\alpha = {} & \int_0^R \xi_r^2 \rho r^2 \mathrm{d}r \iint Y_l^m (\theta, \phi) \, Y_l^{m*} (\theta, \phi) \sin\theta \mathrm{d}\theta \mathrm{d}\phi \\
& + \int_0^R \xi_h^2 \rho r^2 \mathrm{d}r \iint \left(\frac{\partial Y_l^m}{\partial\theta} \frac{\partial Y_l^{m*}}{\partial\theta} + \frac{1}{\sin^2\theta} \frac{\partial Y_l^m}{\partial\phi} \frac{\partial Y_l^{m*}}{\partial\phi} \right) \sin\theta \mathrm{d}\theta \mathrm{d}\phi \\
= {} & \int_0^R \left[\xi_r^2 + l(l+1)\xi_h^2 \right] \rho r^2 \mathrm{d}r.
\end{aligned} \tag{2.3.56}
$$

Equation (2.3.55) implies that a general perturbation ξ can be decomposed as

$$\xi(r, t) = \sum_\alpha q_\alpha(t)\xi_\alpha(r)e^{i\omega_\alpha t}, \tag{2.3.57}$$

where $q_\alpha(t)$ is the *amplitude coefficient* for the mode ξ_α. Furthermore, we may define an energy for the perturbations as (Schenk et al. 2001)

---

[3]Defining the inner product on the space of complex vector functions ξ as $\langle \xi, \xi' \rangle = \int \xi \cdot \xi' \rho \mathrm{d}^3 r$, with the density being the weight function, an operator is Hermitian if, for any ξ, ξ', it satisfies the relation $\langle \xi, \mathcal{C} \cdot \xi' \rangle = \langle \mathcal{C} \cdot \xi, \xi' \rangle$.

$$E_\alpha = 2I_\alpha \omega_\alpha^2 |q_\alpha|^2. \tag{2.3.58}$$

Since linear perturbations yield the linear homogeneous differential equation (2.3.54), the eigenfunctions ξ_α are determined only to within a constant factor. This suggests that, if we were to compare their amplitudes q_α or energies E_α, we would first need to normalise them in the same manner. We set the mode energy at unit amplitude equal to some arbitrary value E_{unit}, namely

$$E_{\text{unit}} = 2I_\alpha \omega_\alpha^2. \tag{2.3.59}$$

Then, Eq. (2.3.58) becomes

$$E_\alpha = |q_\alpha|^2 E_{\text{unit}}. \tag{2.3.60}$$

Now, mode eigenfunctions can be compared against the same standards. Since the mode energy should not depend on the normalisation, a different normalisation choice E'_{unit} would just rescale the amplitudes, according to the relation

$$|q'_\alpha|^2 E'_{\text{unit}} = |q_\alpha|^2 E_{\text{unit}}. \tag{2.3.61}$$

In principle, mode amplitudes are regulated by initial conditions. However, as we will see in Chap. 4, in some cases they can also be uniquely determined by saturation mechanisms, due to nonlinear mode coupling.

2.4   Classes of Modes

According to the analysis presented in Sect. 2.3, an oscillation mode in a nonrotating star is uniquely described by a single spherical harmonic Y_l^m. The *degree l* denotes the total number of nodal lines on the stellar surface along which no motion occurs, with the fluid on each side oscillating in antiphase. The *order m* characterises the azimuthal dependence of the spherical harmonic, $e^{im\phi}$, so $|m|$ determines the number of longitudinal nodal lines. The number of latitudinal nodal lines is, accordingly, $l - |m|$ (see Fig. 1.1).

Expanding an arbitrary perturbation in vector spherical harmonics, we get

$$\boldsymbol{\xi}(r, \theta, \phi) = \sum_l \sum_{m=-l}^{l} \left[W_l^m(r) Y_l^m(\theta, \phi) \boldsymbol{e}_r + V_l^m(r) \nabla Y_l^m(\theta, \phi) \right.$$
$$\left. + U_l^m(r) \nabla Y_l^m(\theta, \phi) \times \boldsymbol{e}_r \right]. \tag{2.4.1}$$

We can distinguish two types of modes in nonrotating stars:

1. *Polar*, or *spheroidal*, modes, for which $U_l^m = 0$.
2. *Axial*, or *toroidal*, modes, for which $W_l^m = V_l^m = 0$.

2.4.1 Polar Modes

Polar modes constitute the "regular" mode spectrum of the star. Their eigenfrequencies are finite in the nonrotating limit and their eigenfunctions are given by Eq. (2.3.19). Apart from the degree l and order m, they are also assigned an *overtone* n, describing the number of radial nodes in their eigenfunctions. In a spherically symmetric star, these are concentric spherical shells within the star, with the fluid oscillating in antiphase on each side. We have three types of polar modes:

I. *Fundamental*, or f-modes: Fundamental oscillations of the fluid, with no nodes in their eigenfunctions ($n = 0$). Their behaviour resembles that of surface gravity waves, like, for instance, sea waves, caused by the density discontinuity at the surface.
II. *Pressure*, or p-modes: Acoustic waves, where the pressure gradient acts as the restoring force. They are the high-frequency overtones of the f-mode ($n \geq 1$), with their eigenfrequency increasing as the overtone increases.
III. *Gravity*, or g-modes: Gravity waves, where the restoring force is buoyancy. They are the low-frequency overtones of the f-mode ($n \geq 1$), with their eigenfrequency decreasing and approaching zero as the overtone increases. Henceforth, to avoid confusion with gravitational waves, we will refer to them either as buoyancy modes or just g-modes.

Parametrising deviations from isentropy or stratification in the star through the variable μ (see Sect. 2.1), then buoyancy effects arise when the Lagrangian perturbation of μ is zero, namely, when the oscillation of a fluid element occurs adiabatically or its composition is frozen, respectively (see Sect. 2.2). This leads to a density difference between the displaced fluid element and its surroundings, which drives it back to its equilibrium position (Finn 1986, 1987; see also McDermott et al. 1983, McDermott 1990, Strohmayer 1993, and Miniutti et al. 2003). This behaviour can be described by the Schwarzschild discriminant A, defined by Eq. (2.3.8), which determines the degree of convective stability ($A < 0$) or instability ($A > 0$). Replacing the equation of state (2.1.4) in Eq. (2.3.8), we get

$$A = -\frac{1}{\Gamma_1}\left(\frac{\partial \ln p}{\partial \ln \mu}\right)_\rho \frac{\mathrm{d}\ln \mu}{\mathrm{d}r}.$$

Usually, it is convenient to parametrise our ignorance about the equation of state using a *polytrope*, namely an equation of state of the form

$$p = K\rho^\Gamma, \tag{2.4.2}$$

where K is the polytropic constant and Γ is the *polytropic exponent* (see Appendix A for a review). Replacing Eq. (2.4.2) in Eq. (2.3.8), we obtain

$$A = \frac{\Gamma_1 - \Gamma}{\Gamma_1}\frac{\mathrm{d}\ln \rho}{\mathrm{d}r}. \tag{2.4.3}$$

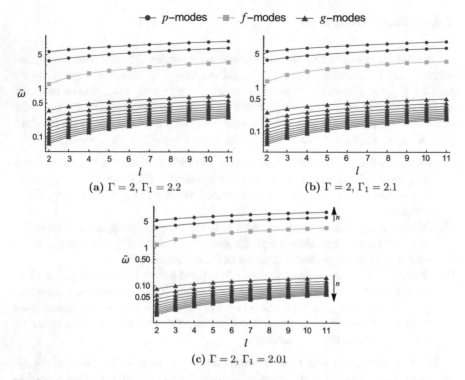

Fig. 2.2 Polar mode spectrum of a star obeying a polytropic equation of state with a polytropic exponent $\Gamma = 2$. Mode eigenfrequencies, normalised as $\tilde{\omega} = \omega/\sqrt{GM/R^3}$, are plotted against the mode degree l. The directions of increasing overtone n are also shown. The adiabatic exponent Γ_1 is decreased from **a** $\Gamma_1 = 2.2$, to **b** $\Gamma_1 = 2.1$, to **c** $\Gamma_1 = 2.01$, thus pushing g-mode eigenfrequencies closer to zero

Then, $\Gamma < \Gamma_1$ denotes convective stability, thus we get the g-modes as described above. If $\Gamma > \Gamma_1$, the star is unstable to convective phenomena, in which case the buoyancy force tends to increase the displacement of a fluid element. Such displacements are often called g^--modes —as opposed to the convectively stable g^+-modes, or simply g-modes.

A sample of the polar mode spectrum of a star obeying various polytropic equations of state is presented in Table 2.1. The eigenfrequencies of a polytrope with $\Gamma = 2$ are plotted in Fig. 2.2, for three different values of Γ_1. We notice that, as the difference $\Gamma_1 - \Gamma$ is decreased, g-mode eigenfrequencies are pushed towards zero, because convective phenomena become less pronounced. The eigenfunctions of a few polar modes with $l = 2$ are presented in Fig. 2.3, for a polytropic star with $\Gamma = 2$ and $\Gamma_1 = 2.1$, where both the radial and horizontal components of Eq. (2.3.19), $\xi_r(r)$ and $\xi_h(r)$, are plotted, with the modes normalised according to Eq. (2.3.59), so that their energy at unit amplitude equals the rest mass energy of the star, namely $E_{\text{unit}} = Mc^2$, c being the speed of light.

Table 2.1 Eigenfrequencies of polar modes, normalised as $\tilde{\omega} = \omega/\sqrt{GM/R^3}$, given for different values of the mode degree l, for a star obeying various polytropic equations of state. The subscript in p- and g-modes denotes their overtone n

(a) $\Gamma = 2$, $\Gamma_1 = 2.2$

l	2	3	4	5	6	7	8	9	10	11
p_2	5.7300	6.2994	6.8045	7.2620	7.6831	8.0752	8.4437	8.7927	9.1249	9.4430
p_1	3.6928	4.2095	4.6508	5.0410	5.3943	5.7196	6.0228	6.3081	6.5784	6.8356
f	1.2284	1.7008	2.0396	2.3135	2.5488	2.7584	2.9495	3.1258	3.2699	3.4472
g_1	0.3515	0.4236	0.4792	0.5247	0.5636	0.5978	0.6286	0.6568	0.6830	0.7071
g_2	0.2434	0.3058	0.3564	0.3990	0.4357	0.4680	0.4969	0.5231	0.5470	0.5690
g_3	0.1875	0.2413	0.2864	0.3254	0.3595	0.3899	0.4172	0.4421	0.4649	0.4860
g_4	0.1529	0.1999	0.2403	0.2758	0.3073	0.3358	0.3616	0.3852	0.4070	0.4272
g_5	0.1293	0.1708	0.2073	0.2397	0.2690	0.2956	0.3199	0.3423	0.3631	0.3824
g_6	0.1120	0.1493	0.1824	0.2123	0.2394	0.2643	0.2872	0.3085	0.3283	0.3468
g_7	0.0989	0.1326	0.1630	0.1906	0.2159	0.2392	0.2609	0.2810	0.2999	0.3176
g_8	0.0885	0.1194	0.1473	0.1730	0.1967	0.2186	0.2391	0.2583	0.2763	0.2932
g_9	0.0802	0.1085	0.1345	0.1584	0.1806	0.2014	0.2208	0.2390	0.2562	0.2725
g_{10}	0.0732	0.0995	0.1237	0.1461	0.1671	0.1867	0.2051	0.2225	0.2389	0.2545

(b) $\Gamma = 2$, $\Gamma_1 = 2.1$

l	2	3	4	5	6	7	8	9	10	11
p_2	5.5743	6.1357	6.6331	7.0832	7.4970	7.8820	8.2437	8.5860	8.9118	9.2233
p_1	3.5785	4.0932	4.5309	4.9168	5.2654	5.5860	5.8845	6.1652	6.4310	6.6833
f	1.2277	1.6996	2.0381	2.3120	2.5474	2.7570	2.9481	3.1245	3.2862	3.3853
g_1	0.2566	0.3081	0.3479	0.3805	0.4083	0.4329	0.4550	0.4752	0.4940	0.5119
g_2	0.1770	0.2220	0.2586	0.2893	0.3158	0.3391	0.3599	0.3788	0.3961	0.4121
g_3	0.1361	0.1750	0.2077	0.2358	0.2605	0.2824	0.3022	0.3202	0.3367	0.3519
g_4	0.1109	0.1449	0.1741	0.1998	0.2226	0.2432	0.2619	0.2790	0.2947	0.3093
g_5	0.0937	0.1238	0.1502	0.1737	0.1948	0.2140	0.2316	0.2479	0.2629	0.2769
g_6	0.0812	0.1081	0.1321	0.1537	0.1734	0.1914	0.2080	0.2234	0.2377	0.2511
g_7	0.0716	0.0961	0.1180	0.1380	0.1563	0.1732	0.1889	0.2035	0.2171	0.2300
g_8	0.0641	0.0864	0.1067	0.1253	0.1424	0.1583	0.1731	0.1870	0.2000	0.2123
g_9	0.0580	0.0786	0.0974	0.1147	0.1308	0.1458	0.1598	0.1730	0.1855	0.1972
g_{10}	0.0530	0.0721	0.0896	0.1058	0.1209	0.1351	0.1485	0.1611	0.1730	0.1843

(c) $\Gamma = 2$, $\Gamma_1 = 2.01$

l	2	3	4	5	6	7	8	9	10	11
p_2	5.4305	5.9849	6.4754	6.9187	7.3259	7.7046	8.0600	8.3963	8.7162	9.0219
p_1	3.4733	3.9866	4.4213	4.8033	5.1479	5.4642	5.7584	6.0349	6.2962	6.5457
f	1.2270	1.6984	2.0367	2.3105	2.5459	2.7556	2.9468	3.1257	3.2894	3.4487
g_1	0.0836	0.1001	0.1128	0.1232	0.1321	0.1400	0.1470	0.1535	0.1596	0.1655

(continued)

Table 2.1 (continued)

g_2	0.0574	0.0720	0.0838	0.0937	0.1022	0.1097	0.1164	0.1225	0.1281	0.1332
g_3	0.0441	0.0567	0.0672	0.0763	0.0843	0.0914	0.0978	0.1036	0.1089	0.1138
g_4	0.0359	0.0469	0.0564	0.0646	0.0720	0.0787	0.0847	0.0902	0.0953	0.1000
g_5	0.0303	0.0401	0.0486	0.0562	0.0630	0.0692	0.0749	0.0802	0.0850	0.0895
g_6	0.0263	0.0350	0.0427	0.0497	0.0561	0.0619	0.0673	0.0722	0.0769	0.0812
g_7	0.0232	0.0311	0.0382	0.0446	0.0506	0.0560	0.0611	0.0658	0.0702	0.0744
g_8	0.0207	0.0280	0.0345	0.0405	0.0460	0.0512	0.0560	0.0605	0.0647	0.0686
g_9	0.0188	0.0254	0.0315	0.0371	0.0423	0.0471	0.0517	0.0559	0.0600	0.0638
g_{10}	0.0171	0.0233	0.0290	0.0342	0.0391	0.0437	0.0480	0.0521	0.0559	0.0596

(d) $\Gamma = 3$, $\Gamma_1 = 3.1$

l	2	3	4	5	6	7	8	9	10	11
p_2	7.0082	7.7283	8.3758	8.9686	9.5185	10.0336	10.5198	10.9815	11.4221	11.8443
p_1	4.3897	5.0078	5.5473	6.0316	6.4747	6.8857	7.2706	7.6341	7.9793	8.3083
f	1.0432	1.4903	1.8267	2.1056	2.3483	2.5657	2.7643	2.9481	3.1209	3.2643
g_1	0.1311	0.1605	0.1844	0.2049	0.2230	0.2394	0.2545	0.2684	0.2815	0.2938
g_2	0.0868	0.1100	0.1294	0.1463	0.1612	0.1747	0.1870	0.1984	0.2090	0.2190
g_3	0.0656	0.0849	0.1014	0.1160	0.1290	0.1408	0.1517	0.1617	0.1711	0.1799
g_4	0.0530	0.0695	0.0839	0.0967	0.1084	0.1190	0.1288	0.1379	0.1464	0.1545
g_5	0.0445	0.0589	0.0717	0.0833	0.0938	0.1035	0.1125	0.1208	0.1287	0.1361
g_6	0.0383	0.0512	0.0627	0.0732	0.0828	0.0917	0.1000	0.1078	0.1152	0.1221
g_7	0.0337	0.0453	0.0558	0.0654	0.0743	0.0825	0.0902	0.0975	0.1044	0.1109
g_8	0.0301	0.0406	0.0502	0.0591	0.0673	0.0750	0.0823	0.0891	0.0956	0.1017
g_9	0.0272	0.0369	0.0457	0.0540	0.0616	0.0688	0.0756	0.0821	0.0882	0.0940
g_{10}	0.0248	0.0337	0.0420	0.0496	0.0568	0.0636	0.0700	0.0761	0.0819	0.0875

(e) $\Gamma = 3$, $\Gamma_1 = 3.01$

l	2	3	4	5	6	7	8	9	10	11
p_2	6.8952	7.6063	8.2456	8.8308	9.3735	9.8817	10.3614	10.8168	11.2514	11.6678
p_1	4.3110	4.9234	5.4573	5.9362	6.3743	6.7804	7.1606	7.5196	7.8604	8.1855
f	1.0432	1.4902	1.8266	2.1055	2.3482	2.5656	2.7642	2.9482	3.1222	3.3075
g_1	0.0422	0.0516	0.0593	0.0658	0.0716	0.0769	0.0817	0.0862	0.0904	0.0944
g_2	0.0279	0.0353	0.0416	0.0470	0.0518	0.0561	0.0600	0.0637	0.0671	0.0703
g_3	0.0211	0.0273	0.0326	0.0372	0.0414	0.0452	0.0487	0.0519	0.0549	0.0578
g_4	0.0170	0.0223	0.0269	0.0311	0.0348	0.0382	0.0413	0.0443	0.0470	0.0496
g_5	0.0143	0.0189	0.0230	0.0267	0.0301	0.0332	0.0361	0.0388	0.0413	0.0437
g_6	0.0123	0.0164	0.0201	0.0235	0.0266	0.0294	0.0321	0.0346	0.0370	0.0392
g_7	0.0108	0.0145	0.0179	0.0210	0.0238	0.0265	0.0290	0.0313	0.0335	0.0356
g_8	0.0097	0.0130	0.0161	0.0190	0.0216	0.0241	0.0264	0.0286	0.0307	0.0326
g_9	0.0087	0.0118	0.0147	0.0173	0.0198	0.0221	0.0243	0.0263	0.0283	0.0302
g_{10}	0.0080	0.0108	0.0135	0.0159	0.0182	0.0204	0.0225	0.0244	0.0263	0.0281

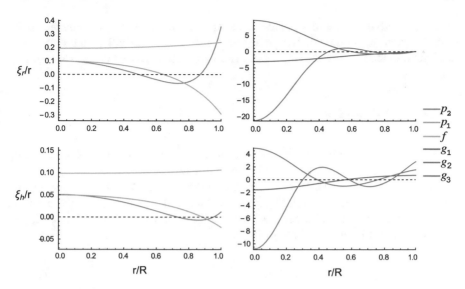

Fig. 2.3 Eigenfunctions of $l = 2$ polar modes, of a star obeying a polytropic equation of state, with a polytropic exponent $\Gamma = 2$ and an adiabatic exponent $\Gamma_1 = 2.1$. The subscript in p- and g-modes denotes their overtone n. The radial and horizontal components of the eigenfunction, $\xi_r(r)$ and $\xi_h(r)$, divided by r, are plotted against the dimensionless radius. The modes have been normalised so that their energy (at unit amplitude) equals the rest mass energy of the star Mc^2

Eigenfrequencies of polar modes in different polytropes can be found in Table I (and Table II, where the Cowling approximation is used; see Sect. 2.3.4) of Robe (1968). Asymptotic representations of g- and p-modes were obtained by Smeyers et al. (1995, 1996) and Smeyers (2003, 2006). For a general description of nonradial oscillations in nonrotating neutron stars, including modes from a solid crust (which we ignore), see McDermott et al. (1988). Stellar pulsation in relativistic stars has also been broadly studied, since the pioneering work of Thorne and Campolattaro (1967). Some interesting studies, among the extensive literature, include Detweiler (1975), Lindblom and Detweiler (1983), Detweiler and Lindblom (1985), Kokkotas and Schutz (1986, 1992), where w-modes (see Sect. 1.4) are introduced, the review by Kokkotas and Schmidt (1999), Ferrari et al. (2003), where the oscillation modes of newborn neutron stars are computed, Gualtieri et al. (2014), where the effects of superfluidity are taken into account, and Krüger et al. (2015), where the influence of internal composition, temperature, and a solid crust is considered.

It should be noted that Eq. (2.3.54) [or, equivalently, the system (2.3.16)–(2.3.18)] does not depend on the azimuthal order m. This is a consequence of the spherical symmetry of the equilibrium state, which demands that the results be independent from the choice of polar axis for the coordinate system. As a result, the eigenvalue ω^2 degenerates $(2l + 1)$-fold with respect to m ($m = -l, -l + 1, \ldots, 0, \ldots, l - 1, l$). The exponential dependence of the mode has the form $e^{i(m\phi + \omega t)}$, which shows that, for nonzero m, the pattern of the oscillation propagates with a phase velocity $-\omega/m$

in the azimuthal direction; a mode with positive m represents a wave travelling in one direction, while a mode with a negative m is a wave travelling in the opposite direction. In the nonrotating limit, there is no distinction between these two modes.

2.4.2   Axial Modes

As shown in Sect. 2.3.6, there is a class of modes which are trivial in a nonrotating star. Their eigenfrequencies are zero in the nonrotating limit and their eigenfunctions are given by Eq. (2.3.53). However, when rotation is introduced, these modes are driven by the Coriolis force and become oscillatory. They are called *rotational*, or *r*-modes, and their eigenfrequencies on the rotating frame, to first order in Ω, are (Papaloizou and Pringle 1978; Saio 1982)

$$\omega = \frac{2m\Omega}{l(l+1)}. \tag{2.4.4}$$

These modes are of the inertial type, like Rossby waves on the Earth's atmosphere and oceans, and comprise horizontal fluid motions, where only the horizontal component of the velocity is perturbed. A schematic of the *r*-modes is shown in Fig. 2.4.

2.4.3   Hybrid Modes

Although not shown by our previous analysis, there is another class of modes, which are also trivial in the nonrotating limit. These modes arise in stars where no buoyancy effects are present. A zero-buoyancy star has to be governed by a barotropic equation of state, i.e., $p = p(\rho)$. If we assume a polytrope, then the Schwarzschild discriminant takes the form (2.4.3) and vanishes if $\Gamma = \Gamma_1$.

In this case, *g*-modes also become trivial and, along with the *r*-modes, form the so-called *hybrid rotational* modes, or *generalised r*-modes. Their eigenfrequencies are confined in the range $[-2\Omega, 2\Omega]$ and their eigenfunctions have both polar and axial components, in the nonrotating limit (Lindblom and Ipser 1999; Lockitch and Friedman 1999). These modes are characterised by their order m, but do not have a fixed degree l. It is worth noting, however, that the purely axial *r*-modes are retrieved in the $l = m$ case and their eigenfrequencies are given by Eq. (2.4.4).

2.5   *The Homogeneous Model

The boundary value problem, given by Eqs. (2.3.16)–(2.3.18) and (2.3.24)–(2.3.27), can only be solved numerically for most equations of state, even for relatively simple

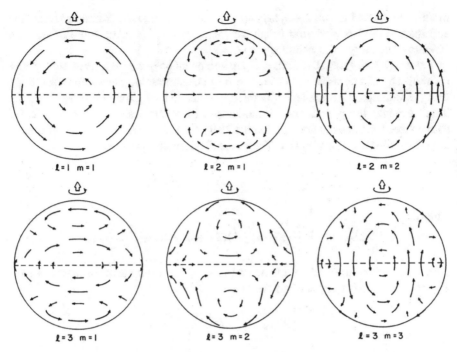

Fig. 2.4 Schematic of r-modes with various degrees l and orders m. Credit: Saio (1982), © American Astronomical Society (AAS). Reproduced with permission

ones, such as polytropes. For the simplified case of a homogeneous star —which technically corresponds to a polytrope with $\Gamma \rightarrow \infty$— there are analytic solutions for the eigenfrequencies and the eigenfunctions, which are thus worth investigating.

It was Thomson (1863) who first solved the problem of nonradial oscillations in a homogeneous, *incompressible* fluid sphere. He found that, in this model, the eigenfrequencies of the f-modes, sometimes also called *Kelvin modes*, are

$$\tilde{\omega}_l^2 = \frac{2l(l-1)}{2l+1}, \tag{2.5.1}$$

where $\tilde{\omega}$ is the dimensionless eigenfrequency, defined by Eq. (2.3.41). Their eigenfunctions are given by (2.3.19), with

$$\xi_r(r) = C_0 r^{l-1} \quad \text{and} \quad \xi_h(r) = \xi_r/l, \tag{2.5.2}$$

C_0 being an arbitrary constant (see Sect. 2.3.7). These are the only polar modes found in this model. The presence of p-modes is prohibited by the absence of compressibility in the medium ($\nabla \cdot \boldsymbol{\xi} = 0$). The latter, combined with uniform density, results in a zero-buoyancy fluid, so g-modes do not exist either (see Smeyers and van Hoolst 2010, Sects. 5.8.3, 8.5, and 10.4.1). When rotation is included, the hybrid rotational modes, discussed in Sect. 2.4.3, also appear and admit analytic solutions (Lindblom and Ipser 1999; Lockitch and Friedman 1999; also, see Kokkotas and Stergioulas 1999 for the purely axial r-modes with $l = m$).

Pekeris (1938) generalised this study to homogeneous, *compressible* fluid spheres. As shown by Chandrasekhar (1963, 1964) and Chandrasekhar and Lebovitz (1964), using a variational principle (subsequently extended by Clement 1964, 1965 for rotating fluids), the f-modes of this model coincide with the Kelvin modes presented above (see also Smeyers and van Hoolst 2010, Sect. 8.5), but now the other classes of polar modes are also allowed. Their eigenfrequencies are given by

$$\tilde{\omega}_{l,n}^2 = D_{n,l} \pm \sqrt{D_{n,l}^2 + l(l+1)}, \qquad (2.5.3)$$

where

$$D_{n,l} = -2 + \frac{\Gamma_1}{2}\left[(n-1)(2n+3+2l) + 3 + 2l\right],$$

with $n \geq 1$. For large values of the overtone n, the asymptotic behaviour of the eigenfrequencies is (Ledoux and Walraven 1958)

$$\tilde{\omega}_{l,n}^2 \approx 2D_{n,l} + \frac{1}{2}\frac{l(l+1)}{D_{n,l}^2}$$

and

$$\tilde{\omega}_{l,n}^2 \approx -\frac{1}{2}\frac{l(l+1)}{D_{n,l}^2}.$$

Thus, for a given l, the oscillation spectrum is divided in two classes: one of positive eigenvalues, approaching infinity for increasing n, and one of negative eigenvalues, approaching zero for increasing n. The first class is the p-modes, while the second class corresponds to g^--modes, i.e., fluid motions associated with convective instabilities (see Sect. 2.4.1). This can be seen from the calculation of the Schwarzschild discriminant (2.3.8) for the homogeneous model, namely

$$A = -\frac{1}{\Gamma_1}\frac{2r}{r^2 - R^2}.$$

For the incompressible model, $\Gamma_1 = \Gamma \to \infty$, which yields $A = 0$ everywhere. For the compressible star, Γ_1 is finite, so $A > 0$ everywhere, with the star exhibiting convective instabilities. This can be interpreted by the fact that, for a compressible fluid, a homogeneous configuration is not a state of minimum potential energy (Ledoux and Walraven 1958).

The eigenfunctions of the polar modes for the compressible model are again given by Eq. (2.3.19) and were found by Sauvenier-Goffin (1951) as

$$\xi_r(r) = \frac{1}{B + 4l + 6} \left\{ \frac{d}{dr} \left[a_{l,n} \left(r^2 - R^2 \right) \right] + \frac{l(l+1)}{r\tilde{\omega}_{l,n}^2} a_{l,n} \left(r^2 - R^2 \right) \right\} \quad (2.5.4)$$

and

$$\xi_h(r) = \frac{1}{\tilde{\omega}_{l,n}^2} \left[\xi_r + \frac{\delta\Phi}{g} - \frac{a_{l,n}}{A} \right], \quad (2.5.5)$$

where

$$\delta\Phi(r) = -\frac{4\pi G\rho}{B + 4l + 6} a_{l,n} \left(r^2 - R^2 \right),$$

$$B = -6 - 4l + \frac{8}{\Gamma_1} + \frac{2\tilde{\omega}_{l,n}^2}{\Gamma_1} - \frac{2l(l+1)}{\Gamma_1 \tilde{\omega}_{l,n}^2},$$

and

$$a_{l,n} \equiv \nabla \cdot \boldsymbol{\xi}(r) = \sum_{j=0}^{n-1} C_{2j} \left(\frac{r}{R} \right)^{2j+l},$$

with

$$C_{2j+2} = C_{2j} \left[\frac{2j(2j+5+2l) - B}{2(j+1)(2j+3+2l)} \right].$$

It should be noted that only one $\delta\Phi$ is associated with a pair of positive and negative eigenvalues, but the same does not apply for ξ_r. Moreover, for the g^--modes, the overtone number n does not coincide with the number of nodes in ξ_r; the g_n^--mode has $n - 1$ nodes.

2.6   Rotating Stars

Taking rotation[4] into account, the situation changes significantly. Due to the Coriolis force and the distortion of the (spherical) equilibrium configuration by the centrifugal force, modes cannot be described by a single spherical harmonic any more. Even though the azimuthal dependence remains of the form $e^{im\phi}$, the latitudinal dependence becomes more complicated. As we will see, rotation lifts the degeneracy of the modes on the azimuthal order m (see Sect. 2.4.1), with every eigenfrequency splitting into $2l + 1$ different values, much like the Zeeman effect of spectral lines

---

[4]In principle, the following discussion applies to magnetic fields as well (e.g., see Gough and Taylor 1984, Dziembowski and Goode 1984, or Unno et al. 1989, § 19), which are ignored here.

when a magnetic field is introduced. As a result, modes with opposite values of m are no longer equivalent: the mode with a positive pattern speed $-\omega/m$ (negative m) travels in the direction of rotation (*prograde* mode), whereas the mode with a negative pattern speed (positive m) travels against the direction of rotation (*retrograde* mode), but their eigenfrequencies are differently affected by rotation. Furthermore, as we discussed in Sects. 2.4.2 and 2.4.3, another class of modes, the inertial modes, becomes relevant in rotating stars, due to the action of the Coriolis force.

We should note that, since we expressed the Euler equation (2.1.2) on the rotating frame, we will refer to perturbations as measured in this frame. If we also assume an inertial frame with coordinates (r', θ', ϕ'), then the transformation between the two frames is simply obtained as

$$\left(r', \theta', \phi'\right) = (r, \theta, \phi + \Omega t) .$$

Hence, a perturbation in the inertial frame has an exponential dependence of the form $e^{i\left(m\phi' + \omega_{\mathrm{in}} t\right)}$, where the eigenfrequency in the inertial frame ω_{in} is given by

$$\omega_{\mathrm{in}} = \omega - m\Omega. \tag{2.6.1}$$

2.6.1 Mode Orthogonality, Decomposition, and Energy

Since operator $\mathcal{B}$ is nonvanishing in this case, the solutions to Eq. (2.2.12) do not obey the orthogonality relation (2.3.55). Even though the operator $\mathcal{L}(\omega) = -\omega^2 - i\omega\mathcal{B} + \mathcal{C}$ is Hermitian for real values of ω (Lynden-Bell and Ostriker 1967), two different solutions of Eq. (2.2.12), (ω_A, ξ_A) and (ω_B, ξ_B), are eigenfunctions of two different operators, $\mathcal{L}(\omega_A)$ and $\mathcal{L}(\omega_B)$, so they are not necessarily orthogonal (Schenk et al. 2001).

However, instead of a configuration space mode expansion, like Eq. (2.3.57), one can use a phase space mode expansion, in which the set of vectors $[\xi_A, i\omega_A\xi_A]$ form a basis on the space of pairs of complex vector functions $[\xi, \xi']$ (Dyson and Schutz 1979). Then, a perturbation can be decomposed as (Schenk et al. 2001)

$$\begin{bmatrix} \xi(r, t) \\ \dot{\xi}(r, t) \end{bmatrix} = \sum_A Q_A(t) \begin{bmatrix} \xi_A(r) \\ i\omega_A\xi_A(r) \end{bmatrix} e^{i\omega_A t},$$

where, as in the case of nonrotating stars, Q_A is the *amplitude coefficient* for the solution ξ_A. We notice that, if (ω_A, ξ_A) is a solution to Eq. (2.2.12), then $(-\omega_A, \xi_A^*)$ is also a solution. Identifying these two solutions by the same index, α, and assuming that $\xi(r, t)$ is real, we get

$$\begin{bmatrix} \boldsymbol{\xi}(r,t) \\ \dot{\boldsymbol{\xi}}(r,t) \end{bmatrix} = \sum_\alpha \left\{ Q_\alpha(t) \begin{bmatrix} \boldsymbol{\xi}_\alpha(r) \\ i\omega_\alpha \boldsymbol{\xi}_\alpha(r) \end{bmatrix} e^{i\omega_\alpha t} + Q_\alpha^*(t) \begin{bmatrix} \boldsymbol{\xi}_\alpha^*(r) \\ -i\omega_\alpha \boldsymbol{\xi}_\alpha^*(r) \end{bmatrix} e^{-i\omega_\alpha t} \right\}.$$

$$(2.6.2)$$

The eigenfunctions $\boldsymbol{\xi}_\alpha$ satisfy a modified orthogonality condition, given by (Schenk et al. 2001)

$$(\omega_\alpha + \omega_\beta)\langle \boldsymbol{\xi}_\alpha, \boldsymbol{\xi}_\beta \rangle - \langle \boldsymbol{\xi}_\alpha, i\boldsymbol{\mathcal{B}}(\boldsymbol{\xi}_\beta) \rangle = b_\alpha \delta_{\alpha\beta}, \qquad (2.6.3)$$

whereas the mode energy, on the rotating frame, is defined as

$$E_\alpha = |Q_\alpha|^2 \omega_\alpha b_\alpha, \qquad (2.6.4)$$

which is reduced to Eq. (2.3.58) for $\Omega = 0$.

The discussion in Sect. 2.3.7 about mode normalisation applies in the case of rotating stars too, with the mode energy at unit amplitude set equal to an arbitrary value E_unit, as

$$E_\text{unit} = \omega_\alpha b_\alpha, \qquad (2.6.5)$$

which makes Eq. (2.6.4)

$$E_\alpha = |Q_\alpha|^2 E_\text{unit}. \qquad (2.6.6)$$

For a different normalisation choice E_unit', the amplitudes Q_α scale as

$$|Q_\alpha'|^2 E_\text{unit}' = |Q_\alpha|^2 E_\text{unit}. \qquad (2.6.7)$$

2.6.2 The Slow-Rotation Approximation

Typically, in order to obtain the eigenfunctions $\boldsymbol{\xi}_\alpha$ for $\Omega \neq 0$, the eigenvalue equation (2.2.12) has to be solved from scratch, which is a far-from-trivial task. However, rotation can also be introduced perturbatively, namely by considering the effects of rotation to the various quantities as perturbations.

To do this, we will expand every quantity in a series with respect to Ω. Eigenfrequencies and eigenfunctions (on the rotating frame) are expanded as

$$\omega_\alpha(\Omega) = \omega_\alpha^{(0)} + \omega_\alpha^{(1)}(\Omega) + \omega_\alpha^{(2)}(\Omega^2) + \mathcal{O}(\Omega^3),$$
$$\boldsymbol{\xi}_\alpha(\Omega) = \boldsymbol{\xi}_\alpha^{(0)} + \boldsymbol{\xi}_\alpha^{(1)}(\Omega) + \boldsymbol{\xi}_\alpha^{(2)}(\Omega^2) + \mathcal{O}(\Omega^3),$$

where the superscript (0) corresponds to the solution in the nonrotating limit, obtained by Eq. (2.3.54), and the rest of the terms are rotational corrections. Equilibrium quantities, i.e., density, pressure, and gravitational potential, are affected only to second order by Ω, because the centrifugal force, which spoils the sphericity of the unperturbed star, is proportional to Ω^2 [see Eq. (2.1.5)]. Thus,

$$\rho(\Omega) = \rho^{(0)} + \rho^{(2)}\left(\Omega^2\right) + \mathcal{O}\left(\Omega^4\right)$$

and likewise for the pressure and the gravitational potential. In a similar manner, operator $\mathcal{C}$ is written as

$$\mathcal{C} = \mathcal{C}^{(0)} + \mathcal{C}^{(2)}\left(\Omega^2\right) + \mathcal{O}\left(\Omega^4\right),$$

whereas operator $\mathcal{B}$ is, by definition,

$$\mathcal{B} = \mathcal{B}^{(1)}\left(\Omega\right).$$

Substituting the above in Eq. (2.2.12) and distinguishing between zeroth-, first-, and second-order terms, we obtain (omitting the subscript α, for simplicity)

$$\omega^{(0)2}\boldsymbol{\xi}^{(0)} = \mathcal{C}^{(0)}\left(\boldsymbol{\xi}^{(0)}\right), \tag{2.6.8}$$

$$-\omega^{(0)2}\boldsymbol{\xi}^{(1)} + \mathcal{C}^{(0)}\left(\boldsymbol{\xi}^{(1)}\right) - 2\omega^{(0)}\omega^{(1)}\boldsymbol{\xi}^{(0)} + i\omega^{(0)}\mathcal{B}^{(1)}\left(\boldsymbol{\xi}^{(0)}\right) = \mathbf{0}, \tag{2.6.9}$$

and

$$-\omega^{(0)2}\boldsymbol{\xi}^{(2)} + \mathcal{C}^{(0)}\left(\boldsymbol{\xi}^{(2)}\right) - 2\omega^{(0)}\omega^{(1)}\boldsymbol{\xi}^{(1)} + i\omega^{(0)}\mathcal{B}^{(1)}\left(\boldsymbol{\xi}^{(1)}\right)$$
$$- 2\omega^{(0)}\omega^{(2)}\boldsymbol{\xi}^{(0)} - \omega^{(1)2}\boldsymbol{\xi}^{(0)} + i\omega^{(1)}\mathcal{B}^{(1)}\left(\boldsymbol{\xi}^{(0)}\right) + \mathcal{C}^{(2)}\left(\boldsymbol{\xi}^{(0)}\right) = \mathbf{0}, \tag{2.6.10}$$

respectively, with Eq. (2.6.8) coinciding with Eq. (2.3.54), as expected. Using Eqs. (2.6.9) and (2.6.10) we find the $\mathcal{O}\left(\Omega\right)$ and $\mathcal{O}\left(\Omega^2\right)$ corrections to the eigenfrequencies and the eigenfunctions.

First- and second-order corrections to polar mode eigenfrequencies are given by (Saio 1981)

$$\omega_\alpha^{(1)} = mC_1\Omega \tag{2.6.11}$$

and

$$\omega_\alpha^{(2)} = \frac{C_2\Omega^2}{\omega_\alpha^{(0)}} \tag{2.6.12}$$

respectively, with their derivation presented in Appendix B. From Eq. (2.6.11) we see that the degeneracy in m is already resolved at first order in Ω. The parameter C_1, given by Eq. (B.1.10), contains first-order effects of the Coriolis force, whereas C_2 includes second-order effects of both the Coriolis and the centrifugal force. A general expression for $\omega_\alpha^{(2)}$ is obtained from Eq. (B.2.8). For the specific case of polytropic stars with the same central density as their nonrotating counterparts (see Appendix B.2.2), it is given by Eq. (B.2.14), which can then be written in the form of Eq. (2.6.12). From Eq. (B.2.16) we see that C_2 can be further decomposed as

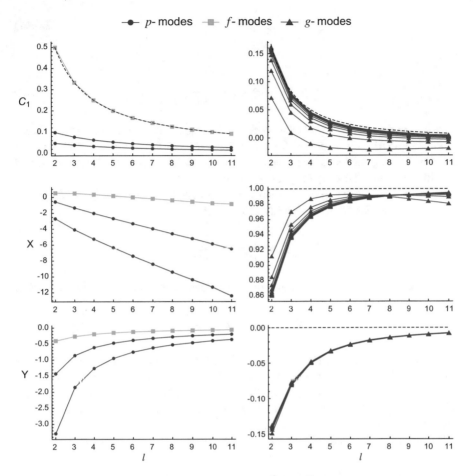

Fig. 2.5 First-order (C_1) and second-order ($C_2 = X + m^2 Y$) correction parameters, plotted against the mode degree l, for the polar modes of a star obeying a polytropic equation of state with a polytropic exponent $\Gamma = 2$ and an adiabatic exponent $\Gamma_1 = 2.1$. In the graphs for f- and p-modes (*left*) the overtone increases downwards, and the dashed line corresponds to $C_1 = 1/l$ (see text). In the graph of C_1 for g-modes (*top right*) the overtone increases upwards and the dashed line corresponds to $C_1 = 1/l(l+1)$ (see text). In the graphs of X and Y for g-modes (*middle right* and *bottom right* respectively) the dashed lines indicate the limiting values of X and Y for increasing l and n

$$C_2 = X + m^2 Y. \tag{2.6.13}$$

In Fig. 2.5 we show the correction parameters C_1, X, and Y, as functions of the degree l, for the polar modes of Fig. 2.2b, namely, for a polytrope with $\Gamma = 2$ and $\Gamma_1 = 2.1$.

From Eq. (B.1.10) we can derive two asymptotic relations for the parameter C_1:

$$\xi_r \gg \xi_h \Rightarrow C_1 \to 0 \tag{2.6.14}$$

and

$$\xi_h \gg \xi_r \Rightarrow C_1 \to \frac{1}{l(l+1)}, \tag{2.6.15}$$

where ξ_r and ξ_h are the radial and horizontal components of the polar mode eigen-function (2.3.19). From Fig. 2.5 we see that Eq. (2.6.14) applies to p-modes and Eq. (2.6.15) to g-modes, as the overtone n increases. The latter is plotted as a dashed line in the graph where C_1 for g-modes is shown.

The maximum value of C_1 can also be obtained, by assuming that $\xi_r = \kappa \xi_h$, where κ is a constant. Replacing in Eq. (B.1.10) we get

$$C_1 = \frac{2\kappa + 1}{\kappa^2 + l(l+1)}.$$

Then, we solve $dC_1/d\kappa = 0$ to find that C_1 achieves its maximum value for $\kappa = l$. So,

$$C_{1,\,\text{max}} = \frac{1}{l} \Leftrightarrow \xi_h = \frac{\xi_r}{l}. \tag{2.6.16}$$

Comparing Eq. (2.6.16) with Eq. (2.5.2) we see that C_1 obtains its maximum value for the case of f-modes in a homogeneous star (Kelvin modes). However, this behaviour seems to apply to f-modes in polytropic stars as well, as it can be seen from Fig. 2.5, where $C_{1,\,\text{max}}$ is plotted as a dashed line. A change of the polytropic exponent Γ does not affect the result.[5]

Hence, the behaviour of the parameter C_1 is summarised as follows:

I. f-modes: $C_1 = \dfrac{1}{l}$

II. p-modes: $C_1 \to 0$ as $n \to \infty$

III. g-modes: $C_1 \to \dfrac{1}{l(l+1)}$ as $n \to \infty$

Equivalent expressions for the asymptotic behaviour of X and Y are hard to derive, due to their complexity. However, from Fig. 2.5 we see that, for the case of g-modes, $X \to 1$ and $Y \to 0$ as the degree l and overtone n are increased.

Changing the adiabatic exponent Γ_1 does not affect the behaviour of the three parameters for f- and p- modes, whereas it induces small changes for g-modes.

The impact of first- and second-order corrections to the eigenfrequency of the $l = 2$ f-mode, in a polytropic star with $\Gamma = 2$, is presented in Fig. 2.6, where the eigenfrequency is plotted against the angular velocity of the star (normalised to the break-up limit, see Sect. 3.2). The degeneracy in m is lifted by rotation and the

---

[5]This is no longer true for smaller values of Γ. As it can be seen in Saio (1981, Table 1), the parameter C_1 is not equal to $1/l$ for the f-modes of a polytrope with $\Gamma = 4/3$.

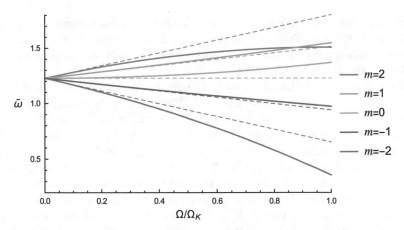

Fig. 2.6 Eigenfrequencies of the $l = 2$ f-mode, in a polytropic star with a polytropic exponent $\Gamma = 2$, plotted against the angular velocity of the star. The $(2l + 1)$-fold degeneracy in m is lifted when rotation is introduced. The eigenfrequencies are normalised as $\tilde{\omega} = \omega/\sqrt{GM/R^3}$ and the angular velocity Ω is normalised to the break-up limit Ω_K. Dashed (solid) lines denote eigenfrequencies corrected up to first (second) order in Ω

eigenfrequency splits into $2l + 1$ different values, one for each m. Eigenfrequencies corrected up to first (second) order in Ω are shown as dashed (solid) lines.

Eigenfunction corrections have the general form

$$\boldsymbol{\xi}_\alpha^{(1)} = \sum_\beta c_{\alpha\beta}^{(1)} \boldsymbol{\xi}_\beta^{(0)} \tag{2.6.17}$$

and

$$\boldsymbol{\xi}_\alpha^{(2)} = \sum_\beta c_{\alpha\beta}^{(2)} \boldsymbol{\xi}_\beta^{(0)}, \tag{2.6.18}$$

namely, they are expanded in terms of the eigenfunctions of the nonrotating star, with $c_{\alpha\beta}^{(1)}$ and $c_{\alpha\beta}^{(2)}$ being the first- and second-order correction coefficients respectively. From Eqs. (2.6.17) and (2.6.18) we see that the rotationally-corrected mode can no longer be described by a single spherical harmonic. First-order correction coefficients are given by Eq. (B.1.12). Alternatively, the components of the first-order eigenfunction correction $\xi_r^{(1)}$, $\xi_\theta^{(1)}$, and $\xi_\phi^{(1)}$ are expressed in closed form in Eqs. (B.1.14)–(B.1.16), respectively.

We should note that higher-order effects should become important for large angular velocities, but the analysis is quite cumbersome already at second order in Ω. A third-order perturbation formalism has been, nevertheless, developed by Soufi et al. (1998; see also Karami 2008). In addition, a non-perturbative method for the computation of modes in rapidly rotating stars was presented by Lignières et al. (2006) and Reese et al. (2006). Tables with results on first-, second-, and third-order corrections to the eigenfrequencies, for various stellar models, can be found in Reese et al. (2006), Karami (2009), and Ballot et al. (2010). A review on the different approaches dealing with the effects of rotation on oscillation modes is given by Reese (2010; see also Dziembowski 2010).

The influence of rotation on the oscillation modes of neutron stars has caught the attention of several authors. Strohmayer (1991) calculated first-order corrections to the eigenfrequencies and eigenfunctions of the nonradial oscillations of a neutron star, including modes due to the presence of a solid crust (not discussed here). The problem has also been studied in the framework of general relativity. Some indicative studies include (Kojima 1992, 1993), where the perturbation equations are derived to first order in Ω (see also Stavridis and Kokkotas 2005, for a more recent study), Ferrari et al. (2004), where the effects of rotation on the oscillation modes of newborn neutron stars are examined, Vavoulidis et al. (2007), where the slow-rotation approximation is applied to crust modes, and Gaertig and Kokkotas (2008), where the oscillation modes of rapidly rotating neutron stars are studied. For a more general investigation of rotating relativistic stars, the reader is referred to Stergioulas (2003) and Friedman and Stergioulas (2013).

References

Abramowitz, M., & Stegun, I. A. (1972). *Handbook of mathematical functions*. New York: Dover. http://adsabs.harvard.edu/abs/1972hmfw.book.....A.

Aerts, C., Christensen-Dalsgaard, J., & Kurtz, D. W. (2010). *Asteroseismology*. Astronomy and Astrophysics Library. New York: Springer. http://adsabs.harvard.edu/abs/2010aste.book.....A.

Ballot, J., Lignières, F., Reese, D. R., & Rieutord, M. (2010). Gravity modes in rapidly rotating stars. Limits of perturbative methods. *Astronomy & Astrophysics, 518*, A30. https://doi.org/10.1051/0004-6361/201014426, arXiv:1005.0275.

Chandrasekhar, S. (1963). Letter to the editor: A general variational principle governing the radial and the non-radial oscillations of gaseous masses. *The Astrophysical Journal, 138*, 896. https://doi.org/10.1086/147694.

Chandrasekhar, S. (1964). A general variational principle governing the radial and the non-radial oscillations of gaseous masses. *The Astrophysical Journal, 139*, 664. https://doi.org/10.1086/147792.

Chandrasekhar, S., & Lebovitz, N. R. (1964). Non-radial oscillations of gaseous masses. *The Astrophysical Journal, 140*, 1517. https://doi.org/10.1086/148056.

Clement, M. J. (1964). A general variational principle governing the oscillations of a rotating gaseous mass. *The Astrophysical Journal, 140*, 1045. https://doi.org/10.1086/148004.

Clement, M. J. (1965). The radial and non-radial oscillations of slowly rotating gaseous masses. *The Astrophysical Journal, 141*, 210. https://doi.org/10.1086/148104.

Cowling, T. G. (1941). The non-radial oscillations of polytropic stars. *Monthly Notices of the Royal Astronomical Society, 101*, 367. https://doi.org/10.1093/mnras/101.8.367.

Cunha, M. S. et al. (2007). Asteroseismology and interferometry. *The Astronomy and Astrophysics Review, 14*, 217–360. https://doi.org/10.1007/s00159-007-0007-0, arXiv:0709.4613.

Detweiler, S. L. (1975). A variational calculation of the fundamental frequencies of quadrupole pulsation of fluid spheres in general relativity. *The Astrophysical Journal, 197*, 203–217. https://doi.org/10.1086/153504.

Detweiler, S., & Lindblom, L. (1985). On the nonradial pulsations of general relativistic stellar models. *The Astrophysical Journal, 292*, 12–15. https://doi.org/10.1086/163127.
Dyson, J., & Schutz, B. F. (1979). Perturbations and stability of rotating stars. I. Completeness of normal modes. *Proceedings of the Royal Society of London. Series A, 368*, 389–410. https://doi.org/10.1098/rspa.1979.0137.
Dziembowski, W. (2010). Asteroseismology of rapidly rotating pulsators. *Highlights of Astronomy, 15*, 360–361. https://doi.org/10.1017/S1743921310009804.
Dziembowski, W., & Goode, P. R. (1984). Simple asymptotic estimates of the fine structure in the spectrum of solar oscillations due to rotation and magnetism. *Memorie della Societa Astronomica Italiana, 55*, 185–213. http://adsabs.harvard.edu/abs/1984MmSAI.55.185D.
Ferrari, V., Miniutti, G., & Pons, J. A. (2003). Gravitational waves from newly born, hot neutron stars. *Monthly Notices of the Royal Astronomical Society, 342*, 629–638. https://doi.org/10.1046/j.1365-8711.2003.06580.x, arXiv:astro-ph/0210581.
Ferrari, V., Gualtieri, L., Pons, J. A., & Stavridis, A. (2004). Rotational effects on the oscillation frequencies of newly born proto-neutron stars. *Monthly Notices of the Royal Astronomical Society, 350*, 763–768. https://doi.org/10.1111/j.1365-2966.2004.07698.x, arXiv:astro-ph/0310896.
Finn, L. S. (1986). g-modes of non-radially pulsating relativistic stars: The slow-motion formalism. *Monthly Notices of the Royal Astronomical Society, 222*, 393–416. http://adsabs.harvard.edu/abs/1986MNRAS.222.393F.
Finn, L. S. (1987). G-modes in zero-temperature neutron stars. *Monthly Notices of the Royal Astronomical Society, 227*, 265–293. http://adsabs.harvard.edu/abs/1987MNRAS.227.265F.
Friedman, J. L., & Stergioulas, N. (2013). *Rotating relativistic stars*. Cambridge, England: Cambridge University Press. http://adsabs.harvard.edu/abs/2013rrs..book.....F.
Gaertig, E., & Kokkotas, K. D. (2008). Oscillations of rapidly rotating relativistic stars. *Physical Review D, 78*, 064063. https://doi.org/10.1103/PhysRevD.78.064063, arXiv:0809.0629.
Gough, D. O., & Taylor, P. P. (1984). Influence of rotation and magnetic fields on stellar oscillation eigenfrequencies. *Memorie della Societa Astronomica Italiana, 55*, 215–226. http://adsabs.harvard.edu/abs/1984MmSAI.55.215G.
Gualtieri, L., Pons, J. A., & Miniutti, G. (2004). Nonadiabatic oscillations of compact stars in general relativity. *Physical Review D, 70*, 084009. https://doi.org/10.1103/PhysRevD.70.084009, arXiv:gr-qc/0405063.
Gualtieri, L., Kantor, E. M., Gusakov, M. E., & Chugunov, A. I. (2014). Quasinormal modes of superfluid neutron stars. *Physical Review D, 90*, 024010. https://doi.org/10.1103/PhysRevD.90.024010, arXiv:1404.7512.
Karami, K. (2008). Third order effect of rotation on stellar oscillations of a B star. *Chinese Journal of Astronomy and Astrophysics, 8*, 285–308. https://doi.org/10.1088/1009-9271/8/3/06, ArXiv:astro-ph/0502194.
Karami, K. (2009). Third order effect of rotation on stellar oscillations of a β-Cephei star. *Astrophysics and Space Science, 319*, 37–44. https://doi.org/10.1007/s10509-008-9937-x, arXiv:0810.5092.
Kojima, Y. (1992). Equations governing the nonradial oscillations of a slowly rotating relativistic star. *Physical Review D, 46*, 4289–4303. https://doi.org/10.1103/PhysRevD.46.4289.
Kojima, Y. (1993). Coupled pulsations between polar and axial modes in a slowly rotating relativistic star. *Progress of Theoretical Physics, 90*, 977–990. https://doi.org/10.1143/PTP.90.977.
Kokkotas, K. D., & Schmidt, B. (1999). Quasi-normal modes of stars and black holes. *Living Reviews in Relativity, 2*, 2. https://doi.org/10.12942/lrr-1999-2, arXiv:gr-qc/9909058.
Kokkotas, K. D., & Schutz, B. F. (1986). Normal modes of a model radiating system. *General Relativity and Gravitation, 18*, 913–921. https://doi.org/10.1007/BF00773556.
Kokkotas, K. D., & Schutz, B. F. (1992). W-modes: A new family of normal modes of pulsating relativistic stars. *Monthly Notices of the Royal Astronomical Society, 255*, 119–128. https://doi.org/10.1093/mnras/255.1.119.

Kokkotas, K. D., & Stergioulas, N. (1999). Analytic description of the r-mode instability in uniform density stars. *Astronomy & Astrophysics, 341*, 110–116. http://adsabs.harvard.edu/abs/1999A %26A...341.110K, arXiv: astro-ph/9805297.

Krüger, C. J., Ho, W. C. G., & Andersson, N. (2015). Seismology of adolescent neutron stars: Accounting for thermal effects and crust elasticity. *Physical Review D, 92*, 063009. https://doi. org/10.1103/PhysRevD.92.063009, arXiv:1402.5656.

Ledoux, P., & Walraven, T. (1958). Variable stars. *Encyclopedia of physics (Handbuch der Physik)* (vol. 51, pp. 353–604). Berlin-Göttingen-Heidelberg: Springer-Verlag. http://adsabs.harvard.edu/ abs/1958HDP....51..353L.

Lignières, F., Rieutord, M., & Reese, D. (2006). Acoustic oscillations of rapidly rotating polytropic stars. I. Effects of the centrifugal distortion. *Astronomy & Astrophysics, 455*, 607–620. https:// doi.org/10.1051/0004-6361:20065015, arXiv:astro-ph/0604312.

Lindblom, L., & Detweiler, S. L. (1983). The quadrupole oscillations of neutron stars. *The Astrophysical Journal Supplement Series, 53*, 73–92. https://doi.org/10.1086/190884.

Lindblom, L., & Ipser, J. R. (1999). Generalized r-modes of the Maclaurin spheroids. *Physical Review D, 59*, 044009. https://doi.org/10.1103/PhysRevD.59.044009, arXiv:gr-qc/9807049.

Lockitch, K. H., & Friedman, J. L. (1999). Where are the r-modes of isentropic stars? *The Astrophysical Journal, 521*, 764–788. https://doi.org/10.1086/307580, arXiv:gr-qc/9812019.

Lynden-Bell, D., & Ostriker, J. P. (1967). On the stability of differentially rotating bodies. *Monthly Notices of the Royal Astronomical Society, 136*, 293. http://adsabs.harvard.edu/abs/ 1967MNRAS.136.293L.

McDermott, P. N. (1990). Density discontinuity g-modes. *Monthly Notices of the Royal Astronomical Society, 245*, 508. http://adsabs.harvard.edu/abs/1990MNRAS.245.508M.

McDermott, P. N., van Horn, H. M., & Scholl, J. F. (1983). Nonradial g-mode oscillations of warm neutron stars. *The Astrophysical Journal, 268*, 837–848. http://dx.doi.org/10.1086/161006.

McDermott, P. N., van Horn, H. M., & Hansen, C. J. (1988). Nonradial oscillations of neutron stars. *The Astrophysical Journal, 325*, 725–748. https://doi.org/10.1086/166044.

Miniutti, G., Pons, J. A., Berti, E., Gualtieri, L., & Ferrari, V. (2003). Non-radial oscillation modes as a probe of density discontinuities in neutron stars. *Monthly Notices of the Royal Astronomical Society, 338*, 389–400. https://doi.org/10.1046/j.1365-8711.2003.06057.x, arXiv:astro-ph/0206142.

Osaki, J. (1975). Nonradial oscillations of a 10 solar mass star in the main-sequence stage. *Publications of the Astronomical Society of Japan, 27*, 237–258. http://adsabs.harvard.edu/abs/ 1975PASJ...27.237O.

Papaloizou, J., & Pringle, J. E. (1978). Non-radial oscillations of rotating stars and their relevance to the short-period oscillations of cataclysmic variables. *Monthly Notices of the Royal Astronomical Society, 182*, 423–442. http://adsabs.harvard.edu/abs/1978MNRAS.182.423P.

Pekeris, C. L. (1938). Nonradial oscillations of stars. *The Astrophysical Journal, 88*, 189. https:// doi.org/10.1086/143971.

Reese, D. R. (2010). Oscillations in rapidly rotating stars. *Astronomische Nachrichten, 331*, 1038. https://doi.org/10.1002/asna.201011452.

Reese, D., Lignières, F., & Rieutord, M. (2006). Acoustic oscillations of rapidly rotating polytropic stars. II. Effects of the Coriolis and centrifugal accelerations. *Astronomy & Astrophysics, 455*, 621–637. arXiv:astro-ph/0605503.

Reisenegger, A., & Goldreich, P. (1992). A new class of g-modes in neutron stars. *The Astrophysical Journal, 395*, 240–249. https://doi.org/10.1086/171645.

Robe, H. (1968). Les oscillations non radiales des polytropes. *Annales d'Astrophysique, 31*, 475. http://adsabs.harvard.edu/abs/1968AnAp...31..475R.

Saio, H. (1981). Rotational and tidal perturbations of nonradial oscillations in a polytropic star. *The Astrophysical Journal, 244*, 299–315. https://doi.org/10.1086/158708.

Saio, H. (1982). R-mode oscillations in uniformly rotating stars. *The Astrophysical Journal, 256*, 717–735. https://doi.org/10.1086/159945.

Sauvenier-Goffin, E. (1951). Note sur les pulsations non-radiale d'une sphère homogène compressible. *Bulletin de la Societe Royale des Sciences de Liège*, *20*, 20–38. http://adsabs.harvard.edu/abs/1951BSRSL.20...20S.

Schenk, A. K., Arras, P., Flanagan, É. É., Teukolsky, S. A., & Wasserman, I. (2001). Nonlinear mode coupling in rotating stars and the r-mode instability in neutron stars. *Physical Review D*, *65*, 024001. https://doi.org/10.1103/PhysRevD.65.024001, arXiv:gr-qc/0101092.

Smeyers, P. (2003). Asymptotic representation of low- and intermediate-degree p-modes in stars. *Astronomy & Astrophysics*, *407*, 643–653. https://doi.org/10.1051/0004-6361:20030744.

Smeyers, P. (2006). The second-order asymptotic representation of higher-order non-radial p-modes in stars revisited. *Astronomy & Astrophysics*. *451*, 237–249. https://doi.org/10.1051/0004-6361:20054546.

Smeyers, P., & van Hoolst, T. (2010). *Linear isentropic oscillations of stars: Theoretical foundations* (vol. 371). Astrophysics and Space Science Library. New York: Springer. http://adsabs.harvard.edu/abs/2010ASSL..371.....S.

Smeyers, P., De Boeck, I., Van Hoolst, T., & Decock, L. (1995). Asymptotic representation of linear, isentropic g-modes of stars. *Astronomy & Astrophysics*, *301*, 105. http://adsabs.harvard.edu/abs/1995A%26A...301.105S.

Smeyers, P., Vansimpsen, T., De Boeck, I., & Van Hoolst, T. (1996). Asymptotic representation of high-frequency, low-degree p-modes in stars and in the Sun. *Astronomy & Astrophysics*, *307*, 105. http://adsabs.harvard.edu/abs/1996A%26A...307.105S.

Soufi, F., Goupil, M. J., & Dziembowski, W. A. (1998). Effects of moderate rotation on stellar pulsation. I. Third order perturbation formalism. *Astronomy & Astrophysics*, *334*, 911–924. http://adsabs.harvard.edu/abs/1998A%26A...334.911S.

Stavridis, A., & Kokkotas, K. D. (2005). Evolution equations for slowly rotating stars. *International Journal of Modern Physics D*, *14*, 543–571. https://doi.org/10.1142/S021827180500592X, arXiv:gr-qc/0411019.

Stergioulas, N. (2003). Rotating stars in relativity. *Living Reviews in Relativity*, *6*, 3. https://doi.org/10.12942/lrr-2003-3, arXiv:gr-qc/0302034.

Strohmayer, T. E. (1991). Oscillations of rotating neutron stars. *The Astrophysical Journal*, *372*, 573–591. https://doi.org/10.1086/170002.

Strohmayer, T. E. (1993). Density discontinuities and the g-mode oscillation spectra of neutron stars. *The Astrophysical Journal*, *417*, 273. https://doi.org/10.1086/173309.

Thomson, W. (1863). Dynamical problems regarding elastic spheroidal shells and spheroids of incompressible liquid. *Philosophical Transactions of the Royal Society of London*, *153*, 583616. https://doi.org/10.1098/rstl.1863.0028.

Thorne, K. S., & Campolattaro, A. (1967). Non-radial pulsation of general-relativistic stellar models. I. Analytic analysis for $l \geq 2$. *The Astrophysical Journal*, *149*, 591. https://doi.org/10.1086/149288.

Unno, W., Osaki, Y., Ando, H., Saio, H., & Shibahashi, H. (1989). *Nonradial oscillations of stars* (2nd ed.). Tokyo: University of Tokyo Press. http://adsabs.harvard.edu/abs/1989nos..book.....U.

Vavoulidis, M., Stavridis, A., Kokkotas, K. D., & Beyer, H. (2007). Torsional oscillations of slowly rotating relativistic stars. *Monthly Notices of the Royal Astronomical Society*, *377*, 1553–1556. https://doi.org/10.1111/j.1365-2966.2007.11706.x, arXiv:gr-qc/0703039.

Chapter 3
The f-mode Instability

As shown in Sects. 2.4 and 2.5 for g^--modes, oscillation modes may not always be stable. Convective instabilities are just a manifestation of *dynamical instabilities*, namely, instabilities associated with the absence of hydrostatic equilibrium in the star, either locally or globally. These instabilities evolve on a dynamical time scale, i.e.,

$$\tau_{\text{dyn}} \sim \frac{R^3}{GM},$$

which is of order the free-fall time of the star.[1] Another interesting class of instabilities is related to the presence of dissipation mechanisms. These are called *secular instabilities* due to their slow evolution, on time scales related to the corresponding dissipative process and will be our main focus on this chapter. Finally, *thermal instabilities* may occur when the star is driven away from thermal equilibrium, e.g., at the commencement of helium burning in the degenerate helium core of a star, resulting in the so-called helium flash. These are not relevant for neutron stars and will not be discussed.

First, we will briefly review the studies about the equilibrium figure of a rotating, self-gravitating body (Sect. 3.1), which revealed that a star is not necessarily deformed by rotation into an oblate spheroid, but may also be shaped like a triaxial ellipsoid, should it rotate sufficiently fast. Depending on the assumed density profile, there is an upper limit on the maximum rotation that a star can support (Sect. 3.2), which determines whether it can actually admit such a state. The possible equilibrium figures were subsequently shown to be related to secular instabilities (Sect. 3.3), associated with damping mechanisms, like viscosity and gravitational radiation. Some intuition on the concept of a secular instability can be gained by means of simplistic mechanical examples (Sect. 3.4). Then, we shall discuss in detail the mechanism behind the gravitational-wave-driven secular instability, known as the Chandrasekhar–Friedman–Schutz (CFS) instability (Sect. 3.5), and how it sets in

---

[1] Namely, the time needed for the star to collapse due to gravity, if pressure were removed.

© Springer Nature Switzerland AG 2018
P. Pnigouras, *Saturation of the f-mode Instability in Neutron Stars*,
Springer Theses, https://doi.org/10.1007/978-3-319-98258-8_3

via various oscillation modes. Finally, we are going to see how viscosity affects the modes and counteracts the instability, giving rise to the so-called instability window (Sect. 3.6).

3.1   *Figures of Equilibrium

The existence of instabilities in a rotating, self-gravitating body was first realised during the study of the possible equilibrium figures that such a body can admit. We are going to briefly present the main contributions to this problem, following Chandrasekhar (1969).

The effect of small rotation on the shape of the Earth was first considered by Isaac Newton (1687, *Philosophiae Naturalis Principia Mathematica*, Book III, Propositions XVIII–XX). In a thought experiment, Newton imagined two wells, one at the equator and the other at the pole, going as deep as the centre of the Earth and filled with a fluid (see Fig. 3.1 for his original illustration). Regarding the Earth as homogeneous and based on the fact that the two fluids must be in equilibrium, he used a series of simple arguments and derived a relation between the oblateness of the Earth f and the centrifugal acceleration at the equator, given by

$$f = \frac{5}{4} \frac{\Omega^2 R_e}{GM/R_e^2},$$

where $f = (R_e - R_p)/R_e$, with R_e and R_p being the equatorial and polar radii respectively. It was known at the time that the relation between the centrifugal acceleration and the gravitational acceleration at the equator is

$$\frac{\Omega^2 R_e}{GM/R_e^2} \approx \frac{1}{290},$$

from which Newton found that, for a slowly rotating, homogeneous Earth,

$$f \approx \frac{1}{230}.$$

This result was experimentally confirmed in 1738 by Maupertuis and Clairaut, ending a long debate about the influence of rotation on the shape of the Earth (see Fig. 3.1). Today, accounting for the Earth's inhomogeneous structure, we know that $f \approx 1/294$.

Extending Newton's arguments to large rotation, Maclaurin (1742) found that the eccentricity of a rotating, homogeneous, incompressible body, defined as

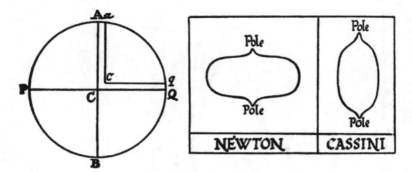

Fig. 3.1 *Left* Illustration from Newton's *Principia*, depicting his Gedankenexperiment about the effect of rotation on the shape of the Earth. *Right* Caricature about Newton's and Cassini's views on the shape of the Earth. Credit: Chandrasekhar (1969, § 1), © 1969 by Yale University

$$e = \sqrt{1 - \frac{R_p^2}{R_e^2}},$$

is related to its angular velocity as

$$\tilde{\Omega}^2 = \frac{\sqrt{1-e^2}}{e^3} 2(3 - 2e^2) \arcsin e - \frac{6}{e^2}(1 - e^2), \qquad (3.1.1)$$

where

$$\tilde{\Omega} = \frac{\Omega}{\sqrt{\pi G \rho}}, \qquad (3.1.2)$$

ρ being the body's density (the derivation of this equation can be also found in Shapiro and Teukolsky 1983, Sect. 7.3). Maclaurin reached the conclusion that rotating bodies in equilibrium must have an oblate shape (*Maclaurin spheroids*), which was disputed later by Jacobi (1834), who explicitly showed that "ellipsoids with three unequal axes can very well be figures of equilibrium" (*Jacobi ellipsoids*)! This remarkable result was subsequently examined by Meyer (1842), who demonstrated that the Jacobi ellipsoidal sequence bifurcates from the Maclaurin spheroidal sequence at the point where $e \approx 0.81267$.

The behaviour of the two sequences with respect to angular velocity and angular momentum is shown in Fig. 3.2. The angular momentum J is defined as

$$J = I\Omega,$$

where I is the moment of inertia of the ellipsoid about the axis of rotation R_p. If R_p, $R_{e,1}$, and $R_{e,2}$ denote the principal axes of the ellipsoid, with $R_p < R_{e,1} \leq R_{e,2}$, then I is obtained as

$$I = \frac{1}{5}M\left(R_{e,1}^2 + R_{e,2}^2\right),$$

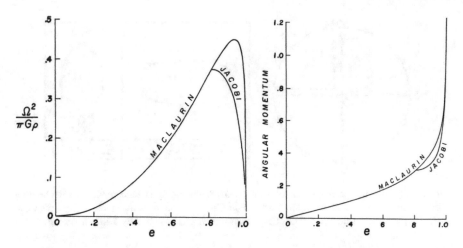

Fig. 3.2 Angular velocity (normalised to $\sqrt{\pi G \rho}$; *left*) and angular momentum (normalised to $\sqrt{G M^3 \bar{R}}$; *right*) versus eccentricity, for a homogeneous, self-gravitating body. The Jacobi ellipsoids branch off the sequence of Maclaurin spheroids at $e \approx 0.81267$. Credit: Chandrasekhar (1969, § 32), © 1969 by Yale University

where the mass M of the ellipsoid is given by

$$M = \frac{4\pi}{3} \rho \bar{R}^3,$$

with $\bar{R} = \left(R_p R_{e,1} R_{e,2} \right)^{1/3}$. The angular momentum is conveniently normalised as

$$\frac{J}{\sqrt{G M^3 \bar{R}}} = \frac{\sqrt{3}}{10} \frac{R_{e,1}^2 + R_{e,2}^2}{\bar{R}^2} \tilde{\Omega}.$$

Also, in this context, the eccentricity of the ellipsoid is calculated with respect to the semimajor axis $R_{e,2}$, namely

$$e = \sqrt{1 - \frac{R_p^2}{R_{e,2}^2}}. \tag{3.1.3}$$

The formulae above apply to Maclaurin spheroids for $R_{e,1} = R_{e,2} \equiv R_e$.

In Fig. 3.2 we see how the Jacobi sequence bifurcates from the Maclaurin sequence at $e \approx 0.81267$, where $\tilde{\Omega}^2 \approx 0.37423$. We can also notice that, although the angular momentum increases monotonically with respect to the eccentricity, the same does not happen with the angular velocity. As the angular momentum increases, Maclaurin spheroids rotate faster, until a maximum angular velocity is reached at $e \approx 0.92995$, where $\tilde{\Omega}_{\text{max}}^2 \approx 0.449331$. From this point on, further increase of the angular momentum results in the spin down of the spheroid, until it becomes an infinitely thin disk. The same happens to the Jacobi ellipsoid as well, whose maximum angular velocity occurs at the bifurcation point and drops thereafter, until the ellipsoid becomes an infinitely long "needle".

Based on the work of Dirichlet (1861), an additional class of homogeneous ellipsoids of equilibrium was found by Dedekind (1861). *Dedekind ellipsoids* are congruent[2] to Jacobi ellipsoids and branch off the Maclaurin spheroidal sequence at the same point as the latter. However, as opposed to Jacobi ellipsoids, they do not rotate uniformly. In fact, they remain stationary in the inertial frame, with their ellipsoidal shape being preserved by internal motions of the fluid, characterised by uniform *vorticity*.

On the inertial frame, the vorticity vector ζ_{in} is defined as

$$\zeta_{in} = \nabla \times v_{in}$$

and serves as a local measure of the fluid's spin. For a fluid body, rotating uniformly with an angular velocity $\mathbf{\Omega}$, the fluid velocity v_{in} can be generally decomposed as

$$v_{in} = v + \mathbf{\Omega} \times r,$$

where v is the fluid velocity measured in the frame rotating with the star, corresponding to internal motions of the fluid, superposed on its rigid rotation. Replacing the velocity in the vorticity equation, we get

$$\zeta_{in} = \zeta + 2\mathbf{\Omega},$$

where ζ is the vorticity, on the rotating frame, of the internal fluid motions.

Another important parameter for measuring the fluid's tendency to rotate is *circulation*, defined as

$$C = \oint_c v_{in} \cdot dl, \tag{3.1.4}$$

where c represents a closed path along the fluid and dl is the differential displacement vector along c. Using Stokes's theorem, we can find a relation between the circulation and the vorticity of the fluid as

$$C = \int_S \zeta_{in} \cdot dS,$$

where S denotes the area whose boundary is the closed curve c and dS is the differential normal area vector of S. In this sense, circulation is interpreted as the flux of vorticity through S.

Based on the above, Jacobi ellipsoids have $\zeta = 0$ and Dedekind ellipsoids have $\Omega = 0$ (henceforth, we will refer to vorticity on the rotating frame). Homogeneous ellipsoids where both ζ and Ω are nonzero constants were found, as a generalisation of

---

[2]Two figures are said to be congruent when they have the same shape and size.

previous work, by Riemann (1861), who proved that ellipsoidal figures of equilibrium are possible in the following cases (Riemann's theorem):

(a) $\Omega = const.$ and $\zeta = 0$, leading to the Maclaurin and Jacobi sequences.
(b) Both Ω and ζ are parallel to a principal axis of the ellipsoid, resulting in ellip-soidal sequences, known as *S-type ellipsoids*, along which the ratio ζ/Ω is con-stant. Thus, Jacobi and Dedekind ellipsoids are S-type ellipsoids with $\zeta/\Omega = 0$ and ∞, respectively.
(c) Both Ω and ζ lie on a principal plane (determined by two principal axes) of the ellipsoid. Although this last case seems to be more general, it cannot encompass the previous one and produces three distinct types of ellipsoids, known as types I, II, and III.

A comprehensive study of these and more results can be found in Chandrasekhar (1969). Moreover, even though these classic studies concern homogeneous, incom-pressible figures of equilibrium, they can be extended to compressible configurations, as we will see below (Lai et al. 1993).

3.2   The Kepler Limit

In Fig. 3.2 we saw that, for a homogeneous, incompressible, uniformly rotating star (Maclaurin spheroid), there exists an upper limit in the angular velocity. Beyond this point, a further increase of the angular momentum results in a decrease of the angular velocity, which is, however, compensated by a (faster) increase of the body's moment of inertia.

Similar studies on uniformly rotating polytropes have shown that they too admit such a maximum (Lai et al. 1993). In these configurations though there is also another important limit, namely the so-called *Kepler limit*, also known as the break-up or mass-shedding limit. This is the point where centrifugal acceleration equals gravi-tational acceleration at the equator, or, equivalently, where the angular velocity of the star equals the angular velocity of a particle in a circular Keplerian orbit at the equator (Jeans 1919, Chap. VII; 1929, Chap. IX). For a homogeneous star, where the polytropic index $n = 0$ (see Appendix A), this occurs when the eccentricity $e = 1$ or the angular momentum $J \rightarrow \infty$, namely at the endpoint of the Maclaurin sequence. As the polytropic index increases, i.e., as the stellar mass is concentrated more and more towards the centre, this point moves to finite angular momenta and, for $n \gtrsim 0.5$, it precedes the point where the angular velocity obtains its maximum (see Fig. 3.5).[3]

If we consider the star as spherical, the Kepler limit Ω_K can be simply calculated as

$$\Omega_K^2 R = \frac{GM}{R^2},$$

---

[3]However, cf. Shapiro et al. (1990), where it is shown that, for a polytropic index very close to 3, the angular velocity again obtains its maximum before the point where mass shedding occurs.

or

$$\Omega_K = \sqrt{\frac{GM}{R^3}}.$$

This approximation ignores the distortion of the equilibrium due to the centrifugal force. A more accurate analytic formula can be obtained using the Roche model (see Appendix A), where

$$\Omega_K = \left(\frac{2}{3}\right)^{3/2} \sqrt{\frac{GM}{R^3}} = \sqrt{2} \left(\frac{2}{3}\right)^2 \sqrt{\pi G \langle \rho \rangle}, \qquad (3.2.1)$$

$\langle \rho \rangle$ being the mean density of the nonrotating star. In this approximation, the equatorial radius R_e is larger by a factor of $3/2$ than the radius R of the nonrotating star, at the Kepler limit. The derivation of Eq. (3.2.1) can be found in Shapiro and Teukolsky (1983, Sect. 7.4). It should be noted of course that this is only an approximate relation, which does not show the different values of Ω_K for different polytropic indices.

The fact that centrally condensed, uniformly rotating objects cannot support much rotation is restricting. For example, in polytropic stars for which $n > 0.808$, mass shedding occurs even before the bifurcation to the Jacobi ellipsoidal sequence (see Fig. 3.2)—more accurately, the polytropic analogue thereof (James 1964). Larger angular momenta could be achieved, however, if we allow for nonuniform rotation, i.e., *differential rotation*. In this case, the angular velocity Ω is not a very useful parameter. A more general way to parametrise rotation is through the ratio of the kinetic energy K to the gravitational potential energy W of the star, namely

$$\beta = \frac{K}{|W|}, \qquad (3.2.2)$$

with

$$K = \frac{1}{2} \int \rho v^2 \mathrm{d}^3 r$$

and

$$W = \frac{1}{2} \int \rho \Phi \mathrm{d}^3 r.$$

Integrating the Euler equation (2.1.2) over the volume of the star, we can obtain the scalar virial equation, i.e.,

$$2K - |W| + 3 \int p \mathrm{d}^3 r = 0$$

(see, for instance, Shapiro and Teukolsky 1983, Sect. 7.1, or Tassoul 2000, Sect. 2.8). Since the volume integral over the pressure is always a non-negative quantity, we conclude that

$$0 \leq \beta \leq \frac{1}{2}.$$
(3.2.3)

In terms of β, Maclaurin spheroids cover the whole range given by Eq. (3.2.3). On the other hand, uniformly rotating, compressible polytropes are much more confined (Lai et al. 1993), but even a small amount of differential rotation may allow β to vary over the full range again (Bodenheimer and Ostriker 1973).

3.3   *Maclaurin, Jacobi, or Dedekind?

As we saw in Sect. 3.1, there is a variety of admissible equilibrium figures for a rotating, self-gravitating fluid body. However, so far we have assumed a perfect fluid. As we will see below, the figure that the star will eventually admit is related to the presence of dissipation mechanisms, which determine its evolution through the parameter space of possible equilibria.

As first shown by Poincaré (1885), the Jacobi ellipsoid corresponds to a lower energy state than the Maclaurin spheroid with the same angular momentum. But, since we have assumed a perfect fluid, the equilibrium energy of the configuration is conserved. If we introduce *viscosity*, energy can then be dissipated, thus allowing the transition to such a lower energy state. This is an underlying consequence of the fact that viscosity conserves angular momentum, but not circulation [Eq. (3.1.4)], in the fluid (e.g., see Lai et al. 1994).

In this sense, Maclaurin spheroids are unstable in the presence of viscous dissipation beyond $e \approx 0.81$, where the bifurcation to the Jacobi sequence occurs. This point signifies the onset of a secular instability. This idea was lingering for decades, ever since Thomson and Tait (1883, § 778″) were explicitly stating that:

> The equilibrium in the revolutional figure is stable, or unstable, according as e is less than or greater than 0.81.

> If there be any viscosity, however slight, in the liquid [...] the equilibrium in any case of energy either a minimax [i.e., a saddle point] or a maximum cannot be secularly stable: and the only secularly stable configurations are those in which the energy is a minimum with a given moment of momentum [i.e., angular momentum].

The fact that this instability was taken for granted can also be seen in the work of Jeans (1919, § 44; 1929, § 196):

> Clearly the Maclaurin spheroids will be [secularly] stable up to the point at which they meet the Jacobian ellipsoids. At this point of bifurcation they lose their [secular] stability, and since the series of Jacobian ellipsoids turns upward at this point, it follows that stability passes to them.[4]

Nevertheless, the viscosity-driven secular instability had not been rigorously proven, until Roberts and Stewartson (1963), using perturbation analysis, showed that viscous

---

[4] Jeans is referring to the fact that a Jacobi ellipsoid with the same angular momentum as a Maclaurin spheroid can always be found beyond the bifurcation point, as seen in Fig. 3.2.

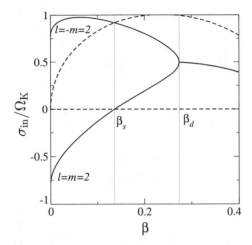

Fig. 3.3 Pattern speed σ_{in}, measured on the inertial frame and normalised to the Kepler limit Ω_K, of the quadrupole f-modes in a Maclaurin spheroid, plotted against the parameter β (solid lines). Also shown are the angular velocity of the star Ω (normalised to the Kepler limit; dashed curve) and the line along which the pattern speed is zero (dashed straight line). The points β_s and β_d indicate the onset of the secular and dynamical instabilities respectively (dotted lines). Credit: Andersson (2003), © IOP Publishing. Reproduced with permission. All rights reserved

Maclaurin spheroids become unstable, beyond the bifurcation point, to quadrupole f-modes (see below) and derived a formula for the instability time scale.

On the other hand, the Dedekind ellipsoid corresponds to a lower energy *and* angular momentum state than the Maclaurin spheroid with the same circulation. Now, viscosity does not allow the transition to the Dedekind sequence, but *gravitational radiation* does, because it emits angular momentum while conserving circulation (e.g., see Miller 1974 and Lai et al. 1994). In this case, as first shown by Chandrasekhar (1970), Maclaurin spheroids suffer a gravitational-wave-driven secular instability to quadrupole f-modes, beyond $e \approx 0.81$.

Quadrupole f-modes, often dubbed "bar modes", are those for which $l = |m| = 2$ and induce an ellipsoidal deformation on the star (see Fig. 1.1). As shown in Sect. 2.6, rotation lifts the degeneracy on the azimuthal order m and, as a result, modes are split into prograde (negative m) and retrograde (positive m), travelling with a pattern speed $\sigma = -\omega/m$ around the star, where ω is the mode eigenfrequency. In Fig. 3.3 we plot, for a Maclaurin spheroid, the pattern speed σ_{in} of the quadrupole f-modes, measured on the *inertial frame* [see Eq. (2.6.1)] and normalised to the Kepler limit Ω_K, as a function of the parameter β (solid lines). Also shown are the angular velocity of the star Ω, normalised to the Kepler limit (dashed curve; compare to Fig. 3.2), and the line along which the pattern speed is zero (dashed straight line). The points β_s and β_d (dotted lines) will be explained below.

At $\beta = \beta_s$ we notice that the prograde ($l = -m = 2$) bar mode has a pattern speed equal to the angular velocity of the star; at this point, an observer on the rotating frame would just see a stationary ellipsoid. Thus, the prograde mode perturbs the star like a

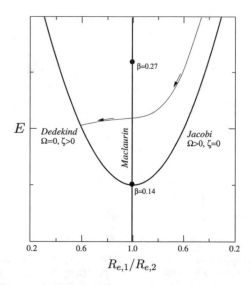

Fig. 3.4 Equilibrium energy E of the Maclaurin, Jacobi, and Dedekind sequences, plotted against the ratio of the equatorial radii $R_{e,1}/R_{e,2}$. The Maclaurin sequence corresponds to $R_{e,1}/R_{e,2} = 1$. The Jacobi ($\zeta = 0$) and Dedekind ($\Omega = 0$) sequences branch off at $\beta \approx 0.14$. At $\beta \approx 0.27$ the dynamical instability sets in. Also shown is a quasi-static evolution of a star, subjected to the gravitational-wave-driven secular instability, along an equilibrium sequence of S-type ellipsoids, where the circulation of the fluid is constant (see text). Credit: Andersson (2003), © IOP Publishing. Reproduced with permission. All rights reserved

Jacobi ellipsoid at β_s. On the other hand, the retrograde ($l = m = 2$) bar mode has a pattern speed equal to zero at $\beta = \beta_s$. On the inertial frame, this looks like a stationary ellipsoid. Hence, the retrograde mode perturbs the star like a Dedekind ellipsoid at β_s. It should have become obvious by now that this is the point of bifurcation to the Jacobi and Dedekind sequences, which occurs at

$$\beta_s \approx 0.1375. \tag{3.3.1}$$

The viscosity- and gravitational-wave-driven secular instabilities set in through the prograde and retrograde modes respectively. Ergo, for $\beta > \beta_s$, the Maclaurin spheroid is secularly unstable to quadrupole f-modes, due to viscous dissipation or emission of gravitational radiation.

A schematic of the Maclaurin sequence, bifurcating to the Jacobi and Dedekind sequences, can be seen in Fig. 3.4, where the equilibrium energy of the configurations is plotted against the ratio of the equatorial radii. Also shown is the evolution of a star towards the Dedekind sequence, due to the emission of gravitational waves. Based on the discussion above, the star evolves along an equilibrium sequence with constant circulation, until it settles on the Dedekind sequence, where it no longer emits gravitational radiation (Lai and Shapiro 1995; see also Lai et al. 1994). Under the influence of viscosity, the evolution would instead occur along an equilibrium sequence with constant angular momentum, which drives the star towards the Jacobi sequence, where viscosity vanishes (Press and Teukolsky 1973). Since the time scale of secular instabilities is much longer than the dynamical time scale, such evolutions can be considered as quasi-static, namely, the star evolves along an equilibrium sequence of S-type ellipsoids with constant circulation or constant angular momentum.

So far, we have discussed the effects of viscosity and gravitational waves separately, i.e., when only one of the two mechanisms is present. In real stars, however, they both ought to be active. In fact, they tend to cancel each other out: viscosity damps differential rotation in a Dedekind ellipsoid and gravitational waves are emitted by a Jacobi ellipsoid. The evolution of a star where both mechanisms are involved depends on the ratio of the time scales associated with each instability (Lindblom and Detweiler 1977; Detweiler and Lindblom 1977; see also Lai and Shapiro 1995). When this ratio is close to unity, namely when the two mechanisms are equally important, the Maclaurin spheroid could remain stable up until the point where the dynamical instability occurs.

The dynamical instability is an "ordinary" fluid instability, not related to any dissipation mechanism. For Maclaurin spheroids, it sets in at $e \approx 0.952887$, where $\tilde{\Omega}^2 \approx 0.440220$ or

$$\beta_d \approx 0.2738. \qquad (3.3.2)$$

This result was first obtained by Riemann (1861), using an energy variational principle. Later, Bryan (1889) showed that, at this point, the Maclaurin spheroid becomes dynamically unstable to quadrupole f-modes. As seen in Fig. 3.3, at the point of

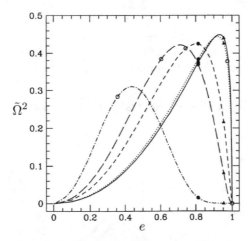

Fig. 3.5 Angular velocity (normalised to $\sqrt{\pi G \langle \rho \rangle}$) versus eccentricity, for polytropic analogues of the Maclaurin sequence. The polytropes shown are: $n = 0$ (original Maclaurin spheroids; solid line), $n = 0.1$ (dotted line), $n = 1$ (short-dashed line), $n = 1.5$ (long-dashed line), and $n = 2.5$ (dotted-dashed line). Also indicated, for each polytrope, is the Kepler limit (open circle), as well as the secular (filled circle) and dynamical (triangle) instability points. Credit: Lai et al. (1993), © American Astronomical Society (AAS). Reproduced with permission

dynamical instability the prograde and retrograde modes merge, i.e., their eigenfrequencies on the rotating frame are exactly opposite (they also acquire an imaginary part, not shown in the figure).

Interestingly enough, β_s and β_d appear strongly independent from the equation of state, or even the angular momentum distribution (if one considers differential rotation; Bodenheimer and Ostriker 1973; Ostriker and Bodenheimer 1973; Lai and Shapiro 1995). The angular velocity of uniformly rotating polytropes, as a function of the eccentricity, is shown in Fig. 3.5, where the secular and dynamical instability points, as well as the Kepler limits, are also indicated. For an extensive review of secular and dynamical instabilities in neutron stars, the reader is referred to Andersson (2003).

3.4   *A Mechanical Example

Studying the circumstances under which a secular instability might take place, Jeans (1919, § 28; 1929, § 181) was stating that:

> It [i.e., secular instability] has reference only to rotating systems or systems in a state of steady motion.

In order to gain some insight, we can look into one of the two mechanical examples devised by Lamb (1908; see also Jeans 1929, § 185), demonstrating the concept of secular instability.

We assume a spherical bowl of radius a, rotating about its vertical axis with an angular velocity Ω, in the inner surface of which a particle with mass m is free to move, as shown in Fig. 3.6. We also define a Cartesian reference frame, rotating with the bowl, with its origin lying at the lowest point of the bowl and the z axis coinciding with the axis of rotation. The Lagrangian $\mathcal{L}$ of the particle is

$$\mathcal{L} = \frac{1}{2}m \left[a^2 \dot{\theta}^2 + a^2 \sin^2 \theta \left(\Omega + \dot{\phi} \right)^2 \right] - mga(1 - \cos \theta),$$

Fig. 3.6 Lamb's mechanical analogue of secular instability, for a particle moving inside a rotating bowl. Credit: Jeans (1929, § 185)

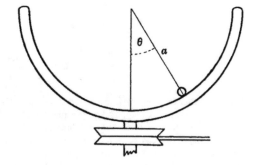

where the angle θ is the colatitude coordinate, so that $\theta = 0$ corresponds to the bottom of the bowl, as shown in Fig. 3.6, ϕ is the azimuthal coordinate, measured on the rotating frame, with $\phi = 0$ corresponding to the x axis, and g is the gravitational acceleration. From this we can obtain the equation of motion along θ, as

$$\ddot{\theta} + \left[\frac{g}{a} - \left(\Omega + \dot{\phi} \right)^2 \cos \theta \right] \sin \theta = 0.$$

Since $\partial \mathcal{L}/\partial t = 0$, the energy is conserved and given by

$$E = \frac{1}{2} ma^2 \left[\dot{\theta}^2 - \left(\Omega^2 - \dot{\phi}^2 \right) \sin^2 \theta \right] + mga(1 - \cos \theta).$$

We can now define the effective potential of the motion (including the effects of centrifugal acceleration; see Sect. 2.1), as

$$V_{\text{eff}}(\theta) = mga(1 - \cos \theta) - \frac{1}{2} ma^2 \Omega^2 \sin^2 \theta,$$

which we can use to obtain the equilibrium points for the particle, namely

$$\frac{dV_{\text{eff}}}{d\theta} = ma \left(g - a\Omega^2 \cos \theta \right) \sin \theta = 0,$$

yielding

$$\theta = 0 \quad \text{or} \quad \theta = \arccos \left(\frac{g}{a\Omega^2} \right).$$

The second solution is allowed only when

$$\Omega \geq \sqrt{\frac{g}{a}}$$

and is simply interpreted as the point where centrifugal acceleration balances gravity. The stability of these equilibrium points can be determined via the second derivative of the effective potential, which gives

$$\left. \frac{d^2 V_{\text{eff}}}{d\theta^2} \right|_{\theta=0} = ma(g - a\Omega^2) > 0 \Leftrightarrow \Omega < \sqrt{\frac{g}{a}}$$

and

$$\left. \frac{d^2 V_{\text{eff}}}{d\theta^2} \right|_{\theta=\arccos\left(\frac{g}{a\Omega^2}\right)} = m \left(a^2 \Omega^2 - \frac{g^2}{\Omega^2} \right) > 0 \Leftrightarrow \Omega > \sqrt{\frac{g}{a}}.$$

Hence, the point $\Omega = \sqrt{g/a}$ is a bifurcation point, where stability is exchanged between the two equilibrium points.

Let us now introduce a small frictional force, of the form $\ddot{\mathbf{r}} = -k\dot{\mathbf{r}}$. The equations of motion, on the rotating Cartesian frame defined above, are then expressed as

$$\ddot{x} - 2\Omega\dot{y} - \Omega^2 x = -k\dot{x} - \frac{gx}{a}$$

and

$$\ddot{y} + 2\Omega\dot{x} - \Omega^2 y = -k\dot{y} - \frac{gy}{a}.$$

It is convenient to define a new variable, $\rho = x + iy$, so that the equations of motion above can be cast into a single equation, namely

$$\ddot{\rho} + (2i\Omega + k)\,\dot{\rho} + \left(\frac{g}{a} - \Omega^2\right)\rho = 0.$$

If we perturb the particle slightly from its lowest equilibrium position, i.e., $(x, y, z) = (0, 0, 0)$, and seek solutions of the form $\rho \propto e^{\lambda t}$, we obtain

$$\rho = c_1 e^{\lambda_1 t} + c_2 e^{\lambda_2 t},$$

where c_1, c_2 are constants and λ_1, λ_2 are given by

$$\lambda_{1,2} = -i\Omega \pm i\sqrt{\frac{g}{a}} - \frac{1}{2}k\left[1 \mp \Omega\sqrt{\frac{a}{g}}\right],$$

neglecting $\mathcal{O}\left(k^2\right)$ terms. To express the solution on an inertial frame, with its axes fixed in space, we can simply multiply ρ with $e^{i\Omega t}$. Then, the solution comprises two circular motions, with angular frequencies $\omega_{1,2} = \pm\sqrt{g/a}$. The counter-rotating (retrograde) motion is always damped by friction. The corotating (prograde) motion is also damped, unless $\Omega > \sqrt{g/a}$, in which case its amplitude increases. Thus, for $\Omega > \sqrt{g/a}$, the particle is driven away from the equilibrium point and ascends the bowl, in a spiral of increasing radius. Based on the stability analysis presented above, the particle will reach a new stable equilibrium at $z = a - g/\Omega^2$.

In this mechanical example we can see how friction may induce a secular instability, with the time scale of the amplitude growth being $\tau_{\text{sec}} \sim 1/k$. To complete this simplistic analogy with the viscosity-driven secular instability, it is worth noting that, at the bifurcation point, the angular frequency of the prograde circular motion ω_1 is equal to the angular velocity of the bowl Ω, just like the pattern speed of the prograde bar mode equals the angular velocity of the Maclaurin spheroid at the point of bifurcation to the Jacobi sequence.

3.5 The CFS Instability

As we saw in Sect. 3.3, quadrupole ($l = m = 2$) f-modes can induce a secular instability in rotating stars, due to gravitational-wave emission (Chandrasekhar 1970). In an attempt to derive a generic stability criterion for rotating stars, Friedman and Schutz (1978a,b) made the astonishing discovery that, for *any* angular velocity Ω, there is always a mode driven unstable by gravitational radiation in an inviscid star!

The onset of the gravitational-wave-driven secular instability occurs when the retrograde quadrupole f-mode obtains a zero pattern speed on the inertial frame (see Fig. 3.3). If the star rotates sufficiently fast, the mode is dragged forwards and appears as prograde on the inertial frame, but is still moving backwards on the rotating frame (its pattern speed is positive, but still smaller than the angular velocity of the star). Hence, gravitational radiation emitted by the mode carries positive angular momentum away from the star, but the angular momentum of the retrograde mode itself is negative (the perturbed fluid is rotating slower than the unperturbed star). As a consequence, the emission of gravitational waves renders the angular momentum of the mode increasingly negative, which in turn makes the energy of the mode grow.[5]

Friedman and Schutz (1978a, b) showed that this instability does not apply only to quadrupole deformations. In fact, higher multipoles become unstable at lower rotation rates (smaller values of β), which implies that the instability is generic, i.e., *all* rotating, inviscid stars are unstable to the emission of gravitational waves. The *Chandrasekhar-Friedman-Schutz (CFS) instability* belongs to a larger class of instabilities, the *rotational dragging instabilities*, emerging in the presence of (i) a radiation mechanism, which deducts angular momentum from the system, and (ii) rotation, which distinguishes modes into prograde and retrograde.

A standard way to introduce (the purely relativistic phenomenon of) gravitational radiation in Newtonian models is by post-Newtonian analysis, where relativistic effects are incorporated in an expansion, whose low- or high-order terms correspond to low- or high-order deviations from Newtonian gravity (see, e.g., Thorne 1969). The power radiated in the form of gravitational waves (GW) by a single mode can be thus expanded in multipoles as (Thorne 1980; Lindblom et al. 1998)

---

[5]In the terminology of Friedman and Schutz (1978a,b), the *canonical angular momentum* J_c of a retrograde mode is negative and is related to its inertial-frame *canonical energy* E_c as

$$E_c = \sigma_{\text{in}} J_c,$$

where σ_{in} is the mode's pattern speed on the inertial frame. Then, considering perturbations described by *canonical displacements*, $E_c \geq 0$ ($E_c < 0$) implies secular stability (instability). It can be shown that the canonical energy of the mode on the rotating frame is given by

$$E_{c,R} = E_c - \Omega J_c,$$

which can be used to derive Eq. (2.6.4).

$$\left(\frac{\mathrm{d}E}{\mathrm{d}t}\right)_{\mathrm{GW}} = -\sum_{l_{\min}}^{\infty} N_l\,\omega\,(\omega - m\Omega)^{2l+1}\left(|\delta D_l^m|^2 + |\delta J_l^m|^2\right). \qquad (3.5.1)$$

First of all, we readily notice that the power emitted is negative (i.e., gravitational radiation damps the mode), unless

$$\omega(\omega - m\Omega) < 0, \qquad (3.5.2)$$

in which case the energy of the mode grows. The onset of the instability occurs when $\omega - m\Omega \equiv \omega_{\mathrm{in}} = 0$, namely when the eigenfrequency of the mode on the inertial frame [see Eq. (2.6.1)] changes sign, as anticipated from the discussion above. The angular velocity at which this happens is called *critical*.

In Eq. (3.5.1), the constant N_l is given by

$$N_l = \frac{4\pi G}{c^{2l+1}}\frac{(l+1)(l+2)}{l(l-1)\left[(2l+1)!!\right]^2}, \qquad (3.5.3)$$

c being the speed of light, whereas δD_l^m and δJ_l^m denote the *mass* and *current multipole moments* respectively, which can be expressed as

$$\delta D_l^m = \int r^l \delta\rho\, Y_l^{*m} \mathrm{d}^3 r \qquad (3.5.4)$$

and

$$\delta J_l^m = \frac{2}{c(l+1)} \int r^l\,(\rho\delta\boldsymbol{v} + \boldsymbol{v}\delta\rho)\cdot\left(\boldsymbol{r}\times\nabla Y_l^{*m}\right)\mathrm{d}^3 r \qquad (3.5.5)$$

(the Eulerian perturbations of density and velocity, $\delta\rho$ and $\delta\boldsymbol{v}$, are functions of all spatial coordinates). Finally, the lower limit of the sum is given by $l_{\min} = \max(2, |m|)$.

Comparing Eqs. (3.5.4), (3.5.5) to (2.4.1) we see that, in the nonrotating limit, polar modes radiate via the mass multipoles, which are associated with density perturbations (i.e., deformations of the star), whereas axial modes radiate via the current multipoles, which correspond to horizontal velocity perturbations (i.e., horizontal fluid motions inducing gravitomagnetic effects). When rotation is introduced, polar (axial) modes also acquire axial (polar) components, in which case they radiate via both mass and current multipoles, although mass (current) multipoles still make the most significant contribution.

The approximate multipole formula (3.5.1) produces accurate results and converges to the actual relativistic gravitational-wave luminosity as the compactness of the star, estimated by GM/Rc^2, decreases, namely, as relativistic effects become less significant. As the compactness approaches typical neutron star values (≈ 0.2), the error of the multipole formula grows (Balbinski and Schutz 1982). However, calculation of gravitational-wave luminosities using a fully relativistic framework is far from trivial and many pieces of neutron star physics are either missing or poorly understood, which often makes the use of Newtonian and post-Newtonian models unavoidable.

Among polar modes, the $l = m$ f-modes are the most susceptible to the CFS instability. As we said before, critical rotation rates are lower for higher multipoles (e.g., see Ipser and Lindblom 1990). However, the higher the multipole, the less efficient the gravitational-wave emission associated with it. This is easy to understand by the fact that large values of the degree l and the order m imply that the stellar surface is divided into many regions oscillating in antiphase (see Sect. 2.4), with the net deviation of the star from axisymmetry being small (incidentally, this is the basis of the Cowling approximation; see Sect. 2.3.4). On the other hand, a low multipole, like the quadrupole, induces a large-scale deformation on the star, thus radiating more power in gravitational waves, but might not get unstable at all; as subtly mentioned in Sect. 3.2, only stiff polytropes, for which $n < 0.808$, can reach the point of secular instability to quadrupole f-modes before shedding mass at the equator (James 1964).[6] The high eigenfrequencies of p-modes prevent them from becoming unstable, as opposed to g-modes, which may be unstable at quite low rotation rates, due to their low eigenfrequencies (Lai 1999; Passamonti et al. 2009; Gaertig and Kokkotas 2009). However, g-modes do not emit gravitational waves as efficiently as f-modes, because they have nodes in their eigenfunctions.

On the other hand, axial modes, i.e., r-modes and $l = m$ hybrid modes, are secularly unstable to the emission of gravitational waves at *all* rotation rates, namely, their critical angular velocity is zero (Andersson 1998; Friedman and Morsink 1998)! Using Eqs. (2.4.4) and (2.6.1), we can calculate the pattern speed of r modes on the rotating and inertial frames as

$$\sigma = -\frac{2\Omega}{l(l+1)}$$

and

$$\sigma_{\text{in}} = \frac{\Omega}{l(l+1)} [l(l+1) - 2]$$

respectively. Hence, r-modes are always retrograde on the rotating frame, but prograde on the inertial frame (for $l \geq 2$). The $l = m$ r-modes radiate gravitational waves more efficiently and are the most important. For $l \neq m$ hybrid modes the CFS instability criterion (3.5.2) is not always satisfied (Lindblom and Ipser 1999; Lockitch and Friedman 1999). For a detailed review of the r-mode instability in neutron stars, the reader is referred to Andersson and Kokkotas (2001).

Finally, a variation of the CFS instability appears for w-modes (Kokkotas et al. 2004), which were briefly mentioned in Sect. 1.4, but are not discussed further in the present study.

---

[6]In relativistic polytropes this limit becomes $n < 1.3$ (Stergioulas and Friedman 1998).

3.6   The Instability Window

Even though the emission of gravitational waves may induce an instability in a rotating star, viscosity acts against it, as we briefly discussed in Sect. 3.3, and tends to stabilise the star.

Viscosity can, in general, be classified into two types: *shear* and *bulk* viscosity. Shear viscosity is the result of momentum transport due to particle scattering in the fluid and describes dissipation along directions which are transverse to the flow. On the other hand, bulk viscosity is associated with dissipation during compressions and rarefactions of the fluid, during which it is driven out of chemical equilibrium. For nuclear matter consisting of neutrons, protons, and electrons, chemical equilibrium is established via the standard neutron-decay and electron-capture reactions

$$n\,(+n) \;\rightarrow\; p\,(+n) + e^- + \bar{\nu}_e$$

and

$$p\,(+n) + e^- \;\rightarrow\; n\,(+n) + \nu_e.$$

Equilibrium implies that

$$\mu_n = \mu_p + \mu_e,$$

where μ denotes the chemical potential of each species. These reactions, often referred to as *Urca processes*, dissipate energy via neutrino emission, which, as opposed to shear viscosity, cools the star down (Gamow and Schoenberg 1941; Haensel 1995).[7] Notice the additional, "spectator" nucleon in these processes. This is necessary for the conservation of energy and momentum in the degenerate neutron star matter (see, e.g., Shapiro and Teukolsky 1983, Sect. 11.2). Such *modified* Urca processes (Chiu and Salpeter 1964) prevail in the star, unless there are dense enough regions where the proton fraction (i.e., the proton number density over the baryon number density) exceeds the critical value $1/9$ (Lattimer et al. 1991), in which case *direct* Urca processes (without the bystander particle) take over.

Following the procedure of Sect. 2.2, but replacing the Euler equation (2.1.2) with the Navier–Stokes equation for a viscous fluid, one can obtain expressions for the damping of the mode due to shear and bulk viscosity (e.g., see Ipser and Lindblom 1991).

Shear viscosity (SV) dissipates the energy of the perturbation at a rate

$$\left(\frac{\mathrm{d}E}{\mathrm{d}t}\right)_{\mathrm{SV}} = -\int 2\eta\, \delta\sigma^{ab}\delta\sigma^*_{ab}\mathrm{d}^3\boldsymbol{r}, \qquad\qquad (3.6.1)$$

---

[7]According to Gamow, who, together with Schoenberg, introduced the term, Urca processes were named after a casino in Rio de Janeiro, which drained gamblers' money just like these processes drain the thermal energy of the star (Haensel 1995)!

where η is the shear viscosity coefficient and $\delta\sigma^{ab}$ is the shear tensor, which, in terms of the contravariant components of the (Eulerian) velocity perturbation, is given by

$$\delta\sigma^{ab} = \frac{1}{2}\left(\nabla^a \delta v^b + \nabla^b \delta v^a - \frac{2}{3}g^{ab}\nabla_c \delta v^c\right), \qquad (3.6.2)$$

with g^{ab} denoting the spatial metric tensor and repeated (dummy) indices implying summation. For normal nuclear matter, comprising (nonsuperfluid) neutrons, (non-superconducting) protons, and electrons, neutron collisions make the most significant contribution to shear viscosity, in which case the parameter η is given by (Flowers and Itoh 1979; Cutler and Lindblom 1987)

$$\eta = 347\,\rho^{9/4}T^{-2}\ \mathrm{g\,cm^{-1}\,s^{-1}}, \qquad (3.6.3)$$

where T is the stellar temperature and the density ρ is measured in cgs units.

If additional physics is added, shear viscosity changes accordingly. Should a solid crust form on the neutron star, shear viscosity on the crust-core interface (*Ekman layer*) is expected to provide an additional source of damping (for instance, see Bildsten and Ushomirsky 2000, for its impact on r-modes). For temperatures below which neutrons become superfluid and protons become superconducting ($\sim 10^9$ K; e.g., see Epstein 1988, Fig. 1) the most significant contribution to shear viscosity comes from electron scattering (Cutler and Lindblom 1987). An additional source of viscosity in the presence of superfluidity is an effect called *mutual friction*, due to electron scattering off of superfluid vortices (Lindblom and Mendell 1995). For the effects of superfluidity on the damping of oscillations of relativistic neutron stars, see Gusakov et al. (2013). Also, for up-to-date results about shear viscosity in neutron stars, see Shternin and Yakovlev (2008).

Bulk viscosity (BV) damps the energy of the mode as

$$\left(\frac{\mathrm{d}E}{\mathrm{d}t}\right)_{\mathrm{BV}} = -\int \zeta \delta\sigma \delta\sigma^* \mathrm{d}^3 r, \qquad (3.6.4)$$

where ζ is the bulk viscosity coefficient and $\delta\sigma$ is the expansion scalar, namely

$$\delta\sigma = \nabla_a \delta v^a. \qquad (3.6.5)$$

For normal nuclear matter, undergoing modified Urca processes, ζ is given by (Sawyer 1989; Cutler et al. 1990)[8]

$$\zeta = 6 \times 10^{-59}\,\rho^2 \omega^{-2}T^6\ \mathrm{g\,cm^{-1}\,s^{-1}}. \qquad (3.6.6)$$

---

[8]The result of Sawyer (1989) was published with a typographical error; see Lindblom (1995).

Direct Urca processes lead to stronger damping (Haensel and Schaeffer 1992). For a recent review on bulk viscosity of dense matter, see Alford et al. (2010).

Typically, the damping rate of the mode can be calculated a priori by taking viscous effects into account, i.e., by obtaining the solutions of the perturbed Navier–Stokes equation. However, like gravitational radiation, viscosity is a secular effect in neutron star oscillations and can be thus incorporated a posteriori into the solutions of the perturbed Euler equation (2.2.9). Assuming a time dependence for the perturbations of the form $e^{i(\omega - i\gamma)}$, with $|\gamma| \ll \omega$, then the energy of the mode (2.6.4), which is a quadratic functional of the displacement $\boldsymbol{\xi}$, implies that

$$\frac{\mathrm{d}E}{\mathrm{d}t} = 2\gamma E, \qquad (3.6.7)$$

with the growth/damping rate γ including the effects of both gravitational radiation and viscosity. Hence, $\gamma > 0$ ($\gamma < 0$) signifies an unstable (stable) mode. A time scale associated with the growth or damping of the mode can also be defined as[9]

$$\tau = -1/\gamma. \qquad (3.6.8)$$

Making use of the formulae for the mode energy (2.6.4) and its rate of change due to gravitational waves (3.5.1), shear viscosity (3.6.1), and bulk viscosity (3.6.4), we can obtain an expression for the mode growth/damping rate γ from Eq. (3.6.7), which depends on the angular velocity Ω and the temperature T of the star. This can be seen in Appendix C, where γ is evaluated for polar modes in the nonrotating limit. Then, solving the equation

$$\gamma \equiv \gamma_{\mathrm{GW}} + \gamma_{\mathrm{SV}} + \gamma_{\mathrm{BV}} = 0, \qquad (3.6.9)$$

we get a curve on the $T - \Omega$ plane where the onset of the CFS instability occurs. This is the *instability window* of the mode and can be seen in Fig. 3.7, for the quadrupole ($l = m = 2$), octupole ($l = m = 3$), and hexadecapole ($l = m = 4$) f-modes of a typical neutron star with $M = 1.4\,M_{\odot}$ and $R = 10\,\mathrm{km}$ ($M_{\odot}$ denotes the solar mass), described by a polytropic equation of state with a polytropic exponent $\Gamma = 2$ and 3 (equivalently, with a polytropic index $n = 1$ and 0.5).[10] It is obvious that, when viscosity is taken into account, the angular velocity where the instability sets in has a strong dependence on the temperature. As a result, there might be modes which become unstable due to the emission of gravitational waves, but the presence of viscosity completely suppresses the instability. Shear and bulk viscosity dominate at temperatures below and above $T \sim 10^9$ K respectively.

---

[9]Traditionally in the literature a negative damping time scale implies an unstable mode, e.g., see Ipser and Lindblom (1991; note that, in this paper, the damping times presented in Table 2 are erroneous, with the correct ones given in Lindblom 1995).

[10]These instability windows were *not* produced using the slow-rotation approximation described in Sect. 2.6.2, because it does not suffice for the f-mode instability to develop (see Sect. 5.1.1).

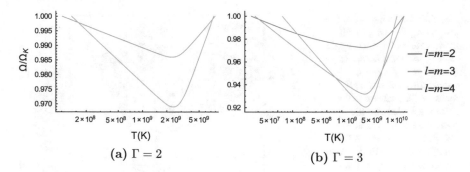

Fig. 3.7 Instability windows of the quadrupole ($l = m = 2$), octupole ($l = m = 3$), and hexade-capole ($l = m = 4$) f-modes, for a polytropic star with $M = 1.4 \, M_\odot$ and $R = 10 \, \text{km}$, with a polytropic exponent **a** $\Gamma = 2$ and **b** $\Gamma = 3$. The angular velocity is normalised to the Kepler limit Ω_K. The $\Gamma = 2$ polytrope is too soft for the quadrupole f-mode to become unstable (James 1964)

From Fig. 3.7 we see that the f-mode instability occurs at quite high rotation rates, with the quadrupole f-mode not becoming unstable for the $\Gamma = 2$ ($n = 1$) polytrope, as expected (James 1964). Relativity has been shown to enhance the instability (Stergioulas and Friedman 1998; Zink et al. 2010; Gaertig et al. 2011; see also Krüger et al. 2010, where differential rotation is included). The same applies to realistic equations of state, which may lead to larger instability windows (Doneva et al. 2013). On the other hand, the r-mode instability window is much deeper; for the quadrupole r-mode, the window minimum lies at a few percent of the mass-shedding limit (Lindblom et al. 1998), but increases for higher multipoles (see Andersson and Kokkotas 2001, Sect. 4.3).

Typically, the size of the instability window increases with the mass of the star (see, e.g., Ipser and Lindblom 1991, Lindblom 1995, and Yoshida and Eriguchi 1995). As briefly mentioned in Sect. 1.4, the most promising sources of the f-mode instability are supramassive neutron stars, namely, rotating stars with a mass larger than the maximum allowed mass of their nonrotating counterparts (see Sect. 1.4), which are supported by their fast rotation against gravitational collapse (Cook et al. 1992, 1994). The origin of these stars is discussed in Sect. 5.4. In such configurations, the f-mode instability can achieve very short growth time scales (Doneva et al. 2015). This is shown in Fig. 3.8, where the gravitational-wave growth time scale of the quadrupole ($l = m = 2$) and octupole ($l = m = 3$) f-modes, defined as $\tau_{GW} = -1/\gamma_{GW}$, is plotted against the (gravitational) mass M and the parameter β, for a neutron star described by the WFF2 equation of state (Wiringa et al. 1988). From this plot, we see that the growth time scale due to gravitational waves depends sensitively on the mass, decreasing by orders of magnitude as the mass increases. Also shown in the graph are the Kepler limit, the gravitational-collapse limit (where centrifugal acceleration on the star is not enough to prevent gravitational collapse to a black hole), and the evolution of a gravitationally radiating star with a *baryon mass*

Fig. 3.8 Contour plots of the gravitational-wave growth time scale for the quadrupole ($l = m = 2$) and octupole ($l = m = 3$) f-modes, versus the (gravitational) mass M (normalised to the solar mass $M_\odot$) and the parameter β, for a neutron star described by the WFF2 equation of state (EOS). Also shown are the Kepler limit, the gravitational-collapse limit (BH-limit), and the evolution of a star with a baryon mass $M_b = 3\,M_\odot$ as it emits gravitational waves (dashed line; see text). Credit: Doneva et al. (2015)

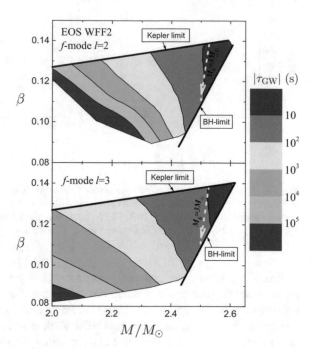

$M_b = 3\,M_\odot$, which is defined as the sum of the masses of the star's constituent particles —as opposed to the gravitational mass M, where the (negative) gravitational binding energy is added (see, e.g., Haensel et al. 2007, Sect. 6.2).

References

Alford, M. G., Mahmoodifar, S., & Schwenzer, K. (2010). Large amplitude behavior of the bulk viscosity of dense matter. *Journal of Physics G, 37*, 125202. https://doi.org/10.1088/0954-3899/37/12/125202, arXiv:1005.3769.

Andersson, N. (1998). A new class of unstable modes of rotating relativistic stars. *The Astrophysical Journal, 502*, 708–713. https://doi.org/10.1086/305919, arXiv:gr-qc/9706075.

Andersson, N. (2003). TOPICAL REVIEW: Gravitational waves from instabilities in relativistic stars. *Classical Quantum Gravity, 20*, 105. http://adsabs.harvard.edu/abs/2003CQGra..20R.105A, arXiv:astro-ph/0211057.

Andersson, N., & Kokkotas, K. D. (2001). The R-mode instability in rotating neutron stars. *International Journal of Modern Physics D, 10*, 381–441. https://doi.org/10.1142/S0218271801001062, arXiv:gr-qc/0010102.

Balbinski, E., & Schutz, B. F. (1982). A puzzle concerning the quadrupole formula for gravitational radiation. *Monthly Notices of the Royal Astronomical Society, 200*, 43–46. https://doi.org/10.1093/mnras/200.1.43P.

Bildsten, L., & Ushomirsky, G. (2000). Viscous boundary-layer damping of r-modes in neutron stars. *The Astrophysical Journal*, *529*, L33–L36. https://doi.org/10.1086/312454, arXiv:astro-ph/9911155.

Bodenheimer, P., & Ostriker, J. P. (1973). Rapidly rotating stars. VIII: Zero-viscosity polytropic sequences. *The Astrophysical Journal*, *180*, 159–170. https://doi.org/10.1086/151951.

Bryan, G. H. (1889). The waves on a rotating liquid spheroid of finite ellipticity. *Philosophical Transactions of the Royal Society of London A*, *180*, 187–219. https://doi.org/10.1098/rsta.1889.0006.

Chandrasekhar, S. (1969). *Ellipsoidal figures of equilibrium*. The Silliman Foundation Lectures. New Haven: Yale University Press. http://adsabs.harvard.edu/abs/1969efe..book.....C.

Chandrasekhar, S. (1970). Solutions of two problems in the theory of gravitational radiation. *Physical Review Letters*, *24*, 611–615. https://doi.org/10.1103/PhysRevLett.24.611.

Chiu, H.-Y., & Salpeter, E. E. (1964). Surface x-ray emission from neutron stars. *Physical Review Letters*, *12*, 413–415. https://doi.org/10.1103/PhysRevLett.12.413.

Cook, G. B., Shapiro, S. L., & Teukolsky, S. A. (1992). Spin-up of a rapidly rotating star by angular momentum loss: Effects of general relativity. *The Astrophysical Journal*, *398*, 203–223. https://doi.org/10.1086/171849.

Cook, G. B., Shapiro, S. L., & Teukolsky, S. A. (1994). Rapidly rotating neutron stars in general relativity: Realistic equations of state. *The Astrophysical Journal*, *424*, 823–845. https://doi.org/10.1086/173934.

Cutler, C., & Lindblom, L. (1987). The effect of viscosity on neutron star oscillations. *The Astrophysical Journal*, *314*, 234–241. https://doi.org/10.1086/165052.

Cutler, C., Lindblom, L., & Splinter, R. J. (1990). Damping times for neutron star oscillations. *The Astrophysical Journal*, *363*, 603–611. https://doi.org/10.1086/169370.

Dedekind, R. (1861). Zusatz zu der vorstehenden Abhandlung. *Journal für die Reine und Angewandte Mathematik*, *58*, 217–228. http://eudml.org/doc/147810.

Detweiler, S. L., & Lindblom, L. (1977). On the evolution of the homogeneous ellipsoidal figures. *The Astrophysical Journal*, *213*, 193–199. https://doi.org/10.1086/155144.

Dirichlet, P. G. L. (1861). Untersuchungen über ein Problem der Hydrodynamik. *Journal für die Reine und Angewandte Mathematik*, *58*, 181–216. http://eudml.org/doc/147809.

Doneva, D. D., Gaertig, E., Kokkotas, K. D., & Krüger, C. (2013). Gravitational wave asteroseismology of fast rotating neutron stars with realistic equations of state. *Physical Review D*, *88*, 044052. https://doi.org/10.1103/PhysRevD.88.044052, arXiv:1305.7197.

Doneva, D. D., Kokkotas, K. D., & Pnigouras, P. (2015). Gravitational wave afterglow in binary neutron star mergers. *Physical Review D*, *92*, 104040. https://doi.org/10.1103/PhysRevD.92.104040, arXiv:1510.00673.

Epstein, R. I. (1988). Acoustic properties of neutron stars. *The Astrophysical Journal*, *333*, 880–894. https://doi.org/10.1086/166797.

Flowers, E., & Itoh, N. (1979). Transport properties of dense matter. II. *The Astrophysical Journal*, *230*, 847–858. https://doi.org/10.1086/157145.

Friedman, J. L., & Morsink, S. M. (1998). Axial instability of rotating relativistic stars. *The Astrophysical Journal*, *502*, 714–720. https://doi.org/10.1086/305920, arXiv:gr-qc/9706073.

Friedman, J. L., & Schutz, B. F. (1978a). Lagrangian perturbation theory of nonrelativistic fluids. *The Astrophysical Journal*, *221*, 937–957. https://doi.org/10.1086/156098.

Friedman, J. L., & Schutz, B. F. (1978b). Secular instability of rotating Newtonian stars. *The Astrophysical Journal*, *222*, 281–296. https://doi.org/10.1086/156143.

Gaertig, E., & Kokkotas, K. D. (2009). Relativistic g-modes in rapidly rotating neutron stars. *Physical Review D*, *80*, 064026. https://doi.org/10.1103/PhysRevD.80.064026, arXiv:0905.0821.

Gaertig, E., Glampedakis, K., Kokkotas, K. D., & Zink, B. (2011). f-mode instability in relativistic neutron stars. *Physical Review Letters*, *107*, 101102. https://doi.org/10.1103/PhysRevLett.107.101102, arXiv:1106.5512.

Gamow, G., & Schoenberg, M. (1941). Neutrino theory of stellar collapse. *Physical Review*, *59*, 539–547. https://doi.org/10.1103/PhysRev.59.539.

Gusakov, M. E., Kantor, E. M., Chugunov, A. I., & Gualtieri, L. (2013). Dissipation in relativistic superfluid neutron stars. *Monthly Notices of the Royal Astronomical Society 428*, 1518–1536. https://doi.org/10.1093/mnras/sts129, arXiv:1211.2452.

Haensel, P. (1995). URCA processes in dense matter and neutron star cooling. *Space Science Reviews, 74*, 427–436. https://doi.org/10.1007/BF00751429.

Haensel, P., & Schaeffer, R. (1992). Bulk viscosity of hot-neutron-star matter from direct URCA processes. *Physical Review D, 45*, 4708–4712. https://doi.org/10.1103/PhysRevD.45.4708.

Haensel, P., Potekhin, A. Y., & Yakovlev, D. G. (2007). *Neutron Stars 1: Equation of State and Structure*, Vol. 326 of Astrophysics and Space Science Library. New York: Springer. http://adsabs.harvard.edu/abs/2007ASSL..326.....H.

Ipser, J. R., & Lindblom, L. (1990). The oscillations of rapidly rotating Newtonian stellar models. *The Astrophysical Journal, 355*, 226–240. https://doi.org/10.1086/168757.

Ipser, J. R., & Lindblom, L. (1991). The oscillations of rapidly rotating Newtonian stellar models. II: Dissipative effects. *The Astrophysical Journal, 373*, 213–221.https://doi.org/10.1086/170039.

Jacobi, C. G. J. (1834). Über die Figur des Gleichgewichts. *Annalen der Physik, 109*, 229–233. https://doi.org/10.1002/andp.18341090808.

James, R. A. (1964). The structure and stability of rotating gas masses. *The Astrophysical Journal, 140*, 552. https://doi.org/10.1086/147949.

Jeans, J. H. (1919). *Problems of cosmogony and stellar dynamics*. Cambridge, England: Cambridge University Press. http://adsabs.harvard.edu/abs/1919pcsd.book.....J.

Jeans, J. H. (1929). *Astronomy and cosmogony* (2nd ed.). Cambridge, England: Cambridge University Press. http://adsabs.harvard.edu/abs/1919pcsd.book.....J.

Kokkotas, K. D., Ruoff, J., & Andersson, N. (2004). w-mode instability of ultracompact relativistic stars. *Physical Review D, 70*, 043003. https://doi.org/10.1103/PhysRevD.70.043003, arXiv:astro-ph/0212429.

Krüger, C., Gaertig, E., & Kokkotas, K. D. (2010). Oscillations and instabilities of fast and differentially rotating relativistic stars. *Physical Review D, 81*, 084019. https://doi.org/10.1103/PhysRevD.81.084019, arXiv:0911.2764.

Lai, D. (1999). Secular instability of g-modes in rotating neutron stars. *Monthly Notices of the Royal Astronomical Society, 307*, 1001–1007. https://doi.org/10.1046/j.1365-8711.1999.02723.x, arXiv:astro-ph/9806378.

Lai, D., & Shapiro, S. L. (1995). Gravitational radiation from rapidly rotating nascent neutron stars. *The Astrophysical Journal, 442*, 259–272. https://doi.org/10.1086/175438, arXiv:astro-ph/9408053.

Lai, D., Rasio, F. A., & Shapiro, S. L. (1993). Ellipsoidal figures of equilibrium: Compressible models. *The Astrophysical Journal Supplement Series, 88*, 205–252. https://doi.org/10.1086/191822.

Lai, D., Rasio, F. A., & Shapiro, S. L. (1994). Hydrodynamics of rotating stars and close binary interactions: Compressible ellipsoid models. *The Astrophysical Journal, 437*, 742–769. https://doi.org/10.1086/175036, arXiv:astro-ph/9404031.

Lamb, H. (1908). On kinetic stability. *Proceedings of the Royal Society of London. Series A, 80*, 168–177. https://doi.org/10.1098/rspa.1908.0013.

Lattimer, J. M., Prakash, M., Pethick, C. J., & Haensel, P. (1991). Direct URCA process in neutron stars. *Physical Review Letters, 66*, 2701–2704. https://doi.org/10.1103/PhysRevLett.66.2701.

Lindblom, L. (1995). Critical angular velocities of rotating neutron stars. *The Astrophysical Journal, 438*, 265–268. https://doi.org/10.1086/175071.

Lindblom, L., & Ipser, J. R. (1999). Generalized r-modes of the Maclaurin spheroids. *Physical Review D, 59*, 044009. https://doi.org/10.1103/PhysRevD.59.044009, arXiv:gr-qc/9807049.

Lindblom, L., & Detweiler, S. L. (1977). On the secular instabilities of the Maclaurin spheroids. *The Astrophysical Journal, 211*, 565–567. https://doi.org/10.1086/154964.

Lindblom, L., Owen, B. J., & Morsink, S. M. (1998). Gravitational radiation instability in hot young neutron stars. *Physical Review Letters, 80*, 4843–4846. https://doi.org/10.1103/PhysRevLett.80.4843, arXiv:gr-qc/9803053.

Lindblom, L., & Mendell, G. (1995). Does gravitational radiation limit the angular velocities of superfluid neutron stars? *The Astrophysical Journal, 444,* 804–809. https://doi.org/10.1086/175653.

Lockitch, K. H., & Friedman, J. L. (1999). Where are the r-modes of isentropic stars? *The Astrophysical Journal, 521,* 764–788. https://doi.org/10.1086/307580, arXiv:gr-qc/9812019.

Maclaurin, C. (1742). *A Treatise of Fluxions.* Edinburgh: T. W. and T. Ruddimans. https://catalog.hathitrust.org/Record/000615723.

Meyer, C. (1842). De Aequilibrii formis Ellipsoidicis. *Journal für die reine und angewandte Mathematik, 24,* 44–59. http://eudml.org/doc/183244.

Miller, B. D. (1974). The effect of gravitational radiation-reaction on the evolution of the Riemann S-type ellipsoids. *The Astrophysical Journal, 187,* 609–620. https://doi.org/10.1086/152671.

Ostriker, J. P., & Bodenheimer, P. (1973). On the oscillations and stability of rapidly rotating stellar models. III: Zero-viscosity polytropic sequences. *The Astrophysical Journal, 180,* 171–180. https://doi.org/10.1086/151952.

Passamonti, A., Haskell, B., Andersson, N., Jones, D. I., & Hawke, I. (2009). Oscillations of rapidly rotating stratified neutron stars. *Monthly Notices of the Royal Astronomical Society, 394,* 730–741. https://doi.org/10.1111/j.1365-2966.2009.14408.x, arXiv:0807.3457.

Poincaré, H. (1885). Sur l'équilibre d'une masse fluide animée d'un mouvement de rotation. *Acta Mathematica, 7,* 259–380. https://doi.org/10.1007/BF02402204.

Press, W. H., & Teukolsky, S. A. (1973). On the evolution of the secularly unstable viscous Maclaurin spheroids. *The Astrophysical Journal, 181,* 513–518. https://doi.org/10.1086/152066.

Riemann, B. (1861). Ein Beitrag zu den Untersuchungen über die Bewegungen eines gleichartigen flüssigen Ellipsoids. *Abhandlungen der Königlichen Gesellschaft der Wissenschaften zu Göttingen, 9,* 3–36. http://eudml.org/doc/135728.

Roberts, P. H., & Stewartson, K. (1963). On the stability of a Maclaurin spheroid of small viscosity. *The Astrophysical Journal, 137,* 777. https://doi.org/10.1086/147555.

Sawyer, R. F. (1989). Bulk viscosity of hot neutron-star matter and the maximum rotation rates of neutron stars. *Physical Review D, 39,* 3804–3806. https://doi.org/10.1103/PhysRevD.39.3804.

Shapiro, S. L., & Teukolsky, S. A. (1983). *Black holes, white dwarfs, and neutron stars: The physics of compact objects.* New York: Wiley. http://adsabs.harvard.edu/abs/1983bhwd.book.....S.

Shapiro, S. L., Teukolsky, S. A., & Nakamura, T. (1990). Spin-up of a rapidly rotating star by angular momentum loss. *The Astrophysical Journal, 357,* L17–L20. http://adsabs.harvard.edu/abs/1928asco.book.....J.

Shternin, P. S., & Yakovlev, D. G. (2008). Shear viscosity in neutron star cores. *Physical Review D, 78,* 063006. https://doi.org/10.1103/PhysRevD.78.063006, arXiv:0808.2018.

Stergioulas, N., & Friedman, J. L. (1998). Nonaxisymmetric neutral modes in rotating relativistic stars. *The Astrophysical Journal, 492,* 301–322. https://doi.org/10.1086/305030, arXiv:gr-qc/9705056.

Tassoul, J.-L. (2000). *Stellar rotation* (vol. 36). Cambridge Astrophysics Series. New York: Cambridge University Press. http://adsabs.harvard.edu/abs/2000stro.book.....T.

Thomson, W., & Tait, P. G. (1883). *Treatise on natural philosophy* (vol. 2, 2nd ed.). Cambridge, England: Cambridge University Press. https://doi.org/10.1017/CBO9780511703935.

Thorne, K. S. (1969). Nonradial pulsation of general-relativistic stellar models. IV: The weak-field limit. *The Astrophysical Journal, 158,* 997. https://doi.org/10.1086/150259.

Thorne, K. S. (1980). Multipole expansions of gravitational radiation. *Reviews of Modern Physics, 52,* 299–340. https://doi.org/10.1103/RevModPhys.52.299.

Wiringa, R. B., Fiks, V., & Fabrocini, A. (1988). Equation of state for dense nucleon matter. *Physical Review C, 38,* 1010–1037. https://doi.org/10.1103/PhysRevC.38.1010.

Yoshida, S., & Eriguchi, Y. (1995). Gravitational radiation driven secular instability of rotating polytropes. *The Astrophysical Journal, 438,* 830–840. https://doi.org/10.1086/175126.

Zink, B., Korobkin, O., Schnetter, E., & Stergioulas, N. (2010). Frequency band of the f-mode Chandrasekhar-Friedman-Schutz instability. *Physical Review D, 81,* 084055. https://doi.org/10.1103/PhysRevD.81.084055, arXiv:1003.0779.

Chapter 4
Mode Coupling: Quadratic Perturbation Scheme

In Chap. 2 we established a linear perturbation formalism, in order to describe the oscillation modes of a star. However, as seen in Chap. 3, unstable modes may grow to large amplitudes, at which the linear approximation is no longer accurate; higher-order perturbative terms are bound to play an important role in the amplitude evolution, since they introduce *mode coupling*. The result of this nonlinear interaction of the unstable mode with other modes is the eventual saturation of the unstable mode's amplitude.

As in previous work about the saturation of the r-mode instability (Schenk et al. 2001; Morsink 2002; Arras et al. 2003; Brink et al. 2004b, a, 2005; Bondarescu et al. 2007, 2009), we will consider quadratic perturbations and study their effects on the evolution of the f-mode, mainly following the pioneering work of Dziembowski (1982; see also the follow-up studies by Dziembowski and Krolikowska 1985 and Dziembowski et al. 1988). Research on mode coupling in stellar oscillations can be traced back to Vandakurov (1979), although it had been already introduced in the study of plasma waves (see, e.g., Stenflo et al. 1970, Wilhelmsson et al. 1970, Verheest 1976, and Anderson 1976). Ever since, the subject has attracted the attention of many investigators (among others, Takeuti and Aikawa 1981, Aikawa 1983, Buchler 1983, Buchler and Regev 1983, Dziembowski and Kovács 1984, Aikawa 1984, Dappen and Perdang 1985, Buchler and Kovacs 1986b, a, Kumar and Goldreich 1989, Verheest 1990, 1993, Wu and Goldreich 2001, Nowakowski 2005, and Passamonti et al. 2007; see also the reviews by Dziembowski 1993 and Smolec 2014).

The process we will follow is similar to that of Chap. 2 for linear perturbations, but now we are also going to consider quadratic terms to define the perturbations (Sect. 4.1). We will show that, in the quadratic-perturbation approximation, modes couple in triplets, which satisfy a resonance condition (Sect. 4.2). Coupling of an unstable mode to other (stable) modes of the star can lead to the saturation of the unstable mode's amplitude, through a mechanism known as parametric resonance instability (Sect. 4.3). For the saturation to be successful, some stability conditions, which determine the amplitude evolution of the coupled triplet, have to be satisfied

© Springer Nature Switzerland AG 2018
P. Pnigouras, *Saturation of the f-mode Instability in Neutron Stars*,
Springer Theses, https://doi.org/10.1007/978-3-319-98258-8_4

(Sect. 4.4), with some interesting behaviours occurring throughout the parameter space, like limit cycles, chaotic orbits, and frequency synchronisation.

Even higher than second-order terms could, in principle, be important at large oscillation amplitudes, but the complexity of the formulation and the requirements of our problem allow us to choose simplicity over accuracy. Work that also includes cubic nonlinearities can be found in Buchler and Goupil (1984), Van Hoolst and Smeyers (1993), and Van Hoolst (1994a, b). Also, for a more general investigation of systems with quadratic and cubic nonlinearities, the reader is referred to Nayfeh and Mook (1979, Chap. 6).

4.1 Quadratic Perturbation Formalism

In order to derive the equation of motion for quadratic perturbations, we have to follow the procedure of Sect. 2.2, used for the derivation of the equation of motion for linear perturbations (2.2.9), except that now we also want to retain second-order perturbative terms.

We decompose the Eulerian perturbation of a quantity f, given by Eq. (2.2.1), as

$$\delta f = \delta_1 f + \delta_2 f, \tag{4.1.1}$$

where the subscripts 1 and 2 denote first- and second-order perturbative terms respectively. The Lagrangian perturbation of f, defined by Eq. (2.2.2), can be decomposed accordingly and is related to the Eulerian, to quadratic order in $\boldsymbol{\xi}$, as

$$\Delta f = \delta_1 f + (\boldsymbol{\xi} \cdot \nabla) f + \delta_2 f + (\boldsymbol{\xi} \cdot \nabla)\delta_1 f + \frac{1}{2}\boldsymbol{\xi} \cdot [(\boldsymbol{\xi} \cdot \nabla) \nabla f] \tag{4.1.2}$$

[compare with Eq. (2.2.3)].

Perturbing the fluid equations (2.1.1)–(2.1.4) about their equilibrium, to quadratic order, we obtain

$$\frac{\partial \delta \rho}{\partial t} + \nabla \cdot (\rho \boldsymbol{v}) + \nabla \cdot (\delta_1 \rho \boldsymbol{v}) = 0, \tag{4.1.3}$$

$$\frac{\partial \boldsymbol{v}}{\partial t} + (\boldsymbol{v} \cdot \nabla)\boldsymbol{v} + 2\boldsymbol{\Omega} \times \boldsymbol{v} = -\frac{\nabla \delta p}{\rho} - \delta\left(\frac{1}{\rho}\right)\nabla p - \delta_1\left(\frac{1}{\rho}\right)\nabla \delta_1 p - \nabla \delta \Phi, \tag{4.1.4}$$

$$\nabla^2 \delta \Phi = 4\pi G \delta \rho, \tag{4.1.5}$$

and

$$\frac{\Delta p}{p} = \Gamma_1 \frac{\Delta \rho}{\rho} + \frac{1}{2}\left[\Gamma_1(\Gamma_1 - 1) + \left(\frac{\partial \Gamma_1}{\partial \ln \rho}\right)_\mu\right]\left(\frac{\Delta_1 \rho}{\rho}\right)^2, \tag{4.1.6}$$

where we set $v \equiv \delta v$ (since we work on the rotating frame, the background velocity is zero). In the perturbed equation of state (4.1.6), μ corresponds to entropy or composition and is assumed to be constant in a perturbed fluid element ($\Delta \mu \approx 0$; see Sect. 2.2), whereas the adiabatic exponent Γ_1 is given by Eq. (2.2.8). Studies where nonadiabatic effects are also considered can be found in Buchler and Goupil (1984) and Van Hoolst (1994a).

Following Dziembowski (1982), we will use the velocity v instead of the displacement ξ to describe the perturbation. The equation of motion for quadratic perturbations can then be written in the form

$$\ddot{v} + \mathcal{B}(\dot{v}) + \mathcal{C}(v) + \mathcal{N} = 0, \tag{4.1.7}$$

where

$$\mathcal{B}(v) = 2\Omega \times v \tag{4.1.8}$$

and

$$\mathcal{C}(v) = \frac{1}{\rho}\nabla\left(\frac{\partial \delta_1 p}{\partial t}\right) - \frac{\nabla p}{\rho^2}\frac{\partial \delta_1 \rho}{\partial t} + \nabla\left(\frac{\partial \delta_1 \Phi}{\partial t}\right), \tag{4.1.9}$$

whereas quadratic terms are collectively denoted by $\mathcal{N}$, which, in terms of first-order quantities, is given by

$$\mathcal{N} = \frac{\partial}{\partial t}\left[\nabla\left(\frac{v \cdot v}{2}\right) - v \times (\nabla \times v) - \frac{\delta_1 \rho}{\rho^2}\nabla \delta_1 p + \frac{(\delta_1 \rho)^2}{2\rho^3}\nabla p\right]$$
$$+ \frac{1}{\rho}\nabla\left[\!\left[\nabla \cdot v \left\{\xi \cdot \nabla(p\Gamma_1) + p\Gamma_1\left[\Gamma_1 + \left(\frac{\partial \ln \Gamma_1}{\partial \ln \rho}\right)_\mu\right]\nabla \cdot \xi\right\} - v \cdot \nabla \delta_1 p\right]\!\right]$$
$$+ \frac{\nabla p}{\rho}(v \cdot \nabla)\left(\frac{\delta_1 \rho}{\rho}\right) + G\nabla\left[\int \frac{\nabla' \cdot (\delta_1 \rho v)}{|r - r'|}d^3 r'\right]. \tag{4.1.10}$$

The details of the derivation of the equations above can be found in Appendix D.1.

With the help of Eq. (2.6.2), we may now expand the velocity in terms of the eigenmodes of the star, i.e.,

$$v(r, t) = \sum_\alpha i\omega_\alpha\left[Q_\alpha(t)\xi_\alpha(r)e^{i\omega_\alpha t} - Q_\alpha^*(t)\xi_\alpha^*(r)e^{-i\omega_\alpha t}\right]. \tag{4.1.11}$$

Starting with the linear terms in Eq. (4.1.7), we get

$$\dot{Q}_\alpha = \frac{1}{\omega_\alpha b_\alpha}\langle \xi_\alpha, \mathcal{N}\rangle e^{-i\omega_\alpha t}, \tag{4.1.12}$$

where b_α is given by Eq. (2.6.3). This is the equation of motion for the amplitude of the mode, Q_α. If quadratic terms are ignored (or, equivalently, if the perturbation is small), then the amplitude Q_α is constant. However, as we will see, a nonzero $\mathcal{N}$ couples the mode denoted by α with other modes, leading to an energy exchange between them.

By further replacing Eq. (4.1.11) in $\mathcal{N}$, we obtain

$$\dot{Q}_\alpha(t) = \frac{i}{b_\alpha} \sum_\beta \sum_\gamma \Big[F_{\alpha\beta\gamma} Q_\beta Q_\gamma e^{i(-\omega_\alpha+\omega_\beta+\omega_\gamma)t} + F_{\alpha\bar{\beta}\gamma} Q_\beta^* Q_\gamma e^{i(-\omega_\alpha-\omega_\beta+\omega_\gamma)t}$$

$$+ F_{\alpha\beta\bar{\gamma}} Q_\beta Q_\gamma^* e^{i(-\omega_\alpha+\omega_\beta-\omega_\gamma)t} + F_{\alpha\bar{\beta}\bar{\gamma}} Q_\beta^* Q_\gamma^* e^{i(-\omega_\alpha-\omega_\beta-\omega_\gamma)t} \Big],$$

$$(4.1.13)$$

where F denotes the *coupling coefficient*, which is generally given by

$$F_{\alpha\beta\gamma} = \frac{1}{i\omega_\alpha} \langle \boldsymbol{\xi}_\alpha, \mathcal{N}(\boldsymbol{\xi}_\beta, \boldsymbol{\xi}_\gamma) \rangle. \qquad (4.1.14)$$

Borrowing the notation of Schenk et al. (2001), a bar over an index means that the corresponding mode eigenfunction in $\mathcal{N}$ has to be complex conjugated and its eigenfrequency sign reversed. The derivation of the amplitude equation of motion can be found in Appendix D.2, where an explicit formula for the coupling coefficient is also given [Eq. (D.2.10)].

4.2 Resonant Mode Coupling

4.2.1 Coupled Triplet Equations of Motion

Observing Eq. (4.1.13), we see that modes couple in triplets, which is a natural consequence of the quadratic-perturbation approximation. This does not, however, restrict the number of couplings for a single mode; if a mode couples to a pair of other modes, it can simultaneously couple to other pairs as well. Also, one can notice that not all terms of Eq. (4.1.13) are equally significant. Rapidly varying terms do not contribute much on long-term dynamics and average to zero, as opposed to slowly oscillating components (see Dziembowski 1982). Hence, couplings which ultimately affect the mode amplitude evolution ought to satisfy a resonance condition, e.g.,

$$\omega_\alpha = \omega_\beta + \omega_\gamma + \Delta\omega, \qquad (4.2.1)$$

where $\Delta\omega$ is a small detuning ($\Delta\omega \ll \omega_k$, $k = \alpha, \beta, \gamma$); this is shown in Sect. 4.2.2, with the help of the multiscale method. Assuming such a relation between the mode eigenfrequencies, we can single out a resonant mode triplet as

$$\dot{Q}_\alpha = \frac{i F_{\alpha\beta\gamma}}{b_\alpha} Q_\beta Q_\gamma e^{-i\Delta\omega t}, \tag{4.2.2a}$$

$$\dot{Q}_\beta = \frac{i F_{\beta\bar{\gamma}\alpha}}{b_\beta} Q_\gamma^* Q_\alpha e^{i\Delta\omega t}, \tag{4.2.2b}$$

$$\dot{Q}_\gamma = \frac{i F_{\gamma\alpha\bar{\beta}}}{b_\gamma} Q_\alpha Q_\beta^* e^{i\Delta\omega t}. \tag{4.2.2c}$$

From the derivation of Eqs. (4.2.2) we see that, in such resonant mode couplings, nonlinear effects develop on a secular time scale, which is large compared to the dynamical time scale associated with mode eigenfrequencies.[1] As we discussed in Chap. 3, the same applies to mechanisms like gravitational waves and viscosity, which usually damp the mode, but may also drive it unstable, by increasing its amplitude. Hence, assuming that these mechanisms act on the same (secular) time scale with nonlinear effects (see Sect. 4.2.2), they can be incorporated manually in Eqs. (4.2.2) as

$$\dot{Q}_\alpha = \gamma_\alpha Q_\alpha + \frac{i\mathcal{H}}{b_\alpha} Q_\beta Q_\gamma e^{-i\Delta\omega t}, \tag{4.2.3a}$$

$$\dot{Q}_\beta = \gamma_\beta Q_\beta + \frac{i\mathcal{H}}{b_\beta} Q_\gamma^* Q_\alpha e^{i\Delta\omega t}, \tag{4.2.3b}$$

$$\dot{Q}_\gamma = \gamma_\gamma Q_\gamma + \frac{i\mathcal{H}}{b_\gamma} Q_\alpha Q_\beta^* e^{i\Delta\omega t}, \tag{4.2.3c}$$

where γ_k represents the linear growth/damping rate of the mode due to gravitational radiation and viscosity, and is given by Eq. (3.6.7). Furthermore, we replaced the coupling coefficients with $\mathcal{H} \equiv \mathcal{F}_{\alpha\beta\gamma} = \mathcal{F}_{\beta\bar{\gamma}\alpha} = \mathcal{F}_{\gamma\alpha\bar{\beta}}$; this has been shown by various authors (e.g., Buchler and Regev 1983), and was explicitly proven by Dziembowski (1982) for coupling coefficients of polar modes in the nonrotating limit. This proof is outlined in Appendix E.

Nonresonant mode coupling has also been considered by various studies (e.g., Buchler and Kovacs 1986a and Verheest 1990, 1993), but second-order amplitude terms make no contribution in this case and nonlinear coupling is introduced at third-order.

4.2.2 *The Multiscale Method

Let us assume that we have an ordinary differential equation which includes a small parameter ϵ. We write the solution to this equation in the form of an asymptotic

---

[1] In fact, this was explicitly assumed in a previous step, namely during the derivation of Eq. (4.1.12) in Appendix D.2 [Eq. (D.2.6)]. Had we not made this assumption, the amplitude equation of motion would be given by the more general Eq. (D.2.5). However, since only resonant triplets contribute to the amplitude evolution, it is, in retrospect, valid.

series, in the sense that

$$y(t) \to \sum_{n=0}^{\infty} y_n(t)\epsilon^n.$$

In the beginning of the evolution, when t is small, low-order terms dominate the solution. However, as t grows larger, the contribution of higher-order terms cannot be neglected. These terms are usually called secular, because their effects become important (compared to low-order terms) at later stages of the evolution. This behaviour appears, for example, in a damped harmonic oscillator, where the zeroth-order solution is simply an undamped harmonic oscillation, with the damping effects occurring at higher orders.

The multiscale method (see, for instance, Nayfeh and Mook 1979, Sect. 2.3.3) is a way to capture such higher-order effects from secular terms and make them appear in the low-order terms. As a result, the low-order approximation of the solution would be valid on secular time scales.

We define the time scales $T_n = \epsilon^n t$ and rewrite the asymptotic solution, so that

$$y(t) \to \sum_{n=0}^{\infty} y_n(T_0, T_1, T_2, \ldots)\epsilon^n.$$

In other words, we let the terms of the series depend on more than one time scale. As we will see, this allows us to "eliminate" secular effects from higher-order terms, thus preventing these terms from becoming significant.

We are going to use this method, in order to study Eqs. (4.2.3). First, we remove the exponential time dependence by setting $C_k = Q_k e^{i\omega_k t}$ ($k = \alpha, \beta, \gamma$) and the equations of motion are written as

$$\dot{C}_\alpha - i\omega_\alpha C_\alpha = \gamma_\alpha C_\alpha + \frac{i\mathcal{H}}{b_\alpha} C_\beta C_\gamma, \qquad (4.2.4a)$$

$$\dot{C}_\beta - i\omega_\beta C_\beta = \gamma_\beta C_\beta + \frac{i\mathcal{H}}{b_\beta} C_\gamma^* C_\alpha, \qquad (4.2.4b)$$

$$\dot{C}_\gamma - i\omega_\gamma C_\gamma = \gamma_\gamma C_\gamma + \frac{i\mathcal{H}}{b_\gamma} C_\alpha C_\beta^*. \qquad (4.2.4c)$$

Now, we seek solutions of the form

$$C_k = \epsilon C_k^{(1)}(T_0, T_1) + \epsilon^2 C_k^{(2)}(T_0, T_1) + \mathcal{O}\left(\epsilon^3\right),$$

where $T_0 = t$ and $T_1 = \epsilon t$. Time derivatives then become

$$\frac{d}{dt} = \frac{\partial}{\partial T_0} + \frac{dT_1}{dT_0} \frac{\partial}{\partial T_1} = \frac{\partial}{\partial T_0} + \epsilon \frac{\partial}{\partial T_1}.$$

Replacing the solutions in Eqs. (4.2.4) and distinguishing between $\mathcal{O}(\epsilon)$ and $\mathcal{O}(\epsilon^2)$ terms, we get

$$\frac{\partial C_\alpha^{(1)}}{\partial T_0} - i\omega_\alpha C_\alpha^{(1)} = 0,$$

$$\frac{\partial C_\beta^{(1)}}{\partial T_0} - i\omega_\beta C_\beta^{(1)} = 0,$$

$$\frac{\partial C_\gamma^{(1)}}{\partial T_0} - i\omega_\gamma C_\gamma^{(1)} = 0,$$

and

$$\frac{\partial C_\alpha^{(1)}}{\partial T_1} + \frac{\partial C_\alpha^{(2)}}{\partial T_0} - i\omega_\alpha C_\alpha^{(2)} = \hat{\gamma}_\alpha C_\alpha^{(1)} + \frac{i\mathcal{H}}{b_\alpha} C_\beta^{(1)} C_\gamma^{(1)},$$

$$\frac{\partial C_\beta^{(1)}}{\partial T_1} + \frac{\partial C_\beta^{(2)}}{\partial T_0} - i\omega_\beta C_\beta^{(2)} = \hat{\gamma}_\beta C_\beta^{(1)} + \frac{i\mathcal{H}}{b_\beta} C_\gamma^{*(1)} C_\alpha^{(1)},$$

$$\frac{\partial C_\gamma^{(1)}}{\partial T_1} + \frac{\partial C_\gamma^{(2)}}{\partial T_0} - i\omega_\gamma C_\gamma^{(2)} = \hat{\gamma}_\gamma C_\gamma^{(1)} + \frac{i\mathcal{H}}{b_\gamma} C_\alpha^{(1)} C_\beta^{*(1)},$$

respectively, where we also set $\gamma_k = \epsilon\hat{\gamma}_k$, so that growth/damping and nonlinear terms appear in the same order.
  The first-order equations have simple solutions of the form

$$C_k^{(1)}(T_0, T_1) = A_k(T_1)e^{i\omega_k T_0}, \tag{4.2.5}$$

which we substitute to the second-order equations, to get

$$\frac{\partial C_\alpha^{(2)}}{\partial T_0} - i\omega_\alpha C_\alpha^{(2)} = \left(\hat{\gamma}_\alpha A_\alpha - \frac{dA_\alpha}{dT_1}\right)e^{i\omega_\alpha T_0} + \frac{i\mathcal{H}}{b_\alpha} A_\beta A_\gamma e^{i(\omega_\beta + \omega_\gamma)T_0}, \tag{4.2.6a}$$

$$\frac{\partial C_\beta^{(2)}}{\partial T_0} - i\omega_\beta C_\beta^{(2)} = \left(\hat{\gamma}_\beta A_\beta - \frac{dA_\beta}{dT_1}\right)e^{i\omega_\beta T_0} + \frac{i\mathcal{H}}{b_\beta} A_\gamma^* A_\alpha e^{i(\omega_\alpha - \omega_\gamma)T_0}, \tag{4.2.6b}$$

$$\frac{\partial C_\gamma^{(2)}}{\partial T_0} - i\omega_\gamma C_\gamma^{(2)} = \left(\hat{\gamma}_\gamma A_\gamma - \frac{dA_\gamma}{dT_1}\right)e^{i\omega_\gamma T_0} + \frac{i\mathcal{H}}{b_\gamma} A_\alpha A_\beta^* e^{i(\omega_\alpha - \omega_\beta)T_0}. \tag{4.2.6c}$$

As we mentioned earlier, the whole point of the multiscale method is to transfer long-term effects from high- to low-order terms. In this case, we want to prevent the second-order terms of the solution, $C_k^{(2)}$, from growing and becoming important. To accomplish this, we have to eliminate the so-called secular terms. In the case of Eqs. (4.2.6), terms that include the factor $e^{i\omega_k T_0}$ have to vanish, because they produce secular terms, causing the solution to grow in time.

Nonresonant Case

If there is no resonance of the form $\omega_\alpha \approx \omega_\beta + \omega_\gamma$ between the modes, then the conditions for the elimination of secular terms from Eqs. (4.2.6) are

$$\frac{dA_k}{dT_1} = \hat{\gamma}_k A_k,$$

or

$$A_k = a_k e^{\hat{\gamma}_k T_1},$$

which makes the first-order solutions (4.2.5)

$$C_k = \epsilon C_k^{(1)} + \mathcal{O}(\epsilon^2) = \epsilon a_k e^{\gamma_k t} e^{i\omega_k t} + \mathcal{O}(\epsilon^2),$$

or, in terms of the original variables Q_k,

$$Q_k = \epsilon a_k e^{\gamma_k t} + \mathcal{O}(\epsilon^2). \tag{4.2.7}$$

Equation (4.2.7) shows that, if there is no resonance between the modes, their amplitudes grow or decrease with time, depending on the sign of γ_k.

Resonant Case

If a resonance of the form $\omega_\alpha = \omega_\beta + \omega_\gamma + \Delta\omega$ exists ($\Delta\omega$ being a small detuning), then the second terms on the right-hand sides of Eqs. (4.2.6) also contribute in the production of secular terms in the solution. Then, the secular-term elimination conditions become

$$\frac{dA_\alpha}{dT_1} = \hat{\gamma}_\alpha A_\alpha + \frac{i\mathcal{H}}{b_\alpha} A_\beta A_\gamma e^{-i\Delta\hat{\omega} T_1}, \tag{4.2.8a}$$

$$\frac{dA_\beta}{dT_1} = \hat{\gamma}_\beta A_\beta + \frac{i\mathcal{H}}{b_\beta} A_\gamma^* A_\alpha e^{i\Delta\hat{\omega} T_1}, \tag{4.2.8b}$$

$$\frac{dA_\gamma}{dT_1} = \hat{\gamma}_\gamma A_\gamma + \frac{i\mathcal{H}}{b_\gamma} A_\alpha A_\beta^* e^{i\Delta\hat{\omega} T_1}, \tag{4.2.8c}$$

where we set $\Delta\omega = \epsilon\Delta\hat{\omega}$. From Eqs. (4.2.8) we obtain our original system (4.2.3).

4.2.3 Coupling Selection Rules

As we already mentioned, the three modes forming the coupled network have to obey a resonance condition, given by Eq. (4.2.1). The structure of the coupling coefficient imposes two more conditions, which have to be met in order for coupling to occur.

As shown in Appendix E.2, the angular dependence of the zeroth-order component (with respect to rotation) of the coupling coefficient has the form

$$Z_{\alpha\beta\gamma} = \iint Y_\alpha^* Y_\beta Y_\gamma \sin\theta d\theta d\phi, \tag{4.2.9}$$

where

$$Y_\alpha \equiv Y_{l_\alpha}^{m_\alpha}.$$

This integral is proportional to the *Clebsch-Gordan coefficients* [see Eq. (E.2.10)] and is nonzero if

$$m_\alpha = m_\beta + m_\gamma \tag{4.2.10}$$

and

$$l_i = l_j + l_k - 2\lambda, \tag{4.2.11}$$

where

$$l_i \geq l_j \geq l_k \quad \text{and} \quad \lambda = 0, 1, \ldots \lambda_{\max} \leq \frac{l_k}{2},$$

with the indices i, j, k taking the values α, β, γ, so that mode i has the largest degree and mode k has the lowest. Equations (4.2.10) and (4.2.11) constitute the selection rules which the coupled mode triplet has to satisfy and restrict the search for possible couplings.[2]

4.2.4 Mode Normalisation

As discussed in Sects. 2.3.7 and 2.6.1, the amplitudes Q depend on mode normalisation. We normalise the modes according to Eq. (2.6.5), by fixing their energy at unit amplitude to some constant value E_{unit}. Then, Eqs. (4.2.3) can be written as

$$\dot{Q}_\alpha = \gamma_\alpha Q_\alpha + i\omega_\alpha \frac{\mathcal{H}}{E_{\text{unit}}} Q_\beta Q_\gamma e^{-i\Delta\omega t}, \tag{4.2.12a}$$

$$\dot{Q}_\beta = \gamma_\beta Q_\beta + i\omega_\beta \frac{\mathcal{H}}{E_{\text{unit}}} Q_\gamma^* Q_\alpha e^{i\Delta\omega t}, \tag{4.2.12b}$$

$$\dot{Q}_\gamma = \gamma_\gamma Q_\gamma + i\omega_\gamma \frac{\mathcal{H}}{E_{\text{unit}}} Q_\alpha Q_\beta^* e^{i\Delta\omega t}. \tag{4.2.12c}$$

From this form of the amplitude equations of motion it is easier to see that the coupling coefficient $\mathcal{H}$ has units of energy.

---

[2]Equations (4.2.10) and (4.2.11) were derived for the coupling coefficient in the nonrotating limit, in which case each mode is described by a single spherical harmonic, thus reducing the angular part of the coupling coefficient to the simple integral (4.2.9). However, they should also be valid when rotation is included, as elegantly shown by Schenk et al. (2001), who also derived some additional, albeit less general, selection rules.

The value of the coupling coefficient is also normalisation-dependent. Since the energy of a mode on the rotating frame, given by Eq. (2.6.6), should be normalisation-independent, Eqs. (4.2.12) may be cast into the alternative form

$$\dot{\mathcal{Q}}_\alpha = \gamma_\alpha \mathcal{Q}_\alpha + i\omega_\alpha \mathcal{H} \mathcal{Q}_\beta \mathcal{Q}_\gamma e^{-i\Delta\omega t}, \tag{4.2.13a}$$

$$\dot{\mathcal{Q}}_\beta = \gamma_\beta \mathcal{Q}_\beta + i\omega_\beta \mathcal{H} \mathcal{Q}_\gamma^* \mathcal{Q}_\alpha e^{i\Delta\omega t}, \tag{4.2.13b}$$

$$\dot{\mathcal{Q}}_\gamma = \gamma_\gamma \mathcal{Q}_\gamma + i\omega_\gamma \mathcal{H} \mathcal{Q}_\alpha \mathcal{Q}_\beta^* e^{i\Delta\omega t}, \tag{4.2.13c}$$

where

$$\mathcal{Q}_i = Q_i E_{\text{unit}}^{1/2}$$

and

$$\mathcal{H} = \frac{\mathcal{H}}{E_{\text{unit}}^{3/2}},$$

which are normalisation-independent quantities. Thus, for a different normalisation choice E'_{unit}, the amplitudes are rescaled according to Eq. (2.6.7) and the coupling coefficient transforms as

$$\frac{\mathcal{H}'}{\mathcal{H}} = \left(\frac{E'_{\text{unit}}}{E_{\text{unit}}} \right)^{3/2}. \tag{4.2.14}$$

Finally, based on the above and Eq. (2.6.6), the energy of the coupled triplet is given by

$$E = \sum_k E_k = E_{\text{unit}} \sum_k |Q_k|^2 = \sum_k |\mathcal{Q}_k|^2, \tag{4.2.15}$$

with $k = \alpha, \beta, \gamma$.

4.3 Parametric Resonance Instability

Having derived the amplitude equations of motion for a resonant coupled triplet (4.2.12), we may now focus on the case where one of the modes (say, mode α) is unstable. Our goal is to check whether coupling of the unstable mode to other modes in the star can stop its growth.

Let us assume that the unstable, or *parent*, mode ($\gamma_\alpha > 0$) couples to two stable, or *daughter*, modes ($\gamma_{\beta,\gamma} < 0$). From the amplitude equations of motion (4.2.12) we see that, in the beginning of the evolution, when the amplitudes are small, linear terms dominate: the amplitude of the parent grows and the amplitudes of the daughters decrease. At some point, nonlinear terms catch up and eventually dominate, thus increasing the daughters' amplitudes. Such an interaction between the modes is an example of a *parametric resonance instability*, i.e., an instability which can occur when the parameters of an oscillator vary in time (see, for example, Landau and Lifshitz 1969, § 27).[3,4] The onset of the parametric instability occurs when the parent exceeds a certain amplitude, called the *parametric instability threshold* (PIT), given by

$$|Q_{\text{PIT}}|^2 = \frac{\gamma_\beta \gamma_\gamma}{\omega_\beta \omega_\gamma} \frac{E_{\text{unit}}^2}{\mathcal{H}^2} \left[1 + \left(\frac{\Delta \omega}{\gamma_\beta + \gamma_\gamma} \right)^2 \right]. \tag{4.3.1}$$

The derivation of Eq. (4.3.1) can be found in Appendix F.1.

Ignoring nonlinear effects in the beginning of the evolution, parent growth is described by $\dot{Q}_\alpha = \gamma_\alpha Q_\alpha$, which means that the parametric instability threshold is crossed at

$$\tau_{\text{PIT}} = \frac{1}{\gamma_\alpha} \ln \left[\frac{Q_{\text{PIT}}}{Q_\alpha(0)} \right], \tag{4.3.2}$$

where $Q_\alpha(0)$ is the parent's initial amplitude. This is of order the growth time scale of the unstable mode, $|\tau_\alpha| = 1/\gamma_\alpha$ [see Eq. (3.6.8)].

The evolution of the parametrically resonant triplet, after the parametric instability threshold is crossed, is characterised by a constant energy exchange between the three modes (see Fig. 4.2a, but ignore the details for now). When the daughters' amplitudes become large enough, the nonlinear term dominates the parent's evolution, causing a drop in its amplitude. As a result, the daughters' nonlinear terms shrink, with their linear terms taking the lead again. Now, the daughters' amplitudes decrease, which means that so does the parent's nonlinear term. Consequently, the linear term dominates the parent's evolution once more and its amplitude increases. This process repeats itself, until some kind of equilibrium is established (as we will see in Sect. 4.4.2, this is not the only possibility).

As shown in Appendix F.2, the amplitude equations of motion (4.2.12) admit such an equilibrium solution, which is given by

$$|Q_\alpha|^2 = \frac{\gamma_\beta \gamma_\gamma}{\omega_\beta \omega_\gamma} \frac{E_{\text{unit}}^2}{\mathcal{H}^2} \left[1 + \left(\frac{\Delta \omega}{\gamma} \right)^2 \right], \tag{4.3.3a}$$

$$|Q_\beta|^2 = -\frac{\gamma_\gamma \gamma_\alpha}{\omega_\gamma \omega_\alpha} \frac{E_{\text{unit}}^2}{\mathcal{H}^2} \left[1 + \left(\frac{\Delta \omega}{\gamma} \right)^2 \right], \tag{4.3.3b}$$

---

[3] Simple examples of parametric instability are pendula in which the length of the string is being varied periodically or the point of support oscillates vertically.

[4] The word "instability", used to describe the phenomenon of parametric resonance, may cause some confusion, because, so far, we were only referring to stability or instability due to the presence of some damping or growth mechanism, like viscosity and/or gravitational radiation (see Sects. 3.5 and 3.6). The parametric resonance discussed here is a consequence of the resonant nonlinear coupling of an unstable mode to two stable modes, resulting in the growth of the latter and, thus, inducing an instability. Hence, to avoid any confusion, we will use phrases like "parametrically unstable", when referring to modes undergoing the parametric resonance instability.

$$|Q_\gamma|^2 = -\frac{\gamma_\alpha \gamma_\beta}{\omega_\alpha \omega_\beta} \frac{E_{\text{unit}}^2}{\mathcal{H}^2} \left[1 + \left(\frac{\Delta\omega}{\gamma} \right)^2 \right],$$ (4.3.3c)

where

$$\gamma = \gamma_\alpha + \gamma_\beta + \gamma_\gamma.$$ (4.3.4)

Notice that, for $|\gamma_\beta + \gamma_\gamma| \gg \gamma_\alpha$, the equilibrium amplitude of the unstable mode (4.3.3a) coincides with the parametric instability threshold (4.3.1).

4.4 Saturation

4.4.1 Stability Conditions

The existence of the equilibrium solution (4.3.3) implies that the two stable (daughter) modes can potentially halt the growth of the unstable (parent) mode, making it saturate at a finite amplitude, given by Eq. (4.3.3a). However, the equilibrium need not always be stable, i.e., saturation may not always be successful.

Performing a linear stability analysis of Eqs. (4.2.12), which is presented in Appendix F.3, we find that the equilibrium solution (4.3.3) is stable if

$$|\gamma_\beta + \gamma_\gamma| > \gamma_\alpha$$ (4.4.1)

and

$$3\left\{ (\zeta_\beta + \zeta_\gamma - 1) \left[(\zeta_\beta - \zeta_\gamma)^2 + 2(\zeta_\beta + \zeta_\gamma) + 1 \right] - 6\zeta_\beta \zeta_\gamma \right\} \left(\frac{\Delta\omega}{\gamma} \right)^4$$
$$+ \left\{ (\zeta_\beta + \zeta_\gamma - 1) \left[(\zeta_\beta - \zeta_\gamma)^2 + (\zeta_\beta + \zeta_\gamma)^2 + 2 \right] - 12\zeta_\beta \zeta_\gamma \right\} \left(\frac{\Delta\omega}{\gamma} \right)^2$$
$$- (\zeta_\beta + \zeta_\gamma - 1)^3 - 2\zeta_\beta \zeta_\gamma > 0,$$ (4.4.2)

where

$$\zeta_{\beta,\gamma} = -\gamma_{\beta,\gamma}/\gamma_\alpha,$$ (4.4.3)

which are the relative damping rates of the daughters, and γ is defined in Eq. (4.3.4).

We can obtain a simpler expression for Eq. (4.4.2) in the following cases:

(a) $\zeta_\beta = \zeta_\gamma \equiv \zeta$

When the damping rates of the daughters $\gamma_{\beta,\gamma}$ are the same, Eqs. (4.4.1) and (4.4.2) give

$$\zeta > \frac{1 + \sqrt{3}}{2} \approx 1.37 \tag{4.4.4}$$

and

$$\Delta^2 > \frac{2\zeta^2 - 2\zeta + 1}{2\zeta^2 - 2\zeta - 1}(1 - 2\zeta)^2, \tag{4.4.5}$$

where

$$\Delta = \Delta\omega/\gamma_\alpha. \tag{4.4.6}$$

The asymptotic behaviour of Δ, for large ζ, is

$$|\Delta| > 2\zeta - 1. \tag{4.4.7}$$

It is interesting to note that Eq. (4.4.4) imposes a stronger constraint on ζ than Eq. (4.4.1).

(b) $\zeta_\gamma \to 0$

If the damping rate of one of the daughters is negligible, Eqs. (4.4.1) and (4.4.2) yield

$$\zeta_\beta > 1 \tag{4.4.8}$$

and

$$\Delta^2 > \left[-\frac{1 + \zeta_\beta^2}{3\left(1 + \zeta_\beta\right)^2} + \frac{2}{3}\sqrt{\frac{\zeta_\beta^4 - \zeta_\beta^2 + 1}{\left(\zeta_\beta + 1\right)^4}} \right] \left(1 - \zeta_\beta\right)^2. \tag{4.4.9}$$

The asymptotic behaviour of Δ, for large ζ_β, is

$$|\Delta| > \frac{\zeta_\beta - 2}{\sqrt{3}}. \tag{4.4.10}$$

Cases (a) and (b), along with the general conditions given by Eqs. (4.4.1) and (4.4.2), are plotted in Fig. 4.1. Since the same arguments apply to both positive and negative values of the detuning $\Delta\omega$, we only consider the case $\Delta\omega > 0$. The two stability conditions (4.4.1) and (4.4.2) [or, equivalently, (4.4.4) and (4.4.5) for case (a), and (4.4.8) and (4.4.9) for case (b)] show that (i) the damping rates of the daughters $\gamma_{\beta,\gamma}$ should be larger (in absolute value) than the growth rate of the parent γ_α, and (ii) the resonance detuning $\Delta\omega$ must have a lower limit, depending on the growth/damping rates of the triplet.

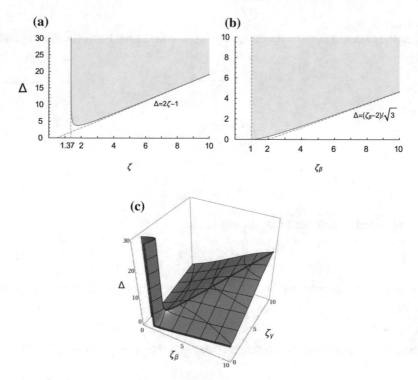

Fig. 4.1 **a** Δ versus $\zeta \equiv \zeta_\beta = \zeta_\gamma$. The stability condition (4.4.5) is satisfied inside the shaded area. The two asymptotes at $\zeta \approx 1.37$ [Eq. (4.4.4)] and $\Delta = 2\zeta - 1$ [Eq. (4.4.7)] are also shown (dashed lines). A global minimum occurs at (1.77, 3.73). **b** Δ versus ζ_β, with $\zeta_\gamma \to 0$. The stability conditions (4.4.8) and (4.4.9) are satisfied inside the shaded area. The asymptote at $\Delta = (\zeta_\beta - 2)/\sqrt{3}$ [Eq. (4.4.10)] is also shown, along with the limit $\zeta_\beta = 1$ posed by Eq. (4.4.8) (dashed lines). **c** Δ versus ζ_β versus ζ_γ. The stability conditions (4.4.1) and (4.4.2) are satisfied inside the region that lies above the plotted surface. The thick lines correspond to cases (a) and (b)

4.4.2   Possible Evolutions

The impact of the stability conditions, given by Eqs. (4.4.1) and (4.4.2), on the parametrically resonant system (4.2.12) has been studied by various authors (e.g., Wersinger et al. 1980b, a, and Dimant 2000; see also Ott 1981), who discovered interesting behaviours throughout the parameter space.

When both conditions are satisfied, the equilibrium solution for the triplet amplitudes (4.3.3) is stable, i.e., the triplet amplitudes converge around their equilibrium solution and saturation is successful, as seen in Fig. 4.2a.

The significance of Eq. (4.4.1) is fairly easy to see: the daughters have to dissipate the incoming energy from the parent faster than the parent grows. Otherwise, the parent's amplitude keeps growing (at a rate lower than γ_α), dragging the daughters along as it does so. The three modes diverge from their equilibria, by constantly

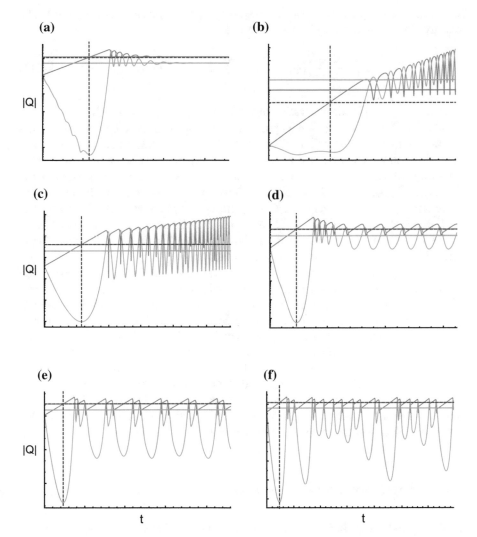

Fig. 4.2 Evolution of a parametrically resonant triplet with two identical daughters [case (a) in Sect. 4.4.1]. Horizontal solid lines denote the equilibrium amplitudes of the modes (4.3.3), whereas the horizontal and vertical dashed lines indicate the parametric instability threshold (4.3.1) and the time it is crossed by the parent (4.3.2) respectively. *Top panel:* **a** Successful saturation, with both stability conditions (4.4.4) and (4.4.5) satisfied ($\zeta = 5$, $\Delta = 20$). **b** Unsuccessful saturation, with condition (4.4.4) being false ($\zeta = 1/3$, $\Delta = 10/3$). **c** Unsuccessful saturation, with condition (4.4.5) being false ($\zeta = 4$, $\Delta = 0.1$). *Bottom panel:* Quasi-successful saturation, with condition (4.4.5) being false. **d** Limit cycle of period 1 ($\zeta = 5$, $\Delta = 5$). **e** Limit cycle of period 2 ($\zeta = 12$, $\Delta = 2$). **f** Chaotic motion ($\zeta = 15$, $\Delta = 2$)

exchanging energy at an increasing frequency, and saturation fails (Fig. 4.2b). The energy of the system (4.2.15) grows at a rate

$$\frac{d}{dt} \sum_k |Q_k|^2 E_{unit} = \sum_k 2\gamma_k |Q_k|^2 E_{unit}, \tag{4.4.11}$$

with $k = \alpha, \beta, \gamma$ [for the derivation of Eq. (4.4.11), see Appendix F.2].

On the other hand, Eq. (4.4.2) poses a surprising constraint, by demanding that the detuning have a lower limit. When this condition is not satisfied [but Eq. (4.4.1) is], a rich variety of evolutions can occur, depending on the values of the parameters. Growing solutions may still appear for small values of the damping rates or the detuning (Fig. 4.2c), but bounded evolutions dominate throughout the rest of the parameter space. Wersinger et al. (1980a, b) report the appearance of limit cycles, with periods[5] ranging from 1 to 32, as well as chaotic orbits, where the amplitudes of the modes oscillate around their equilibrium values. In these cases, saturation is considered "quasi-successful" (Fig. 4.2, bottom panel).

The simplest case, of a limit cycle with period 1, was more thoroughly examined by Moskalik (1985; see also Moskalik 1986) and is shown in Fig. 4.2d. The time scale of the modulation is of order the growth time scale of the parent mode $|\tau_\alpha|$ [see Eq. (3.6.8)] and its peak-to-peak depth mainly depends on the ratio $|\Delta/\zeta|$, for a triplet with two identical daughters [case (a) in Sect. 4.4.1]. The modulation is larger when $|\Delta/\zeta|$ is small, i.e., when (i) the three modes are close to resonance ($\Delta\omega = \gamma_\alpha \Delta \approx 0$), or (ii) the daughters are strongly damped. The latter may sound unexpected, but, as seen from Eqs. (4.2.12), large daughter damping rates "delay" the nonlinear terms from becoming significant, thus allowing the parent to reach higher amplitude values. This characteristic probably explains why triplets with small detunings need to also have small daughter damping rates in order to successfully saturate (see Fig. 4.1).

4.4.3 *Frequency Synchronisation

From the discussion in Sects. 4.4.1 and 4.4.2 we see that the detuning $\Delta\omega$ of the resonant triplet affects the evolution of the system in a notable manner. Interestingly enough though, this mismatch between the mode eigenfrequencies is compensated by nonlinear effects. Applying a procedure presented in detail in Appendix F.2, we split the complex amplitude Q_k into its real amplitude and phase components, as

$$Q_k = |Q_k| e^{i\vartheta_k},$$

with $k = \alpha, \beta, \gamma$. Incorporating the phase in the harmonic time dependence of the mode, we get

$$\xi_k \propto e^{i(\omega_k t + \vartheta_k)},$$

---

[5] A cycle with period n intersects the Poincaré section n times (Wersinger et al. 1980b, a).

suggesting that the eigenfrequency ω_k of the mode is shifted, due to nonlinear coupling, to

$$\omega_k' = \omega_k + \dot{\vartheta}_k, \tag{4.4.12}$$

where the shift $\dot{\vartheta}_k$ is given by

$$\dot{\vartheta}_k = \omega_k \frac{\mathcal{H}}{E_{\text{unit}}} \frac{|Q_\alpha Q_\beta Q_\gamma|}{|Q_k|^2} \cos\varphi, \tag{4.4.13}$$

with

$$\varphi = \vartheta_\alpha - \vartheta_\beta - \vartheta_\gamma + \Delta\omega t.$$

As shown in Appendix F.2, equilibrium implies that $\dot{\varphi} = 0$, or

$$\omega_\alpha' = \omega_\beta' + \omega_\gamma'. \tag{4.4.14}$$

This has been referred to as *frequency synchronisation* (Aikawa 1984) or phase lock (Dziembowski and Kovács 1984), and is an anticipated effect of nonlinear resonance.

Replacing equilibrium values in Eq. (4.4.13), we find that the shifted eigenfrequency is

$$\omega_k' = \omega_k - |\gamma_k| \frac{\Delta\omega}{\gamma}, \tag{4.4.15}$$

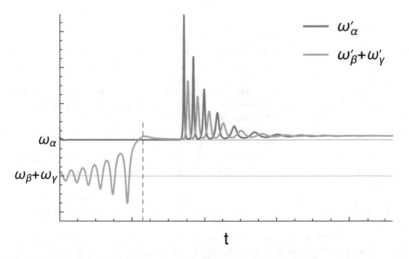

Fig. 4.3 Evolution of the shifted (due to nonlinear coupling) mode eigenfrequencies towards nonlinear resonance, for the triplet of Fig. 4.2a. Horizontal lines denote the eigenfrequencies of the linear system and the vertical dashed line indicates the time when the parametric instability threshold is crossed by the parent. Credit: Pnigouras and Kokkotas (2016)

where γ is given by Eq. (4.3.4). It should be noted that Eqs. (4.4.14) and (4.4.15) are valid only in the case of successful saturation, when the mode amplitudes are constants.[6] The evolution of the shifted mode eigenfrequencies towards nonlinear resonance is shown in Fig. 4.3, for the triplet of Fig. 4.2a.

References

Aikawa, T. (1983). On the secular variation of amplitudes in double-mode Cepheids. *Monthly Notices of the Royal Astronomical Society, 204*, 1193–1202. http://adsabs.harvard.edu/abs/1983MNRAS.204.1193A.

Aikawa, T. (1984). Period shifts and synchronization in resonant mode interactions of non-linear stellar pulsation. *Monthly Notices of the Royal Astronomical Society, 206*, 833–842. http://adsabs.harvard.edu/abs/1984MNRAS.206..833A.

Anderson, D. (1976). Nonresonant wave-coupling and wave-particle interactions. *Physica Scripta, 13*, 117–121. https://doi.org/10.1088/0031-8949/13/2/010.

Arras, P., Flanagan, É. É., Morsink, S. M., Schenk, A. K., Teukolsky, S. A., & Wasserman, I. (2003). Saturation of the r-mode instability. *The Astrophysical Journal, 591*, 1129–1151. https://doi.org/10.1086/374657, arXiv:astro-ph/0202345.

Bondarescu, R., Teukolsky, S. A., & Wasserman, I. (2007). Spin evolution of accreting neutron stars: Nonlinear development of the r-mode instability. *Physical Review D, 76*, 064019. https://doi.org/10.1103/PhysRevD.76.064019, arXiv:0704.0799.

Bondarescu, R., Teukolsky, S. A., & Wasserman, I. (2009). Spinning down newborn neutron stars: Nonlinear development of the r-mode instability. *Physical Review D, 79*, 104003. https://doi.org/10.1103/PhysRevD.79.104003, arXiv:0809.3448.

Brink, J., Teukolsky, S. A., & Wasserman, I. (2004a). Nonlinear coupling network to simulate the development of the r-mode instability in neutron stars. I. Construction. *Physical Review D, 70*, 124017. https://doi.org/10.1103/PhysRevD.70.124017, arXiv:gr-qc/0409048.

Brink, J., Teukolsky, S. A., & Wasserman, I. (2004b). Nonlinear couplings of R-modes: Energy transfer and saturation amplitudes at realistic timescales. *Physical Review D, 70*, 121501. https://doi.org/10.1103/PhysRevD.70.121501, arXiv:gr-qc/0406085.

Brink, J., Teukolsky, S. A., & Wasserman, I. (2005). Nonlinear coupling network to simulate the development of the r mode instability in neutron stars. II. Dynamics. *Physical Review D, 71*, 064029. https://doi.org/10.1103/PhysRevD.71.064029, arXiv:gr-qc/0410072.

Buchler, J. R. (1983). Resonance effects in radial pulsators. *Astronomy & Astrophysics, 118*, 163–165. http://adsabs.harvard.edu/abs/1983A%26A...118..163B.

Buchler, J. R., & Regev, O. (1983). The effects of nonlinearities on radial and nonradial oscillations. *Astronomy & Astrophysics, 123*, 331–342. http://adsabs.harvard.edu/abs/1983A%26A...123..331B.

Buchler, J. R., & Goupil, M.-J. (1984). Amplitude equations for nonadiabatic nonlinear stellar pulsators I. The formalism. *The Astrophysical Journal, 279*, 394–400. https://doi.org/10.1086/161900.

Buchler, J. R., & Kovacs, G. (1986a). On the modal selection of radial stellar pulsators. *The Astrophysical Journal, 308*, 661–668. https://doi.org/10.1086/164537.

Buchler, J. R., & Kovacs, G. (1986b). The effects of a 2:1 resonance in nonlinear radial stellar pulsations. *The Astrophysical Journal, 303*, 749–765. https://doi.org/10.1086/164122.

---

[6]In principle, they could also apply to the quasi-stable equilibria discussed in Sect. 4.4.2, as average-value relations (Moskalik 1985).

Dappen, W., & Perdang, J. (1985). Non-linear stellar oscillations. Non-radial mode interactions. *Astronomy & Astrophysics*, *151*, 174–188. http://adsabs.harvard.edu/abs/1985A %26A...151..174D.

Dimant, Y. S. (2000). Nonlinearly saturated dynamical state of a three-wave mode-coupled dissipative system with linear instability. *Physical Review Letters*, *84*, 622. https://doi.org/10.1103/ PhysRevLett.84.622.

Dziembowski, W. (1982). Nonlinear mode coupling in oscillating stars. I. Second order theory of the coherent mode coupling. *Acta Astronomica*, *32*, 147–171. http://adsabs.harvard.edu/abs/ 1982AcA....32..147D.

Dziembowski, W. (1993). Mode selection and other nonlinear phenomena in stellar oscillations. In W. W. Weiss and A. Baglin, (Eds.), *IAU Colloquia 137: Inside the Stars* (vol. 40). Astronomical Society of the Pacific Conference Series. http://adsabs.harvard.edu/abs/1993ASPC...40..521D.

Dziembowski, W., & Kovács, G. (1984). On the role of resonances in double-mode pulsation. *Monthly Notices of the Royal Astronomical Society*, *206*, 497–519. http://adsabs.harvard.edu/ abs/1984MNRAS.206..497D.

Dziembowski, W., & Krolikowska, M. (1985). Nonlinear mode coupling in oscillating stars. II. Limiting amplitude effect of the parametric resonance in main sequence stars. *Acta Astronomica*, *35*, 5–28. http://adsabs.harvard.edu/abs/1985AcA....35....5D.

Dziembowski, W., Krolikowska, M., & Kosovichev, A. (1988). Nonlinear mode coupling in oscillating stars. III. Amplitude limiting effect of the rotation in the Delta Scuti stars. *Acta Astronomica*, *38*, 61–75. http://adsabs.harvard.edu/abs/1988AcA....38...61D.

Kumar, P., & Goldreich, P. (1989). Nonlinear interactions among solar acoustic modes. *The Astrophysical Journal*, *342*, 558–575. https://doi.org/10.1086/167616.

Landau, L. D., & Lifshitz, E. M. (1969). *Mechanics* (vol. 1, 2nd ed.). Course of Theoretical Physics. New York: Pergamon Press. http://adsabs.harvard.edu/abs/1969mech.book.....L.

Morsink, S. M. (2002). Nonlinear Couplings between r-Modes of Rotating Neutron Stars. *The Astrophysical Journal*, *571*, 435–446. https://doi.org/10.1086/339858, arXiv:astro-ph/0202051.

Moskalik, P. (1985). Modulation of amplitudes in oscillating stars due to resonant mode coupling. *Acta Astronomica*, *35*, 229–254. http://adsabs.harvard.edu/abs/1985AcA....35..229M.

Moskalik, P. (1986). Amplitude modulation due to internal resonances as a possible explanation of the Blazhko effect in RR Lyrae stars. *Acta Astronomica*, *36*, 333–353. http://adsabs.harvard.edu/ abs/1986AcA....36..333M.

Nayfeh, A. H., & Mook, D. T. (1979). *Nonlinear Oscillations*. Pure & Applied Mathematics. New York: Wiley. http://adsabs.harvard.edu/abs/1979noos.book.....N.

Nowakowski, R. M. (2005). Multimode resonant coupling in pulsating stars. *Acta Astronomica*, *55*, 1–41. http://adsabs.harvard.edu/abs/2005AcA....55....1N, arXiv:astro-ph/0501510.

Ott, E. (1981). Strange attractors and chaotic motions of dynamical systems. *Reviews of Modern Physics*, *53*, 655–671. https://doi.org/10.1103/RevModPhys.53.655.

Passamonti, A., Stergioulas, N., & Nagar, A. (2007). Gravitational waves from nonlinear couplings of radial and polar nonradial modes in relativistic stars. *Physical Review D*, *75*, 084038. https:// doi.org/10.1103/PhysRevD.75.084038, arXiv:gr-qc/0702099.

Pnigouras, P., & Kokkotas, K. D. (2016). Saturation of the f-mode instability in neutron stars. II. Applications and results. *Physical Review D*, *94*, 024053. https://doi.org/10.1103/PhysRevD.94. 024053, arXiv:1607.03059.

Schenk, A. K., Arras, P., Flanagan, É. É., Teukolsky, S. A., & Wasserman, I. (2001). Nonlinear mode coupling in rotating stars and the r-mode instability in neutron stars. *Physical Review D*, *65*, 024001. https://doi.org/10.1103/PhysRevD.65.024001, arXiv:gr-qc/0101092.

Smolec, R. (2014). Mode selection in pulsating stars. In J. A. Guzik, W. J. Chaplin, G. Handler and A. Pigulski (Eds.), *Precision Asteroseismology* (vol. 9). Proceedings of the IAU Symposium 301. Wroclaw, Poland. https://doi.org/10.1017/S1743921313014439, arXiv:1309.5959.

Stenflo, L., Weiland, J., & Wilhelmsson, H. (1970). A solution of equations describing explosive instabilities. *Physica Scripta*, *1*, 46. https://doi.org/10.1088/0031-8949/1/1/008.

Takeuti, M., & Aikawa, T. (1981). Resonance phenomenon in classical cepheids. *Science reports of the Tohoku University, Eighth Series*, *2*, 106–129. http://adsabs.harvard.edu/abs/1981SRToh... 2..106T.

Van Hoolst, T. (1994a). Coupled-mode equations and amplitude equations for nonadiabatic, non-radial oscillations of stars. *Astronomy & Astrophysics*, *292*, 471–480. http://adsabs.harvard.edu/abs/1994A%26A...292..471V.

Van Hoolst, T. (1994b). Nonlinear, nonradial, isentropic oscillations of stars: Hamiltonian formalism. *Astronomy & Astrophysics*, *286*, 879–889. http://adsabs.harvard.edu/abs/1994A%26A... 286..879V.

Van Hoolst, T., & Smeyers, P. (1993). Non-linear, non-radial, isentropic oscillations of stars: Third-order coupled-mode equations. *Astronomy & Astrophysics*, *279*, 417–430. http://adsabs.harvard. edu/abs/1993A%26A...279..417V.

Vandakurov, Y. V. (1979). Nonlinear Coupling of Stellar Pulsations. *Soviet Astronomy*, *23*, 421. http://adsabs.harvard.edu/abs/1979SvA....23..421V.

Verheest, F. (1976). Possible nonlinear wave-wave coupling between three or four waves in plasmas. *Plasma Physics*, *18*, 225–234. https://doi.org/10.1088/0032-1028/18/3/008.

Verheest, F. (1990). Nonresonant three-mode coupling as a model for double-mode pulsators. *Astrophysics and Space Science*, *166*, 77–91. https://doi.org/10.1007/BF00655609.

Verheest, F. (1993). Nonresonant mode coupling in double-mode pulsators. *Astrophysics and Space Science*, *200*, 325–330. https://doi.org/10.1007/BF00627139.

Wersinger, J.-M., Finn, J. M., & Ott, E. (1980a). Bifurcation and "strange" behavior in instability saturation by nonlinear three-wave mode coupling. *Physics of Fluids*, *23*, 1142–1154. https://doi. org/10.1063/1.863116.

Wersinger, J.-M., Finn, J. M., & Ott, E. (1980b). Bifurcations and strange behavior in instability saturation by nonlinear mode coupling. *Physical Review Letters*, *44*, 453–456. https://doi.org/10. 1103/PhysRevLett.44.453.

Wilhelmsson, H., Stenflo, L., & Engelmann, F. (1970). Explosive instabilities in the well-defined phase description. *Journal of Mathematical Physics*, *11*, 1738–1742. https://doi.org/10.1063/1. 1665320.

Wu, Y., & Goldreich, P. (2001). Gravity Modes in ZZ Ceti Stars. IV. Amplitude Saturation by Parametric Instability. *The Astrophysical Journal*, *546*, 469–483. https://doi.org/10.1086/318234, arXiv:astro-ph/0003163.

Chapter 5
Results

As opposed to the linear perturbation scheme, presented in Chap. 2, which gives rise to the oscillation spectrum of the star, quadratic perturbations form networks of interacting, resonantly coupled, mode triplets, which can potentially saturate unstable modes by means of the parametric resonance instability mechanism, as shown in Chap. 4. Building on the foundations laid in these chapters, we are going to investigate the saturation of f-modes, driven unstable by the emission of gravitational waves via the CFS mechanism, as seen in Chap. 3, in neutron stars modelled as Newtonian polytropes.

After setting up the problem at hand (Sect. 5.1), we derive some helpful approximate relations for the parametric instability threshold (Sect. 5.2), used later during the discussion of the results. We will present results both for typical (Sect. 5.3) and supramassive (Sect. 5.4) neutron stars, for various polytropic and adiabatic exponents. In each section, we review the evolution of the instability and discuss the results. We also present estimations about the contribution of the f-mode instability to the stochastic gravitational-wave background, both from typical and supramassive neutron stars (Sect. 5.5). Finally, we review the studies on the saturation of the r-mode instability via mode coupling and compare them to our results (Sect. 5.6).

5.1 Setup

We can obtain the saturation amplitude of an unstable f-mode, for a specific temperature T and angular velocity Ω of the star, by following these steps:

1. Calculate mode eigenfrequencies ω and eigenfunctions $\boldsymbol{\xi}$. We first obtain the modes in the nonrotating limit (Sect. 2.3) and then find first- and second-order rotational corrections to the eigenfrequencies, using the slow-rotation approximation (Sect. 2.6.2). Due to their simple spherical harmonic dependence, eigenfunctions are only evaluated in the nonrotating limit. Our study focuses on polar modes, with degrees $l \leq 11$ and overtones $n \leq 10$.

© Springer Nature Switzerland AG 2018
P. Pnigouras, *Saturation of the f-mode Instability in Neutron Stars*,
Springer Theses, https://doi.org/10.1007/978-3-319-98258-8_5

2. Calculate mode growth/damping rates γ, due to the effects of gravitational waves, shear viscosity, and bulk viscosity (Sects. 3.5 and 3.6).
3. Find all possible couplings among the unstable $l = m = 2, 3, 4$ f-modes and the rest of the polar modes considered (Sect. 4.2). Calculate their parametric instability thresholds (Sect. 4.3).
4. Locate the triplet with the lowest parametric instability threshold and check whether saturation is successful (Sect. 4.4).

Repeating this for a grid of (T, Ω) pairs, we get the unstable mode's saturation amplitude throughout the instability window (Sect. 3.6). Below in this section we are going to review every step in more detail.

5.1.1 *Eigenfrequencies and Eigenfunctions

As described in Sect. 2.6.2, we use the slow-rotation approximation in order to determine the mode eigenfrequencies. The main reason for this is that we want as many modes as possible to be available for coupling, and solving Eq. (2.2.12) for the exact eigenfrequencies and eigenfunctions of a rotating star can be very cumbersome, if one wants to obtain many modes. Details about the validity of this approximation are discussed below.

All polar modes with degrees $l \leq 11$ and overtones $n \leq 10$ are acquired in the nonrotating limit, by solving the boundary value problem defined by Eqs. (2.3.16)–(2.3.18) and (2.3.24)–(2.3.27) [or Eqs. (2.3.33)–(2.3.36) and (2.3.43)–(2.3.46)], using a shooting-to-a-fitting-point method (Press et al. 1992, Sect. 17.2; 1996, Chap. B17). Higher overtones and multipoles were harder to obtain, due to numerical issues. Whether these modes are enough is going to be addressed in retrospect, in Sect. 5.3.3.

Subsequently, the eigenfrequencies are corrected due to rotation as

$$\omega = \omega^{(0)} + mC_1\Omega + \frac{C_2}{\omega^{(0)}}\Omega^2 + \mathcal{O}(\Omega^3), \qquad (5.1.1)$$

$\omega^{(0)}$ being the eigenfrequency in the nonrotating limit, whereas C_1 and C_2 are parameters that depend on the equation of state and mode properties. Although we do calculate the rotationally corrected (to first order) eigenfunctions [$\xi = \xi^{(0)} + \xi^{(1)} + \mathcal{O}(\Omega^2)$] in order to obtain C_2 (see Appendix B.2), we are using the nonrotating solutions $\xi^{(0)}$ to evaluate the various mode quantities, like growth/damping rates and coupling coefficients (more details about this are discussed later on in this section).

Higher than second-order corrections to the eigenfrequencies should become important at large angular velocities of the star. In fact, $\mathcal{O}(\Omega^2)$ corrections for g-modes are divergent as their overtone increases (i.e., as $\omega^{(0)} \to 0$). As we saw in Sect. 2.6.2, the parameter C_2 can be decomposed as $X + m^2 Y$ and, for g-modes with increasing l and n, $X \to 1$ and $Y \to 0$ (see Fig. 2.5). So, in this case, second-order corrections scale as $1/\omega^{(0)}$. This behaviour seems to be independent from the (polytropic) equation of state in use.

The validity of the slow-rotation approximation for g-modes can be seen in much detail in Ballot et al. (2010; see also Ballot et al. 2012), where simulations were performed for a $\Gamma = 4/3$ polytrope and rotational corrections up to third order were calculated.[1] Then, the corrected eigenfrequencies were compared to the "exact" ones, obtained from complete simulations. What they found is that second-order corrections are satisfactory for the high-frequency g-modes (low overtones), but even third-order corrections are insufficient for the low-frequency ones. They attribute this result to the fact that, in the subinertial regime ($\omega < 2\Omega$), the modes acquire a mixed gravito-inertial character (see Unno et al. 1989, Chap. VI), which significantly changes their propagation zone (Dintrans and Rieutord 2000), a property which is not considered by perturbative methods.

Similar calculations for p-modes have also been performed by Lignières et al. (2006) and Reese et al. (2006), where it was shown that the slow-rotation approximation, even at third-order, fails at relatively low rotation rates. However, p-modes reside at a frequency range which is too high for our resonance condition (4.2.1) to be satisfied. As thoroughly explained in Sect. 4.2, nonlinear coupling is relevant for the amplitude evolution only if the parent mode eigenfrequency ω_α nearly equals the daughter mode eigenfrequencies $\omega_\beta + \omega_\gamma$. As a result, only modes with eigenfrequencies lower than the f-mode eigenfrequency can become suitable daughters, hence g-modes and CFS-stable f-modes (more than 1500 modes in our study).

Given the approximations we have applied, we immediately see that the couplings among the modes can significantly change if we consider their "correct" eigenfrequencies and eigenfunctions. A daughter pair that resonates with the parent in the slow-rotation approximation might not do so in the complete solution. Furthermore, the strength of the various couplings should be affected by the form of the eigenfunctions, which, at the large rotation rates considered here, are expected to differ from their nonrotating counterparts (e.g., see Kastaun et al. 2010). However, the nature of the problem is such that a precise evaluation of the coupled triplet network is not the important point. What we are looking for is a low-order estimation of the value of the lowest parametric instability threshold, around which the parent saturates (Sect. 4.3). Since the daughter pair responsible for the saturation of the parent is chosen from a "sea" of available modes, this a highly statistical process, from which the triplet that minimizes Eq. (4.3.1) is always picked. Besides, even if we could have used the exact eigenfrequency and eigenfunction solutions, we would, at best, have calculated the correct couplings of a very simple neutron star model, rife with other simplifications and approximations.

While the slow-rotation approximation is used for daughter modes, the same cannot be done for parent modes. The f-mode instability becomes active at large angular velocities, close to the Kepler limit Ω_K (Sects. 3.2 and 3.6) and, as a result, even

---

[1]Based on the results given there, we cannot verify the behaviour of Fig. 2.5 for the $\Gamma = 4/3$ polytrope, because they only consider low-degree modes.

second-order rotational corrections do not suffice for the f-modes to become unstable. To fix this, we manually introduced "higher-order" corrections to their eigenfrequencies, based on the exact solutions provided in Ipser and Lindblom (1990)[2] —alternatively, one could use the empirical eigenfrequency relations in Doneva et al. (2013) and Doneva and Kokkotas (2015).

We should note that, in principle, coupling of the f-mode to inertial modes (Sects. 2.4.2 and 2.4.3) can be possible, as it is not forbidden by any coupling selection rule (Schenk et al. 2001). The main reason we only considered polar modes is that the coupling coefficient for polar mode coupling has a relatively simple, known form, which was derived in Appendix E [Eq. (E.3.16)]. By considering a stratified or finite-temperature star, where $\Gamma \neq \Gamma_1$ (see Sect. 2.4), we get the low-frequency modes that will play the role of daughter modes, i.e., g-modes (we take the simplest case, where $\Gamma_1 \approx const.$). We could have cases where an r-mode is one of the daughter modes but, since r-modes are purely axial (to zeroth order in Ω) in stars with nonzero buoyancy, this would make the coupling less efficient. If the coupling coefficient were evaluated to higher orders in Ω, coupling to r-modes could become significant. Had we considered a zero-buoyancy star, where $\Gamma = \Gamma_1$, then no g-modes would be present —more precisely, they would become trivial (see Sect. 2.4.3). In this case, the daughters would have to be CFS-stable f-modes and generalised r-modes. The latter have both polar and axial components in the nonrotating limit, which could make them more suitable daughters than the purely axial r-modes, but would also require modifications in the form of the coupling coefficient, accounting for the additional axial components of the daughters.

As briefly mentioned before though, g-modes are also driven by rotation, together with buoyancy, in rotating stars (Dintrans and Rieutord 2000; Yoshida and Lee 2000) and have been shown to approach the hybrid rotational modes of zero-buoyancy stars for large angular velocities (Passamonti 2009; Passamonti et al. 2009; Gaertig and Kokkotas 2009). Given that the f-mode instability operates at high rotation rates, this suggests that either studying zero- or nonzero-buoyancy stars would not affect the coupling, since, for both cases, all the daughter modes (except for the CFS-stable f-modes) would be of the inertial type. In practice, however, the slow-rotation approximation that we use does not take into account this inertial-led behaviour of g-modes, even though their eigenfrequencies are dominated by the correction terms, for large Ω.

5.1.2   *Growth/Damping Rates

For the growth ($\gamma > 0$) or damping ($\gamma < 0$) rates of the modes, defined in Eq. (3.6.7), we consider the basic mechanisms for the dissipation of fluid oscillations, of a neutron star consisting of normal nuclear matter, namely, (nonsuperfluid) neutrons,

---

[2]In particular, we introduced third- and fourth-order terms in Eq. (5.1.1), so that parent mode eigenfrequencies fit the curves and values given in Figs. 2–4 and Table 1 of Ipser and Lindblom (1990).

(nonsuperconducting) protons, and electrons. These are gravitational waves (GW), shear viscosity (SV), and bulk viscosity (BV; Sects. 3.5 and 3.6), the contributions of which are given by Eqs. (3.5.1), (3.6.1) and (3.6.4), respectively.

The growth/damping rates due to the mechanisms above are evaluated, for polar modes in the nonrotating limit, in Appendix C. Since the CFS instability arises only when rotation is present and, specifically, when the inertial-frame eigenfrequency of the mode changes sign [Eq. (3.5.2)], the factor $\omega(\omega - m\Omega)$ in Eq. (3.5.1) is calculated using the rotationally corrected eigenfrequencies. On the other hand, use of the rotationally corrected eigenfunctions would spoil the direct spherical harmonic dependence of the mode, making the evaluation of the various integrals harder to follow.

Hence, only γ_{GW} changes with Ω, whereas γ_{SV} and γ_{BV} depend solely on the temperature, scaling as T^{-2} and T^6 respectively. The instability window of the mode can then be obtained by solving Eq. (3.6.9), which, based on the above, is simplified as

$$\gamma_{GW}\left(\Omega\right) + \left(\frac{10^9\,\mathrm{K}}{T}\right)^2 \gamma_{SV}\left(T = 10^9\,\mathrm{K}\right) + \left(\frac{T}{10^9\,\mathrm{K}}\right)^6 \gamma_{BV}\left(T = 10^9\,\mathrm{K}\right) = 0,$$

where the damping rates due to shear and bulk viscosity need only be evaluated once for (say) $T = 10^9\,\mathrm{K}$. Then, for a chosen value of Ω, the equation above can be easily solved as a quartic equation for T^2.

5.1.3 *Couplings

Having calculated all the quantities associated with every mode, namely, eigenfrequencies, eigenfunctions, and growth/damping rates, we proceed with finding all the possible couplings among unstable f-modes and the rest of the polar modes considered (Sect. 4.2). We choose the $l = m = 2, 3, 4$ f-modes to be the parent modes, because they have the best "instability window size/growth time scale" ratio among all the unstable polar modes (Sects. 3.5 and 3.6).

We subject all the possible parent-daughter-daughter triplets to a screening process, using the coupling selection rules as criteria (Sect. 4.2.3). Although the selection rules for the orders m and the degrees l are either satisfied or not, there is an inherent freedom in the resonance condition (4.2.1), stemming from the detuning parameter $\Delta\omega$. Thus, we define a cut-off parameter $\Delta\omega_{max}$ such that, if $|\Delta\omega| \le \Delta\omega_{max}$, then the modes are considered resonant. The actual value of this parameter is chosen after a few trial runs, so that the triplets with the lowest parametric instability thresholds (see Sect. 5.1.4 below) do not change by further increasing it.

Then, we proceed with the calculation of the coupling coefficient $\mathcal{H}$ for every coupled triplet. We are evaluating the coupling coefficient in the nonrotating limit, with rotational corrections taken into account only through the eigenfrequencies, as derived in Appendix E [Eq. (E.3.16)]. The angular dependence of the coupling coefficient is thus reduced to the simple spherical harmonic integral (4.2.9), which does not happen if one considers the rotationally corrected eigenfunctions instead.

We have now obtained all the parameters needed to calculate every coupled triplet's parametric instability threshold [Eq. (4.3.1)].

5.1.4 Saturation

As we saw in Sect. 4.2, a tacit consequence of quadratic nonlinearities is that modes couple in triplets. This means that individual couplings consist of three modes only, with the daughter modes impeding the growth of the parent mode, via the parametric resonance instability mechanism (Sect. 4.3). Of course, the same parent can couple to more than one pairs of daughters. However, not all couplings become important. Remember that, until the parametric instability threshold is crossed [Eqs. (4.3.1) and (4.3.2)], the parent does not really "feel" the presence of the daughters. Since each coupled triplet has its own threshold, only the couplings with the lowest thresholds will affect the parent's evolution, because they will be crossed first.

Thus, for the last step, all the coupled triplets are sorted in ascending order, according to their parametric instability thresholds. Starting with the triplet that has the lowest threshold, we can examine whether it leads to saturation or not (Sect. 4.4.1), by examining the validity of the stability conditions (4.4.1) and (4.4.2), which may be roughly approximated as

$$|\gamma_\beta + \gamma_\gamma| \gtrsim \gamma_\alpha \tag{5.1.2}$$

and

$$|\Delta\omega| \gtrsim |\gamma_\alpha + \gamma_\beta + \gamma_\gamma|. \tag{5.1.3}$$

If these conditions are met for the daughter pair with the lowest parametric instability threshold, the triplet's amplitudes successfully converge towards their equilibrium solution (4.3.3) and saturate (Fig. 4.2a). If not, the daughter pair with the second lowest threshold is checked, and so on, until the first stable equilibrium is found. The lowest parametric instability threshold that leads to successful saturation will be called *stable* and approximately equals the parent's saturation amplitude [compare Eq. (4.3.1) with Eq. (4.3.3a), in conjunction with the stability condition (4.4.1)].

*The Rigorous Approach

In Sect. 4.4.1, where the stability conditions (4.4.1) and (4.4.2) were presented, we considered the special cases a) $\gamma_\beta = \gamma_\gamma$ and b) $\gamma_\gamma/\gamma_\alpha \to 0$, for which the complicated form of Eq. (4.4.2) is simplified. There, we saw that the stability condition (4.4.2) places a lower limit on the detuning, and may also impose a stronger constraint on the growth/damping rates than the stability condition (4.4.1). This is expressed by the approximate relations (5.1.2) and (5.1.3) above, which, from now on, will be referenced, instead of the exact ones (4.4.1) and (4.4.2), because it will thus be more clear whether we refer to the constraint on the growth/damping rates or the one on the detuning.

Usually, in our study, the triplet with the lowest parametric instability threshold *does* satisfy the stability conditions (5.1.2) and (5.1.3), i.e., the lowest threshold is usually stable. In the few cases when it is not, however, the iterating process described before, used to locate the lowest stable threshold, can be defective.

In Sect. 4.4.2 we saw the impact of the stability conditions on the evolution of the parametrically resonant triplet. If Eq. (5.1.2) is violated, saturation cannot be achieved; the daughters do not dissipate the incoming energy from the parent quickly enough and, consequently, the three modes grow and diverge from their equilibrium solution (Fig. 4.2b). However, the parent now grows at a rate lower than γ_α, due to the presence of the daughters. If another unstable threshold is crossed next, the growth rate will be reduced even more by the newly excited daughter pair, and so on, until the parent's growth rate approaches zero and the parent saturates. Additionally, since the daughters grow together with the parent, they should excite more modes themselves (*daughter-daughter couplings*), thus speeding up the saturation process, because the energy available to the parent is distributed to even more modes. Hence, in this alternative scenario, saturation can be achieved even by daughters that do not satisfy the stability condition (5.1.2).

The same behaviour may also occur if condition (5.1.3) is false (Fig. 4.2c), but the most frequent outcome in this case is a state of quasi-successful saturation, where the modes oscillate around their equilibrium solution (Fig. 4.2, bottom panel). Thus, since this condition is not, in most cases, necessary for the system to saturate, one may ask why did we always take it into account.

With the exception of certain cases [e.g., cases (a) and (b) above], the approximate relations (5.1.2) and (5.1.3) are hard to disentangle from the more general stability condition (4.4.2). Whenever the latter is violated, it is not easy to systematically check if this happens because of small daughter damping rates or a low detuning. The only secure way to determine whether saturation is achieved at the lowest parametric instability threshold is to solve the equations of motion (4.2.12) and inspect the result, like we did in Fig. 4.2. However, our goal is to calculate the saturation amplitude *throughout* the instability window, which means that we would have to integrate Eqs. (4.2.12) for the lowest threshold of every single (T, Ω) pair (up to ≈ 2600 grid points for some models). This is beyond the scope of our approach, but also quite unnecessary, because, in their vast majority, the lowest thresholds are stable. As we shall see in Sect. 5.6, this does not happen for the couplings of the unstable r-mode.

5.2 Approximate Relations for the Parametric Instability Threshold

For later use, we are going to examine the parametric instability threshold (4.3.1) for two limiting cases:

1. One daughter mode is damped much more quickly than the other.
2. The daughter modes are equally damped.

For each case, we will further consider two additional limits:

a. The detuning approximately equals the daughters' damping rates.
b. The detuning is much larger than the daughters' damping rates.

The limit in which the detuning is much smaller than the daughters' damping rates
is inconsistent with the stability condition (5.1.3).[3]

1. $|\gamma_\beta| \gg |\gamma_\gamma|$

If one daughter's damping rate is much larger than the other's, Eq. (4.3.1) becomes

$$|Q_{\text{PIT}}|^2 \approx \frac{\gamma_\beta \gamma_\gamma}{\omega_\beta \omega_\gamma} \frac{E_{\text{unit}}^2}{\mathcal{H}^2} \left[1 + \left(\frac{\Delta\omega}{\gamma_\beta} \right)^2 \right]. \qquad (5.2.1)$$

We then take the two subcases:

1a. $|\Delta\omega| \approx |\gamma_\beta|$

$$|Q_{\text{PIT}}|^2 \approx 2 \frac{\gamma_\beta \gamma_\gamma}{\omega_\beta \omega_\gamma} \frac{E_{\text{unit}}^2}{\mathcal{H}^2}, \qquad (5.2.2)$$

1b. $|\Delta\omega| \gg |\gamma_\beta|$

$$|Q_{\text{PIT}}|^2 \approx \frac{\gamma_\gamma}{\gamma_\beta} \frac{\Delta\omega^2}{\omega_\beta \omega_\gamma} \frac{E_{\text{unit}}^2}{\mathcal{H}^2}. \qquad (5.2.3)$$

The case $|\Delta\omega| \approx |\gamma_\gamma|$ is skipped, because this would mean $|\Delta\omega| \ll |\gamma_\beta|$, thus vio-
lating the stability condition (5.1.3).[4]

2. $\gamma_\beta \approx \gamma_\gamma$

In cases when the daughter damping rates are the same, Eq. (4.3.1) becomes

$$|Q_{\text{PIT}}|^2 \approx \frac{\gamma_\beta^2}{\omega_\beta \omega_\gamma} \frac{E_{\text{unit}}^2}{\mathcal{H}^2} \left[1 + \left(\frac{\Delta\omega}{2\gamma_\beta} \right)^2 \right]. \qquad (5.2.4)$$

The two subcases additionally give:

2a. $|\Delta\omega| \approx |\gamma_\beta|$

$$|Q_{\text{PIT}}|^2 \approx \frac{5}{4} \frac{\gamma_\beta^2}{\omega_\beta \omega_\gamma} \frac{E_{\text{unit}}^2}{\mathcal{H}^2}, \qquad (5.2.5)$$

2b. $|\Delta\omega| \gg |\gamma_\beta|$

---

[3]Based on the discussion in Sect. 5.1.4, one could also consider this limit, because it usually leads
to quasi-successful saturation, but the parametric instability threshold approximation for this case
differs at most by a factor of 2 with case (a).
[4]Cf. Footnote 3.

$$|Q_{\text{PIT}}|^2 \approx \frac{\Delta\omega^2}{4\,\omega_\beta\omega_\gamma}\frac{E_{\text{unit}}^2}{\mathcal{H}^2}.\qquad(5.2.6)$$

5.3   Typical Neutron Stars

5.3.1   Instability Evolution

The f-mode instability is expected to be important in newborn neutron stars (Sect. 1.4). After a core-collapse supernova explosion, a very hot ($T \sim 10^{11}$ K) proto-neutron star is formed (Burrows and Lattimer 1986; Burrows 1990), which subsequently cools down due to neutrino emission, powered by Urca processes (Sect. 3.6; for a review on neutron star evolution, the reader is referred to Prakash et al. 2001). Depending on its angular velocity, the star might enter the f-mode instability window. For the Newtonian polytropes that we use, this means that we require initial angular velocities $\Omega > 0.9\,\Omega_K$ (see Fig. 3.7), which, as we discussed in Sect. 1.4, are theoretically feasible.

As soon as the star enters the instability window, the unstable mode grows exponentially until it saturates. Shear viscosity, triggered by the oscillation, heats up the star and balances neutrino cooling, establishing thermal equilibrium.[5] Gravitational waves emitted from the perturbed star carry off angular momentum and the star descends the instability window along a thermal equilibrium curve ($T \approx const.$), until it finally exits the window. This process applies to both the f-mode (Passamonti et al. 2013) and the r-mode instability (Owen et al. 1998; Bondarescu et al. 2009; for the r-mode instability in strange stars, see also Andersson et al. 2002).

The saturation amplitude of the unstable f-mode determines the gravitational wave strain associated with the perturbation. The detectability of the signal also depends on the competition of the f-mode instability with other spin-down mechanisms, such as the r-mode instability and magnetic braking: should the r-mode saturation amplitude be larger than (or even comparable to) the f-mode one, or the magnetic field be greater than some critical value, then one or both of these mechanisms will dominate the spin evolution of the neutron star (Passamonti et al. 2013).

5.3.2   Models

To study the f-mode saturation, we applied the quadratic perturbation scheme, presented in Chap. 4, in polytropic stars. We used two polytropic configurations, with polytropic exponents $\Gamma = 2$ and 3, and varied the adiabatic exponent Γ_1, leading

---

[5] As opposed to shear viscosity, bulk viscosity cools down the star by neutrino emission. However, its contribution to the star's cooling is negligible (Passamonti et al. 2013).

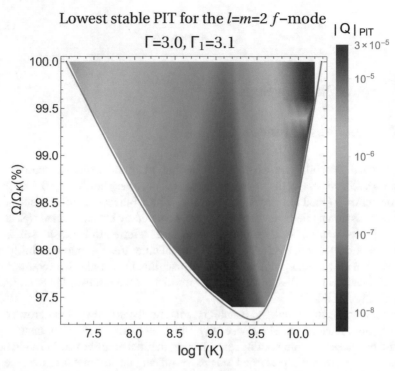

Fig. 5.1 Contour plot of the lowest stable parametric instability threshold (PIT) inside the instability window of the $l = m = 2$ f-mode, for a typical neutron star with $M \approx 1.4\,M_\odot$ and $R \approx 10$ km. The angular velocity Ω, normalised to the Kepler limit Ω_K, is drawn on the vertical axis, and the (decimal) logarithm of the temperature T on the horizontal axis. The star is described by a polytrope with a polytropic exponent $\Gamma = 3$ and an adiabatic exponent $\Gamma_1 = 3.1$. The mode amplitude is given by the relation $|Q| = \sqrt{E/E_{\text{unit}}}$, with $E_{\text{unit}} = Mc^2$. Credit: Pnigouras and Kokkotas (2016)

to stronger or weaker buoyancy effects (the smaller the difference between Γ_1 and Γ, the closer to zero g-mode eigenfrequencies are pushed in the nonrotating limit, see Sect. 2.4.1). Because of the complications described in Sect. 5.1.1 regarding g-mode eigenfrequencies, models in which $\Gamma_1 - \Gamma$ was very small exhibited divergent behaviour and were thus ignored.

The results for three models are presented in Figs. 5.1, 5.2 and 5.3, where we plot the lowest stable parametric instability threshold ($\approx$ saturation amplitude, see Sect. 5.1.4) throughout the instability window. In the first two models, $\Gamma = 2$, and $\Gamma_1 = 2.2$ and 2.1, whereas in the third one $\Gamma = 3$ and $\Gamma_1 = 3.1$. The unstable f-modes we consider are the quadrupole ($l = m = 2$), the octupole ($l = m = 3$), and the hexadecapole ($l = m = 4$). All three of these modes become unstable in the $\Gamma = 3$ polytrope, but only the last two in the $\Gamma = 2$ polytrope (see Sects. 3.5 and 3.6).

In all three models we consider typical neutron stars, with $M \approx 1.4\,M_\odot$ and $R \approx 10$ km, where $M_\odot$ is the solar mass. The exact parameters of the models are

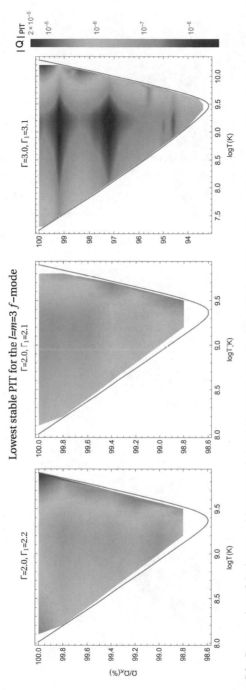

Fig. 5.2 Contour plots of the lowest stable parametric instability threshold (PIT) inside the instability window of the $l = m = 3$ f-mode, for a typical neutron star with $M \approx 1.4 M_\odot$ and $R \approx 10$ km. The angular velocity Ω, normalised to the Kepler limit Ω_K, is drawn on the vertical axis, and the (decimal) logarithm of the temperature T on the horizontal axis. The star is described by a polytrope with a polytropic exponent $\Gamma = 2$, and an adiabatic exponent $\Gamma_1 = 2.2$ and 2.1, as well as a polytrope with $\Gamma = 3$ and $\Gamma_1 = 3.1$. The mode amplitude is given by the relation $|Q| = \sqrt{E/E_{\text{unit}}}$, with $E_{\text{unit}} = Mc^2$. Credit: Pnigouras and Kokkotas (2016)

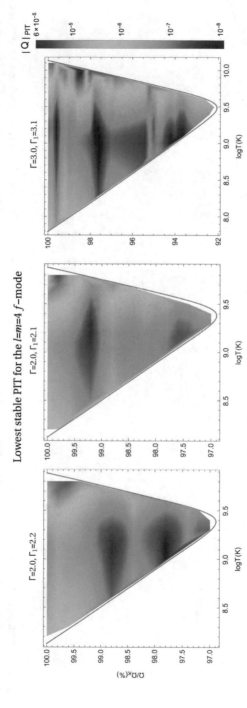

Fig. 5.3 Contour plots of the lowest stable parametric instability threshold (PIT) inside the instability window of the $l = m = 4$ f-mode, for a typical neutron star with $M \approx 1.4\,M_\odot$ and $R \approx 10$ km. The angular velocity Ω, normalised to the Kepler limit Ω_K, is drawn on the vertical axis, and the (decimal) logarithm of the temperature T on the horizontal axis. The star is described by a polytrope with a polytropic exponent $\Gamma = 2$, and an adiabatic exponent $\Gamma_1 = 2.2$ and 2.1, as well as a polytrope with $\Gamma = 3$ and $\Gamma_1 = 3.1$. The mode amplitude is given by the relation $|Q| = \sqrt{E/E_{\mathrm{unit}}}$, with $E_{\mathrm{unit}} = Mc^2$. Credit: Pnigouras and Kokkotas (2016)

Table 5.1 Typical neutron star models, used for the calculation of the f-mode saturation amplitude. The following parameters are presented in the table, by column: mass M, radius R, polytropic exponent Γ, adiabatic exponent Γ_1, polytropic constant K, and central density ρ_c

M ($M_\odot$)	R (km)	Γ	Γ_1	K $\left(\text{gr}^{1-\Gamma}\,\text{cm}^{3\Gamma-1}\,\text{s}^{-2}\right)$	ρ_c $\left(\text{gr}\,\text{cm}^{-3}\right)$
1.378	10.1	2	2.2	4.34481×10^4	2.08697×10^{15}
1.378	10.1	2	2.1	4.34481×10^4	2.08697×10^{15}
1.354	10.0	3	3.1	6.26042×10^{-11}	1.18038×10^{15}

presented in Table 5.1. The normalisation used is $E_{\text{unit}} = Mc^2$ (c being the speed of light), which is the mode energy at unit amplitude ($E = |Q|^2 E_{\text{unit}}$; Sect. 2.6.1).

In addition, a hypothetical (based on the results of Passamonti et al. 2013) evolution of a star inside the instability window of the octupole f-mode, together with the variation of the mode's saturation amplitude, can be seen in Fig. 5.4.

5.3.3 Discussion

Two main features can be observed in Figs. 5.1, 5.2 and 5.3: (a) the decrease of the saturation amplitude from the edge to the interior of the instability window, and (b) horizontal bands where the amplitude behaves differently from the "background" —although the bands themselves also follow the modulation of the first feature.

(a) First Feature

The first feature can be clearly seen in Fig. 5.1, where the second feature is absent. This decline of the saturation amplitude can easily be explained if one looks at the *coupling spectrum*, i.e., the daughter pairs responsible for the saturation of the parent throughout the instability window (see Appendix G, where the coupling spectrum of the octupole f-mode is presented, for the model with $\Gamma = 3$ and $\Gamma_1 = 3.1$).

As a rule, we have two types of daughter pairs: either a CFS-stable f-mode and a g-mode (f-g coupling) or two g-modes (g-g coupling). Depending on the coupling type and the daughters' parameters, we can simplify the formula for the parametric instability threshold (4.3.1), as shown in Sect. 5.2.

In the case of f-g couplings, which is the most common, the f-mode damping rate γ_β is much larger (in absolute value) than the g-mode damping rate γ_γ. Then, Eq. (4.3.1) is approximated by Eq. (5.2.2) or (5.2.3). Since the f-mode damping is mainly due to gravitational waves, it does not change much with temperature, which makes it roughly constant for some angular velocity Ω. On the other hand, for this type of coupling, the g-mode daughter is predominantly damped by viscosity. From Eqs. (5.2.2) and (5.2.3), this means that

$$|Q_{\text{PIT}}| \propto \sqrt{|\gamma_\gamma|} \propto \begin{cases} T^{-1}, & T \lesssim 10^9\,\text{K} \\ T^3, & T \gtrsim 10^9\,\text{K} \end{cases} \quad \text{for} \quad \Omega = const. \qquad (5.3.1)$$

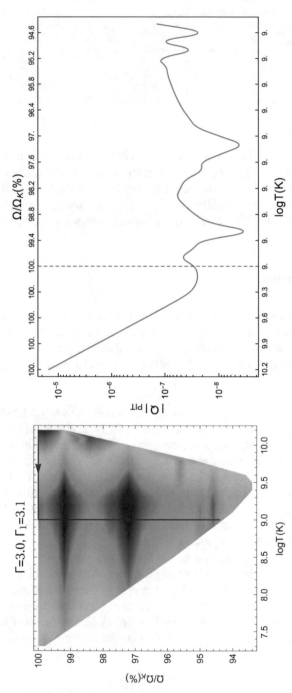

Fig. 5.4 Hypothetical evolution of a typical neutron star with $M \approx 1.4\,M_\odot$ and $R \approx 10$ km, through the instability window of the $l = m = 3$ f-mode (*left*), and the corresponding evolution of the lowest stable parametric instability threshold (PIT; *right*). In the *left* (*right*) graph, the angular velocity Ω, normalised to the Kepler limit Ω_K, is drawn on the vertical (top horizontal) axis, and the logarithm of the temperature T on the horizontal (bottom horizontal) axis. The star obeys a polytropic equation of state with a polytropic exponent $\Gamma = 3$ and an adiabatic exponent $\Gamma_1 = 3.1$. In this example, the star enters the window during its cooling phase, rotating at its maximum angular velocity, until thermal equilibrium is established (indicated by the vertical dashed line), at which point it descends the window at $T = 10^9$ K (Passamonti et al. 2013). Credit: Pnigouras and Kokkotas (2016)

In other words, along an $\Omega = const.$ line, the saturation amplitude follows the behaviour of the g-mode daughter damping rate.[6] The temperature dependence in Eq. (5.3.1) is a result of shear and bulk viscosity, dominating the damping at low and high temperatures, and scaling as T^{-2} and T^6, respectively (see Sect. 3.6).

When a g-g coupling prevails, the relation between the daughters' damping rates can vary. If gravitational waves dominate the damping for one of them, everything is reduced to the f-g coupling case. This happens when one of them is a low g-mode multipole. If viscosity dominates for both, it is better to start with the observation that the detuning is usually much larger (in absolute value) than the damping rates. Then, the relevant approximate formulae for the parametric instability threshold are Eqs. (5.2.3) and (5.2.6). Both equations show that the saturation amplitude should not change along constant-angular-velocity lines. This is obvious in Eq. (5.2.6), but can also be seen in Eq. (5.2.3), because $|Q_{PIT}| \propto \sqrt{\gamma_\gamma/\gamma_\beta}$ and the damping rates follow the same temperature scaling. Nevertheless, this situation is not observed much (see paragraph d below).

(b) Second Feature

For the second feature to be understood, we need to look at constant temperature lines instead. The difficulty here is that all quantities that appear in Eq. (4.3.1) change as Ω is varied. This makes the modulation of the saturation amplitude along a $T = const.$ line harder to follow.

Looking at the coupling spectrum, we see that the same daughter pair is usually responsible for the saturation of the parent along an $\Omega = const.$ line. After all, this is the basis of the reasoning that led to Eq. (5.3.1). This is no longer true along constant temperature lines: the daughter pair which gives the lowest stable parametric instability threshold may change many times. Occasionally, this change might be abrupt, making the saturation amplitude higher or lower, compared to neighbouring angular velocities. As a result, these characteristic horizontal bands appear, which, however, individually still follow the behaviour of the first feature.

Although the effect is highly statistical, given the number of variables and available modes, we can single out two main reasons for it: The first is the occurrence of a very fine resonance between the parent and some daughter pair, which only appears for a specific angular velocity. Such a resonance has a very low detuning $|\Delta\omega|$, which can lead to the drop of the saturation amplitude. The second, less frequent, reason is related to the validity of the stability conditions (5.1.2) and (5.1.3). If the saturating triplet satisfies one of these conditions marginally for some value of the angular velocity, it will not be long before it cannot saturate the parent any more, and some other daughter pair will take its place.

(c)* Varying the Adiabatic Exponent

As mentioned before, using different values for the difference between the adiabatic and polytropic exponents, $\Gamma_1 - \Gamma$, shifts the nonrotating-limit g-mode eigenfrequencies closer to or further away from the f-mode eigenfrequency (the latter depends

---

[6]The daughters' damping rates γ_β and γ_γ are the only quantities in Eq. (4.3.1) that can change along a constant-angular-velocity line.

mainly on Γ and is highly unaffected by any change in Γ_1, see Table 2.1 and Fig. 2.2). However, since we are interested in fast-rotating stars, rotational corrections to g-mode eigenfrequencies will prevail, causing g-modes to become rotationally-driven, rather than buoyancy-driven (see Sect. 5.1.1).

Consequently, fast-rotating models with different adiabatic exponents (but the same polytropic exponent) should be nearly indistinguishable —at least as far as f- and g-modes are concerned. This means that the couplings and the saturation amplitudes should not change much if a different value of Γ_1 is chosen for some polytrope. This can indeed be seen, to some extent, in Figs. 5.2 and 5.3. In practice, however, as discussed in Sect. 5.1.1, g-modes do not exhibit inertial behaviour in the slow-rotation approximation. Hence, in principle, there should be differences in the results if one considers coupling to inertial modes; for instance, the inertial mode damping rates reported by Lockitch and Friedman (1999) are larger (in absolute value) than our g-mode damping rates, which, according to Eq. (5.3.1), should systematically *increase* the saturation amplitude.

(d) Is the Number of Modes Enough?

In our models, we searched for couplings of unstable f-modes to more than 1500 polar modes and obtained many triplets with fine resonances, meaning that our frequency spectrum was dense enough for the parent to always resonate with daughter pairs. These fine resonances could probably become even finer and/or more frequent, had we included more modes in the calculation. However, a small detuning alone does not necessarily lead to smaller amplitudes. This can be seen in many g-g couplings, where even though better resonances were achieved compared to f-g couplings, the latter were much more abundant in the coupling spectrum (see paragraph *a* above). This shows that triplets with larger detunings might give the lowest parametric instability thresholds instead, depending on how their parameters are tuned and which of the cases shown in Sect. 5.2 they fall under.

5.4   Supramassive Neutron Stars

5.4.1   Instability Evolution

In Sects. 1.4 and 3.6 we discussed supramassive neutron stars as promising sources of the f-mode instability. These are stars whose mass exceeds the maximum mass allowed by their equation of state in the nonrotating limit (see Sect. 1.3) and are supported against gravitational collapse by their rapid rotation (Cook et al. 1992, 1994).[7] According to Fig. 3.8, the growth time scale of the mode [Eq. (3.6.8)] due to the emission of gravitational waves, $|\tau_{GW}|$, can be as short as $10-100\,\mathrm{s}$ for $M > 2.4\,M_\odot$, increasing by orders of magnitude as the mass of the star is

---

[7]If the star is differentially rotating, it may be able to support even larger masses (Baumgarte et al. 2000).

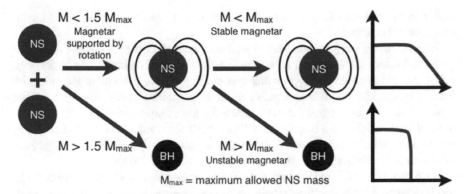

Fig. 5.5 Illustration of the "magnetar model" for SGRBs. A possible outcome of the binary neutron star (NS) merger is a rapidly rotating neutron star, powering the afterglow emission. The remnant spins down and, depending on its mass, either remains stable, with the emission decaying slowly, or collapses to a black hole (BH), with the emission ending abruptly. Credit: Rowlinson (2013)

decreased, thus rendering such configurations prospective targets for gravitational-wave-asteroseismology studies.

Supramassive neutron stars are often associated with short γ-ray bursts (SGRBs), namely, γ-ray bursts (GRBs) with short durations ($\sim$1 s) and hard (i.e., high-energy) spectra. It is generally believed that GRBs originate from the coalescence of compact binaries (either neutron-star binaries or neutron-star–black-hole binaries; see Lee and Ramirez-Ruiz 2007 for a review), leading to the formation of a neutron star or a black hole. Observations show that many SGRBs are accompanied by a persistent X-ray afterglow emission, thus indicating the presence of a long-lived central engine, with the X-ray light curve featuring a plateau, which may be followed by a shallow or a steep decay (see Rowlinson et al. 2013, Fig. 8). According to a popular scenario, the former could be attributed to the spin-down of a stable neutron star and the latter to the gravitational collapse of a supramassive neutron star (Rowlinson et al. 2013; Rowlinson 2013; cf. also Ciolfi and Siegel 2015 and Rezzolla and Kumar 2015). This is illustrated in Fig. 5.5.

The formation of a rapidly rotating neutron star has long been associated with GRBs (Duncan and Thompson 1992; Usov 1992; Dai and Lu 1998; Zhang and Mészáros 2001). As suggested above, supramassive post-merger neutron star remnants could be formed shortly after the binary coalescence (see also Fryer et al. 2015, as well as simulations by Duez et al. 2006, Hotokezaka et al. 2013, Kastaun and Galeazzi 2015). If the massive remnant rotates rapidly enough, it will not promptly collapse to a black hole, but will remain stable until some spin-down mechanism drains its angular momentum and gravitational collapse can no longer be halted. Other possible formation channels include an accretion-induced collapse of a white dwarf (Usov 1992), as well as the "collapsar" model (MacFadyen et al. 2001; see also Woosley 1993 and Heger et al. 2003), where a weak supernova explosion is followed by fallback of material on the newly formed proto-neutron star, inducing gravitational collapse.

Recent studies suggest that these objects may remain stable for up to $\approx 4 \times 10^4$ s (Ravi and Lasky 2014), which is enough time for the f-mode instability to develop. After the initial differentially rotating and cooling phase, the star might enter the instability window and follow a path similar to the one described in Sect. 5.3.1 for typical neutron stars, but, since the window for a supramassive star is much larger, the star may collapse to a black hole before it exits the window (Doneva et al. 2015; for an alternative scenario, where gravitational waves originate from a magnetic-field-induced deformation of the star, see Dall'Osso et al. 2015).

As in the case of typical neutron stars, the saturation amplitude of the unstable f-mode determines whether the associated gravitational-wave signal can be detected, based on the competition among the f-mode instability, the r-mode instability, and magnetic braking (Doneva et al. 2015). This can be seen in Fig. 5.6, where we plot the signal-to-noise ratio (e.g., see Owen et al. 1998) of a gravitational-wave signal emitted by the $l = m = 2$ f-mode, for second- and third-generation detectors, versus the dipole component of the magnetic field on the stellar surface and the saturation amplitude of the $l = m = 2$ r-mode. The graphs were made for relativistic supramassive neutron stars, under the Cowling approximation (Sect. 2.3.4), with baryon masses $M_b \approx 3 M_\odot$ (see Sect. 3.6), governed by the WFF2 (Wiringa et al. 1988) and APR (Akmal et al. 1998) equations of state. The assumed distance to the sources is $d = 20$ Mpc and the saturation amplitude of the f-mode is taken equal to 10^{-3}, meaning a saturation *energy* $E = 10^{-6} E_{\text{unit}}$, where $E_{\text{unit}} = Mc^2$.

5.4.2 Models

The main complication regarding the application of our formalism in supramassive stars is that they do not admit a Newtonian limit (Cook et al. 1992). Hence, we applied the quadratic perturbation scheme, presented in Chap. 4, in configurations which *emulate* supramassive stars, namely in Newtonian models with artificially large masses.

We considered a star with $M = 2.5 M_\odot$ and $R = 12$ km, obeying a polytropic equation of state with a polytropic exponent $\Gamma = 3$, and an adiabatic exponent $\Gamma_1 = 3.2$ and 3.1. In order to achieve the instability growth time scales of Doneva et al. (2015), shown in Fig. 3.8, we also had to manually enhance the Kepler limit of our models, thus allowing for rotation rates beyond the Newtonian value of the mass-shedding limit [roughly given by Eq. (3.2.1)]. This way, the factor $\omega(\omega - m\Omega)$,

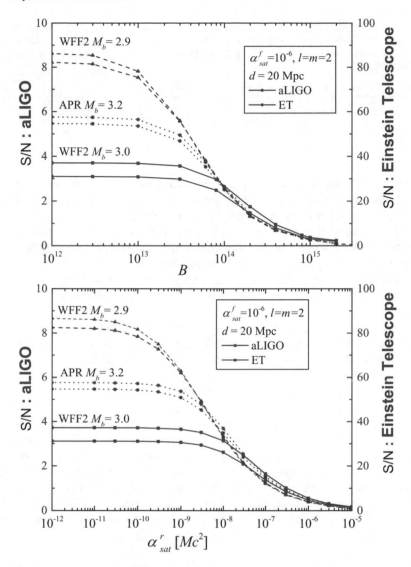

Fig. 5.6 Signal-to-noise ratio (S/N) of a gravitational-wave signal emitted by the $l = m = 2$ f-mode, plotted both for second- and third-generation detectors, like Advanced LIGO (aLIGO; left vertical axis, blue lines) and the Einstein Telescope (ET; right vertical axis, red lines) respectively, against the dipole component of the magnetic field on the stellar surface B (*top*) and the saturation energy (normalised to Mc^2) of the $l = m = 2$ r-mode $\alpha^r_{sat} \equiv |Q|^2 = E/Mc^2$ (*bottom*). The three models used have baryon masses $M_b \approx 3\,M_\odot$ and are described by the WFF2 and APR equations of state. The distance to the sources is $d = 20\,\mathrm{Mpc}$ and the assumed saturation energy (normalised to Mc^2) of the f-mode is $\alpha^f_{\mathrm{sat}} = 10^{-6}$. Credit: Doneva et al. (2015)

Table 5.2 Supramassive neutron star models, used for the calculation of the f-mode saturation amplitude. The following parameters are presented in the table, by column: mass M, radius R, polytropic exponent Γ, adiabatic exponent Γ_1, polytropic constant K, central density ρ_c, and (enhanced) Kepler limit Ω_K

$M\ (M_\odot)$	R (km)	Γ	Γ_1	$K\ \left(\text{gr}^{1-\Gamma}\,\text{cm}^{3\Gamma-1}\,\text{s}^{-2}\right)$	$\rho_c\ \left(\text{gr cm}^{-3}\right)$	$\Omega_K/\sqrt{GM/R^3}$
2.5	12	3	3.2	8.42051×10^{-11}	1.26372×10^{15}	0.74
2.5	12	3	3.1	8.42051×10^{-11}	1.26372×10^{15}	0.74

appearing in the gravitational-wave growth rate formula (3.5.1), can obtain smaller values,[8] leading to shorter growth time scales. These time scales cannot be obtained legitimately by our Newtonian polytropes, because the models used in Doneva et al. (2015) are relativistic (employing the Cowling approximation) and governed by realistic equations of state. The exact parameters of our models are shown in Table 5.2.

Given the assumptions above and the simplicity of our approach, our models should be merely considered as toy models, used to demonstrate the impact of larger masses and shorter instability growth times on the saturation amplitude of the unstable modes.

The results for the models described above are presented in Figs. 5.7 and 5.8, where the lowest stable parametric instability threshold ($\approx$ saturation amplitude, see Sect. 5.1.4) is plotted inside the instability windows of the quadrupole ($l = m = 2$) and the octupole ($l = m = 3$) f-modes. The normalisation used for the mode energy is $E_{\text{unit}} = Mc^2$. Since the left part of the windows is not expected to be significant for the evolution of a newborn neutron star (see Fig. 5.4), and given their considerably larger size, compared to the corresponding windows from typical neutron stars, we restricted our calculations to $T \geq 10^8$ K. Furthermore, the models of Doneva et al. (2015) become unstable to gravitational-collapse when the star radiates away up to 20% of its angular momentum, so we considered rotation rates greater than $0.8\,\Omega_K$. A model without an enhanced Kepler limit is also shown, for comparison.

Finally, a hypothetical (based on the results of Doneva et al. 2015) evolution of a star inside the instability window of the quadrupole f-mode is shown in Fig. 5.9, along with the variation of the mode's saturation amplitude.

5.4.3 Discussion

(a) Features

The same features that were discussed in the previous section for typical neutron stars can also be seen in Figs. 5.7 and 5.8. The same reasoning can be used to explain the characteristic decrease of the saturation amplitude from the edge to the interior of

---

[8]Remember that the instability implies that the factor $\omega(\omega - m\Omega)$ is negative [Eq. (3.5.2)], so, in this sense, "smaller values" means "more negative".

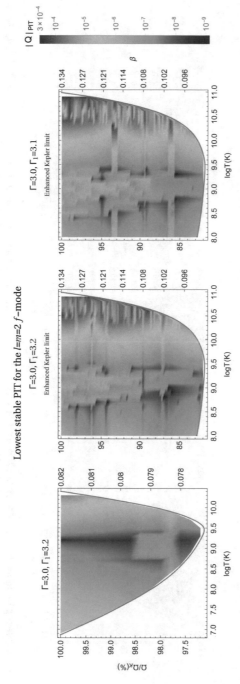

Fig. 5.7 Contour plots of the lowest stable parametric instability threshold (PIT) inside (part of) the instability window of the $l = m = 2$ f-mode, for our toy model of a supramassive neutron star with $M = 2.5\, M_\odot$ and $R = 12$ km. The angular velocity Ω, normalised to the Kepler limit Ω_K, is drawn on the left vertical axis, the ratio of kinetic to gravitational potential energy β on the right vertical axis, and the (decimal) logarithm of the temperature T on the horizontal axis. The star is described by a polytrope with a polytropic exponent $\Gamma = 3$, and an adiabatic exponent $\Gamma_1 = 3.2$ and 3.1. The Kepler limit has been enhanced, to imitate the behaviour of the models used in Doneva et al. (2015). A model with its actual Kepler limit is also shown. The mode amplitude is given by the relation $|Q| = \sqrt{E/E_{\text{unit}}}$, with $E_{\text{unit}} = Mc^2$. Credit: Pnigouras and Kokkotas (2016)

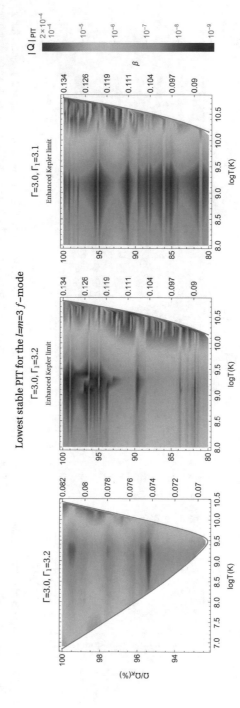

Fig. 5.8 Contour plots of the lowest stable parametric instability threshold (PIT) inside (part of) the instability window of the $l = m = 3$ f-mode, for our toy model of a supramassive neutron star with $M = 2.5$ $M_\odot$ and $R = 12$ km. The angular velocity Ω, normalised to the Kepler limit Ω_K, is drawn on the left vertical axis, the ratio of kinetic to gravitational potential energy β on the right vertical axis, and the (decimal) logarithm of the temperature T on the horizontal axis. The star is described by a polytrope with a polytropic exponent $\Gamma = 3$, and an adiabatic exponent $\Gamma_1 = 3.2$ and 3.1. The Kepler limit has been enhanced, to imitate the behaviour of the models used in Doneva et al. (2015). A model with its actual Kepler limit is also shown. The mode amplitude is given by the relation $|Q| = \sqrt{E/E_{unit}}$, with $E_{unit} = Mc^2$. Credit: Pnigouras and Kokkotas (2016)

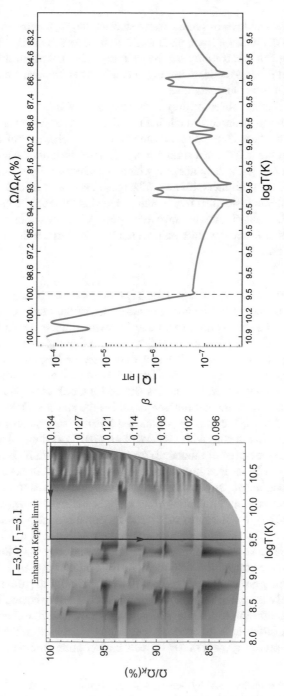

Fig. 5.9 Hypothetical evolution of our toy model of a supramassive neutron star with $M = 2.5\,M_\odot$ and $R = 12\,$km, through the instability window of the $l = m = 2\ f$-mode (*left*), and the corresponding evolution of the lowest stable parametric instability threshold (PIT; *right*). In the *left* (*right*) graph, the angular velocity Ω, normalised to the Kepler limit Ω_K, is drawn on the left vertical (top horizontal) axis, the ratio of kinetic to gravitational potential energy β on the right vertical axis, and the logarithm of the temperature T on the horizontal (bottom horizontal) axis. The star obeys a polytropic equation of state with a polytropic exponent $\Gamma = 3$ and an adiabatic exponent $\Gamma_1 = 3.1$. In this example, the star enters the window during its cooling phase, rotating at its maximum angular velocity, until thermal equilibrium is established (indicated by the vertical dashed line), at which point it descends the window at $T \approx 3 \times 10^9\,$K (Doneva et al. 2015). Credit: Pnigouras and Kokkotas (2016)

the instability window, as well as the horizontal bands that appear at certain angular velocities.

The fact that the windows of supramassive stars are larger, compared to their counterparts from typical neutron stars, justifies the increase of the maximum value that the saturation amplitude can attain: according to Eq. (5.3.1), the saturation amplitude scales with the daughter g-mode's damping rate, which can achieve greater (absolute) values at higher or lower temperatures.

An additional feature, observed only in the case of supramassive stars (and mainly in the results for the quadrupole f-mode in Fig. 5.7), is this vertical "brushstroke-like" structure at intermediate temperatures. The anomalous behaviour of the saturation amplitude in this area occurs, in a similar manner to the horizontal band feature, due to daughter pair changes: a daughter pair which can successfully saturate the parent fails to do so once the star enters this area. The reason is that some daughter g-modes become CFS-unstable inside this region and can no longer stop the parent's growth (remember that the daughter modes have to be stable). The unstable parent will then get saturated by a different daughter pair, which may lead to a sudden change of the saturation amplitude.

(b) Properties

As previously mentioned, we enhanced the Kepler limit of our Newtonian polytropic models, in order to reproduce the growth time scales of the models used in Doneva et al. (2015). The factors leading to so short growth time scales in the latter are relativity[9] (Stergioulas and Friedman 1998; see also Gaertig et al. 2011), realistic equations of state (Doneva et al. 2013), and, of course, the large masses and angular momenta of these supramassive stars. The behaviour of the angular momentum is an important subtlety of such stars. As shown by Cook et al. (1992), there are regions where the loss of angular momentum spins the star *up*. This feature, which we discussed in Sect. 3.2, cannot be mimicked by our Newtonian polytropes. As a result, we get the same τ_{GW} with Doneva et al. (2015) at the (enhanced) Kepler limit, but not for lower angular momenta, for which our growth time scale becomes significantly longer. This does not happen in the models of Doneva et al. (2015) which, as described above, spin up instead when angular momentum is lost due to gravitational wave emission, inducing only a slight increase in τ_{GW}. This is why Doneva et al. (2015) use the ratio of kinetic to gravitational potential energy β, introduced in Sect. 3.2, instead of Ω, to parametrise rotation. In Figs. 5.7 and 5.8 we use both parameters: Ω, as a reminder of the Newtonian origin of the calculation, and β, to indicate the connection with the realistic model.

The fact that the gravitational wave growth time scale of the parent mode should stay approximately the same throughout the window is not expected to change the results about the saturation amplitude. Although in our models τ_{GW} changes by orders of magnitude for different angular momenta, it is not included in the evaluation of Eq. (4.3.1). The parent's growth rate γ_α affects the couplings indirectly, through the

---

[9]If the Cowling approximation, used by Doneva et al. (2015), were dropped, the instability should be amplified even more, because it sets in at smaller rotation rates (Zink et al. 2010).

stability conditions (5.1.2) and (5.1.3). This means that even if τ_{GW} had its Keplerian value everywhere, then, since the parent is successfully saturated at the Kepler limit by some daughter pairs, there will always be daughters with similar properties which will saturate it at lower angular momenta as well.

(c)* Nonlinear Eigenfrequency Shift

In Sect. 4.4.3, we saw that nonlinear coupling induces a shift to the mode eigenfrequencies which, when saturation is achieved, are then given by Eq. (4.4.15). For the models used in this chapter we find that, in general, the eigenfrequency shift is negligible for the parent mode. Among triplets which give the lowest parametric instability thresholds throughout the instability windows, we found a maximum shift of ~ 0.1 Hz for typical neutron stars. For supramassive stars, some couplings induce an eigenfrequency shift as high as ~ 0.1 kHz. Given that the $l = m$ f-modes have eigenfrequencies of a few kHz, the latter could be significant. However, these couplings are very few, since one needs the parameters in Eq. (4.4.15) finely tuned (large γ_α and $|\Delta\omega|$; $|\gamma_{\beta,\gamma}|$ as small as possible) to produce a considerable eigenfrequency shift. For the daughter modes, the shift is larger, since they also have to cover the mismatch $\Delta\omega$ (at most ~ 0.1 kHz) to catch up with the parent (see Fig. 4.3).[10]

5.5 Stochastic Background

In addition to gravitational-wave signals from individual sources, the superposition of unresolved and uncorrelated gravitational-wave signals from many sources throughout the Universe is expected to give a *stochastic background* of gravitational waves. Oscillation modes which are prone to the CFS instability in neutron stars should, in principle, contribute to this stochastic background (see Ferrari et al. 1999 and Zhu et al. 2011 for studies on unstable r-modes; for a review on the astrophysical stochastic gravitational-wave background, the reader is referred to Regimbau 2011).

Mainly based on the work of Ferrari et al. (1999), Surace et al. (2016) provided an estimation of the contribution of CFS-unstable f-modes to the stochastic gravitational-wave background, both for the case of typical, supernova-derived, neutron stars (Sect. 5.3), as well as for supramassive, merger-derived, neutron stars (Sect. 5.4). The result is shown in Fig. 5.10, where the energy density of the stochastic gravitational-wave background (normalised to the critical density of the Universe[11]) Ω_{gw} is plotted against the observed gravitational-wave frequency. The f-modes considered are the same as in Sects. 5.3 and 5.4, namely, the $l = m = 2$, 3 and 4 for typical neutron stars, and the $l = m = 2$ and 3 for supramassive neutron stars. The instability windows of the modes were obtained from Doneva et al. (2013) and

---

[10]From Eqs. (4.4.1) and (4.4.15) we see that the eigenfrequency shift always has the same sign with the detuning $\Delta\omega$, meaning that the parent eigenfrequency is always shifted away from the daughter eigenfrequencies.

[11]Namely, the density for which the geometry of the Universe is flat.

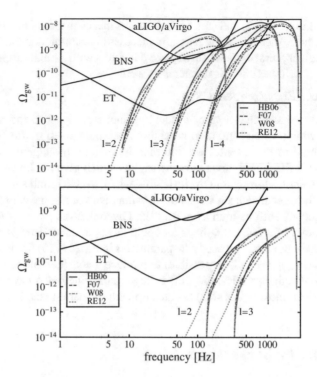

Fig. 5.10 Energy density of the stochastic gravitational-wave background (normalised to the critical density of the Universe) $\Omega_{\rm gw}$, due to CFS-unstable f-modes, versus the observed gravitational-wave frequency, for four different cosmic star formation rate models (denoted as HB06, F07, W08, RE12). The individual contributions of the $l = m = 2$, 3 and 4 f-modes is presented for typical, supernova-derived, neutron stars (*top*), and of the $l = m = 2$ and 3 f-modes for supramassive, merger-derived, neutron stars (*bottom*). Also shown are the stochastic background due to binary neutron-star mergers (BNS) and the sensitivity curves of second- and third-generation detectors, like Advanced LIGO/Advanced Virgo (aLIGO/aVirgo) and the Einstein Telescope (ET) respectively. Credit: Surace et al. (2016), reproduced with permission, © European Southern Observatory (ESO)

Doneva et al. (2015) for typical and supramassive neutron stars respectively, where the WFF2 equation of state (Wiringa et al. 1988) was used. For the evolution of a typical neutron star through the windows, the work of Passamonti et al. (2013) was also consulted, where a relativistic polytrope with $\Gamma \approx 2.61$, resembling the WFF2 equation of state, is considered. The results presented in Fig. 5.10 assume that 10% (100%) of the population of newborn typical (supramassive) neutron stars rotate with initial angular velocities close to the Kepler limit and, hence, may have unstable f-modes. Furthermore, the effects of other spin-down mechanisms, like the r-mode instability and magnetic braking, are ignored, which implies that gravitational radiation from the unstable f-modes is solely responsible for the energy and angular momentum loss of the star. The energy density is calculated for four different cosmic star formation rate models, which can be found in Surace et al. (2016). Also shown in the graphs are the stochastic background due to binary neutron-star mergers, as well as the sensitivity curves of second- and third-generation detectors.

From Fig. 5.10 we see that, for typical neutron stars, the $l = m = 2$ background might be detectable with the current, second-generation, detectors, because it peaks at low frequencies ($\approx 50-200$ Hz), where the sensitivity of the detectors is better, and at an amplitude which is higher than the one from coalescing binary neutron stars. The $l = m = 3$ background is too close to the sensitivity limit and thus unlikely to be detected with second-generation detectors, but lies well inside the sensitivity curve of third-generation detectors. Finally, although it obtains the highest amplitudes, the $l = m = 4$ background occupies the high-frequency regime of third-generation detectors and will be hard to detect. We should additionally note that the $l = m = 3$ and 4 f-mode backgrounds peak at frequencies (~ 1 kHz) and amplitudes ($\Omega_{gw} \sim 10^{-8}$) where many other astrophysical backgrounds also reside (see Fig. 6 in Regimbau 2011), which will thus make it more difficult to discriminate between them.

On the other hand, stochastic backgrounds from unstable f-modes in supramassive neutron stars should be undetectable, even with third-generation detectors. This can be mainly attributed to the supramassive neutron star formation rate from binary mergers, which is lower than the typical neutron star formation rate from supernovae, as well as the fact that the backgrounds peak at high frequencies (~ 1 kHz), where the detectors are not sensitive enough.

5.6 *Comparison with r-modes

Instantly after its discovery (Andersson 1998; Friedman and Morsink 1998), the r-mode instability drew much attention, due to its short growth time scale and large instability window (see Sect. 3.5). Unstable r-modes are not only considered a promising gravitational-wave source, with important implications for gravitational-wave asteroseismology, but have also been proposed as an explanation for the observed neutron star spin periods, as discussed in Sect. 1.4.

Apart from determining the strength (and hence the detectability) of the generated gravitational-wave signal, the saturation amplitude of unstable r-modes is also important for the evolution of the neutron star inside the instability window (a process which presumably sets its final spin rate), because it affects the spin-down time scale. During initial studies (e.g., Owen et al. 1998), the saturation energy of the r-mode was taken to be of order the rotational energy of the star, i.e., $E_{unit} = MR^2\Omega^2/2$. Later studies, which will be reviewed below, showed that nonlinear coupling saturates the r-mode at much lower amplitudes.

After Schenk et al. (2001) laid the groundwork, by deriving a consistent mode coupling formalism, Morsink (2002) calculated nonlinear couplings among r-modes in a stratified polytrope. Then, Arras et al. (2003) provided the first analytic estimate of the saturation amplitude, by coupling the $l = m = 2$ r-mode to other inertial modes, in

a star with negligible buoyancy. After calculating the modes in the Wentzel–Kramers–
Brillouin (WKB) limit, they considered two cases: (i) the discrete (or "weak-driving")
limit and (ii) the continuum (or "strong-driving") limit.

The first case is identical to the one we study here: the unstable mode grows and,
after crossing the lowest parametric instability threshold, excites the corresponding
daughter pair and saturates. Thus, the number of modes involved is small and the
mode spectrum can be regarded as discrete. For saturation to occur, the stability conditions (5.1.2) and (5.1.3) have to be satisfied for the coupled triplet.[12] In this sense,
according to Eq. (5.1.2), the parent is weakly driven, compared to the daughters'
damping rates. Although their estimations about the saturation amplitude are really
low (see Sect. 6 of their paper), Arras et al. conclude that the discrete limit is not
a good approximation, neither for nascent nor for accreting neutron stars, because
a large number of daughter modes is expected to be excited. This brings us to the
continuum limit.

In the second case, the modes are treated as a continuum, since a large number
of daughter modes is excited. This happens when the coupled triplets fail to satisfy
the stability condition (5.1.2), so the driving rate of the parent is larger than the
damping rates of the daughters. As the parent grows, it crosses many parametric
instability thresholds but cannot reach a stable equilibrium. However, as it couples
with more and more daughters, a turbulent energy cascade is formed and its growth
rate decreases, until it finally settles and saturates (see discussion in Sect. 5.1.4).
Arras et al. find that the saturation energy is given by $E/E_{unit} = 10^{-6}(\alpha_e/0.1)\nu_{kHz}^5$,
where $E_{unit} = MR^2\Omega^2/2$, ν is the spin rate of the star (measured in kHz) and α_e
parametrizes some "uncertainties" of their approach. Although taken to be ≈ 0.1, α_e
could be as low as $\sim 10^{-4}$ (see Sect. 2 of their paper), which would make the saturation
energy even lower. With the help of Eq. (2.6.7) we find that, for $E_{unit} = Mc^2$, the
saturation energy above is approximately two orders of magnitude lower.

Our study shows that the f-mode falls into the discrete limit case: the lowest
parametric instability threshold is usually stable and successfully saturates the mode.
Comparing our Eq. (5.3.1) to Arras et al.'s Eqs. (78) and (80) we see that the temperature scalings do not change: if bulk viscosity dominates the damping, their saturation
amplitude scales as T^6, whereas, if shear viscosity dominates, it scales as T^{-2} (their
A is equivalent to our $|Q|$). Although our scalings are different by a factor of $1/2$
in power, this should not be a surprise. As explained in Sect. 5.3.3, the temperature
dependence in Eq. (5.3.1) is due to the g-mode daughter damping rate changing with
temperature (as opposed to the f-mode daughter damping rate). On the other hand,
Arras et al. extract their Eqs. (78) and (80) assuming two identical daughters, in which
case the parametric instability threshold is approximated by Eq. (5.2.4), where the
daughter damping rate is squared.[13]

The analytical work of Arras et al. was followed by the simulations of Brink et al.
(2004a, b, 2005), where the saturation amplitude of the unstable r-mode was found

---

[12]Arras et al. are relaxing the saturation conditions by allowing the second one [Eq. (5.1.3)] not to
be true. The reason behind this was explained in Sect. 5.1.4.

[13]They also ignore the term $\Delta\omega/(\gamma_\beta + \gamma_\gamma)$, see Footnote 12.

"experimentally". Brink et al. used an incompressible, homogeneous star (Maclaurin spheroid; see Sect. 3.1), which permits the analytical calculation of eigenfrequencies, eigenfunctions, and all the quantities that involve them (growth/damping rates, coupling coefficients). As opposed to Arras et al., they dropped the Cowling approximation (Sect. 2.3.4), which, as they concluded, would otherwise neglect important terms in the couplings coefficients.

Their simulations included inertial modes with $l \leq 29$ ($\approx$5000 modes), which resulted in $\approx 1.5 \times 10^5$ direct couplings to the $l = m = 2$ r-mode, plus a large number of couplings among daughters themselves (daughter-daughter couplings; see Sect. 5.1.4). Starting with the integration of the triplet with the lowest parametric instability threshold, via Eqs. (4.2.12), they gradually added more couplings, as the r-mode kept growing and more modes were rising above the noise level. With this technique, i.e., following the evolution of the mode amplitudes, one can achieve much longer integration times than ordinary hydrodynamic simulations, where the integration time step is set by the oscillation periods of the modes.

In their work, they studied three types of large (i.e., involving all modes) systems: the conservative (Hamiltonian) system, as well as the strongly- and weakly-damped nonconservative system.

In the conservative system, the growth/damping rates are zero and, hence, the modes simply interact nonlinearly. This is known as the Fermi–Pasta–Ulam (FPU) problem (Fermi et al. 1955; Fermi 1965; see also Dauxois 2008) and its extensive study has shown that, after the initial excitation of a large-scale mode, a state of energy equipartitioning is reached, should the initial amplitude be larger than some threshold. This was indeed observed by Brink et al., with the equipartition time scale decreasing as the initial r-mode amplitude was increased.

The strongly-damped system corresponds to Arras et al.'s weak-driving limit. Brink et al. showed that, in this case, the mode amplitude evolution resembles that of a single triplet: the daughters' damping rates are large enough to halt the parent's growth, after the lowest parametric instability threshold is crossed.[14]

In the weakly-damped system, which in turn corresponds to the strong-driving limit of Arras et al., the situation is significantly more complex. The daughters are now not damped enough to stop the growth of the r-mode. However, daughter-daughter couplings distribute the incoming energy to many modes, thus preventing the r-mode from growing far beyond the second lowest parametric instability threshold (see Figs. 12 and 13 in Brink et al. 2005). In fact, at this amplitude the rate to equipartition is similar to the r-mode's growth rate, making the contribution of the FPU mechanism to the damping of the instability quite important.

The simulations above were performed for a star rotating at $\approx 0.6\,\Omega_K$. The strongly-damped system resides at low temperatures ($T \sim 10^6 - 10^7$ K), whereas the weakly-damped system at intermediate temperatures ($T \sim 10^8 - 10^9$ K). At these

---

[14]Like Arras et al., Brink et al. also ignore the stability condition (5.1.3) (see Footnote 12). This condition is not satisfied by the triplets with the lowest parametric instability thresholds in their simulations and, as a result, saturation occurs in the form of limit cycles or aperiodic motions (see Sect. 4.4.2).

temperatures, shear viscosity is the dominant damping mechanism (bulk viscosity is zero for an incompressible star).

Although Arras et al.'s cascade picture was mostly confirmed in the weakly-damped regime, the saturation amplitudes reported by Brink et al. are lower. The r-mode saturation energy was found to be $E/E_{unit} \approx 10^{-10} - 10^{-8}$, where $E_{unit} = MR^2\Omega^2/2$, with the higher values occurring at the weakly-damped system and the low-temperature end of the strongly-damped system. Since our system is strongly-damped at all temperatures, this is in agreement with our results, where the saturation amplitude obtains larger values at low and high temperatures.

Based on the work of Brink et al., a series of simulations were performed by Bondarescu et al. (2007, 2009), where the evolution of the neutron star through the $l = m = 2$ r-mode instability window was studied. Since the r-mode instability is relevant both for accreting and nascent neutron stars, both cases were examined.

Bondarescu et al. used a $\Gamma = 2$ polytrope, assuming the couplings of Brink et al.'s incompressible model. In particular, they incorporated the coupling with the lowest parametric instability threshold (as obtained from Brink et al.'s simulations) into the neutron star's spin and temperature evolution equations. This makes the neutron star evolution more "dynamical", as opposed to previous work, where the r-mode saturation amplitude was treated as an arbitrary constant. In most cases, the coupling occurring at the lowest threshold was enough to stop the r-mode from growing, which places the star in Brink et al.'s strongly-damped regime. It should be noted that, in addition to shear viscosity, Bondarescu et al. also considered viscosity at the crust-core boundary layer (see Sect. 3.6) and hyperon bulk viscosity, thus "enhancing" the dissipation effects.

By varying certain properties of the star, like the hyperon superfluid transition temperature, the fraction of the star above the threshold for direct Urca reactions (see Sect. 3.6), and the crust-core slippage factor, a number of interesting scenarios occur. Essentially, these quantities parametrise the strength of viscous and cooling effects. Except for their runaway evolutions, where the r-mode grows beyond the lowest parametric instability threshold and the three-mode system fails, saturation always occurs at the lowest threshold. The latter slowly changes during the evolution, due to the temperature dependence of the daughters' damping rates. For the rest of the quantities comprising the parametric instability threshold [Eq. (4.3.1)], Bondarescu et al. assume the values of the triplet giving the lowest threshold as "statistically relevant" constants.

Taking a step further, we calculated the saturation amplitude of the unstable f-modes, due to the three-mode coupling mechanism, *throughout* their instability windows. By doing this, we extract the whole coupling spectrum of the f-modes, which enables us to follow the different kinds of couplings and their strengths. Doing this for the r-mode will, most probably, not change the main results dramatically, because inertial modes are confined in a relatively small frequency range $[-2\Omega, 2\Omega]$ and fine resonances are fairly easy to obtain. On the other hand, in the case of the f-mode there are different kinds of daughter pairs (f-g or g-g, see Sect. 5.3.3), which makes it more subtle to be described by use of "effective" values for parameters like the detuning or the coupling coefficient.

References

Akmal, A., Pandharipande, V. R., & Ravenhall, D. G. (1998). Equation of state of nucleon matter and neutron star structure. *Physical Review C, 58*, 1804–1828. https://doi.org/10.1103/PhysRevC.58. 1804, arXiv:nucl-th/9804027.

Andersson, N. (1998). A new class of unstable modes of rotating relativistic stars. *The Astrophysical Journal, 502*, 708–713. https://doi.org/10.1086/305919, arXiv:gr-qc/9706075.

Andersson, N., Jones, D. I., & Kokkotas, K. D. (2002). Strange stars as persistent sources of gravitational waves. *Monthly Notices of the Royal Astronomical Society, 337*, 1224–1232. https://doi.org/10.1046/j.1365-8711.2002.05837.x, arXiv:astro-ph/0111582.

Arras, P., Flanagan, É. É., Morsink, S. M., Schenk, A. K., Teukolsky, S. A., & Wasserman, I. (2003). Saturation of the r-mode instability. *The Astrophysical Journal, 591*, 1129–1151. https://doi.org/10.1086/374657, arXiv:astro-ph/0202345.

Ballot, J., Lignières, F., Prat, V., Reese, D. R., & Rieutord, M. (2012). 2D Computations of g-modes in Fast Rotating Stars. In H. Shibahashi, M. Takata, & A. E. Lynas-Gray (Eds.), *Progress in solar/stellar physics with helio- and asteroseismology* (vol. 462). Astronomical Society of the Pacific Conference Series.http://adsabs.harvard.edu/abs/2012ASPC..462..389B.

Ballot, J., Lignières, F., Reese, D. R., & Rieutord, M. (2010). Gravity modes in rapidly rotating stars. Limits of perturbative methods. *Astronomy & Astrophysics, 518*, A30. https://doi.org/10.1051/0004-6361/201014426, arXiv:1005.0275.

Baumgarte, T. W., Shapiro, S. L. & Shibata, M. (2000). On the maximum mass of differentially rotating neutron stars. *The Astrophysical Journal, 528*, L29–L32. https://doi.org/10.1086/312425, arXiv:astro-ph/9910565.

Bondarescu, R., Teukolsky, S. A., & Wasserman, I. (2007). Spin evolution of accreting neutron stars: Nonlinear development of the r-mode instability. *Physical Review D, 76*, 064019. https://doi.org/10.1103/PhysRevD.76.064019, arXiv:0704.0799.

Bondarescu, R., Teukolsky, S. A., & Wasserman, I. (2009). Spinning down newborn neutron stars: Nonlinear development of the r-mode instability. *Physical Review D, 79*, 104003. https://doi.org/10.1103/PhysRevD.79.104003, arXiv:0809.3448.

Brink, J., Teukolsky, S. A., & Wasserman, I. (2004a). Nonlinear coupling network to simulate the development of the r-mode instability in neutron stars. I. Construction. *Physical Review D, 70*, 124017. https://doi.org/10.1103/PhysRevD.70.124017, arXiv:gr-qc/0409048.

Brink, J., Teukolsky, S. A., & Wasserman, I. (2004b). Nonlinear couplings of R-modes: Energy transfer and saturation amplitudes at realistic timescales. *Physical Review D, 70*, 121501. https://doi.org/10.1103/PhysRevD.70.121501, arXiv:gr-qc/0406085.

Brink, J., Teukolsky, S. A., & Wasserman, I. (2005). Nonlinear coupling network to simulate the development of the r mode instability in neutron stars. II. Dynamics. *Physical Review D, 71*, 064029. https://doi.org/10.1103/PhysRevD.71.064029, arXiv:gr-qc/0410072.

Burrows, A. (1990). Neutrinos from supernova explosions. *Annual Review of Nuclear and Particle Science, 40*, 181–212. https://doi.org/10.1146/annurev.ns.40.120190.001145.

Burrows, A., & Lattimer, J. M. (1986). The birth of neutron stars. *The Astrophysical Journal, 307*, 178–196. https://doi.org/10.1086/164405.

Ciolfi, R., & Siegel, D. M. (2015). Short gamma-ray bursts in the "time-reversal" scenario. *The Astrophysical Journal, 798*, L36. https://doi.org/10.1088/2041-8205/798/2/L36, arXiv:1411.2015.

Cook, G. B., Shapiro, S. L., & Teukolsky, S. A. (1992). Spin-up of a rapidly rotating star by angular momentum loss: Effects of general relativity. *The Astrophysical Journal, 398*, 203–223. https://doi.org/10.1086/171849.

Cook, G. B., Shapiro, S. L., & Teukolsky, S. A. (1994). Rapidly rotating neutron stars in general relativity: Realistic equations of state. *The Astrophysical Journal, 424*, 823–845. https://doi.org/10.1086/173934.

Dai, Z. G., & Lu, T. (1998). Gamma-ray burst afterglows and evolution of postburst fireballs with energy injection from strongly magnetic millisecond pulsars. *Astronomy & Astrophysics, 333*, L87–L90. http://adsabs.harvard.edu/abs/1998A%26A...333L..87D, arXiv:astro-ph/9810402.

Dall'Osso, S., Giacomazzo, B., Perna, R., & Stella, L. (2015). Gravitational waves from massive magnetars formed in binary neutron star mergers. *The Astrophysical Journal*, *798*, 25. https://doi.org/10.1088/0004-637X/798/1/25, arXiv:1408.0013.

Dauxois, T. (2008). Fermi, Pasta, Ulam, and a mysterious lady. *Physics Today*, *61*, 55. https://doi.org/10.1063/1.2835154, arXiv:0801.1590.

Dintrans, B., & Rieutord, M. (2000). Oscillations of a rotating star: A non-perturbative theory. *Astronomy & Astrophysics*, *354*, 86–98. http://adsabs.harvard.edu/abs/2000A%26A...354...86D.

Doneva, D. D., & Kokkotas, K. D. (2015). Asteroseismology of rapidly rotating neutron stars: An alternative approach. *Physical Review D*, *92*, 124004. https://doi.org/10.1103/PhysRevD.92.124004, arXiv:1507.06606.

Doneva, D. D., Gaertig, E., Kokkotas, K. D., & Krüger, C. (2013). Gravitational wave asteroseismology of fast rotating neutron stars with realistic equations of state. *Physical Review D*, *88*, 044052. https://doi.org/10.1103/PhysRevD.88.044052, arXiv:1305.7197.

Doneva, D. D., Kokkotas, K. D., & Pnigouras, P. (2015). Gravitational wave afterglow in binary neutron star mergers. *Physical Review D*, *92*, 104040. https://doi.org/10.1103/PhysRevD.92.104040, arXiv:1510.00673.

Duez, M. D., Liu, Y. T., Shapiro, S. L., Shibata, M., & Stephens, B. C. (2006). Evolution of magnetized, differentially rotating neutron stars: Simulations in full general relativity. *Physical Review D*, *73*, 104015. https://doi.org/10.1103/PhysRevD.73.104015, arXiv:astro-ph/0605331.

Duncan, R. C., & Thompson, C. (1992). Formation of very strongly magnetized neutron stars: Implications for gamma-ray bursts. *The Astrophysical Journal*, *392*, L9–L13. https://doi.org/10.1086/186413.

Fermi, E., Pasta, J., Ulam, S., & Tsingou, M. (1955). *Studies of nonlinear problems*. Report LA-1940, Los Alamos Scientific Laboratory, Los Alamos.

Fermi, E. (1965). *Collected papers* (vol. II, pp. 977–988). Chicago: The University of Chicago Press.

Ferrari, V., Matarrese, S., & Schneider, R. (1999). Stochastic background of gravitational waves generated by a cosmological population of young, rapidly rotating neutron stars. *Monthly Notices of the Royal Astronomical Society*, *303*, 258–264. https://doi.org/10.1046/j.1365-8711.1999.02207.x, arXiv:astro-ph/9806357.

Friedman, J. L., & Morsink, S. M. (1998). Axial instability of rotating relativistic stars. *The Astrophysical Journal*, *502*, 714–720. https://doi.org/10.1086/305920, arXiv:gr-qc/9706073.

Fryer, C. L., Belczynski, K., Ramirez-Ruiz, E., Rosswog, S., Shen, G., & Steiner, A. W. (2015). The fate of the compact Remnant in neutron star mergers. *The Astrophysical Journal*, *812*, 24. https://doi.org/10.1088/0004-637X/812/1/24, arXiv:1504.07605.

Gaertig, E., & Kokkotas, K. D. (2009). Relativistic g-modes in rapidly rotating neutron stars. *Physical Review D*, *80*, 064026. https://doi.org/10.1103/PhysRevD.80.064026, arXiv:0905.0821.

Gaertig, E., Glampedakis, K., Kokkotas, K. D., & Zink, B. (2011). F-mode instability in relativistic neutron stars. *Physical Review Letters*, *107*, 101102. https://doi.org/10.1103/PhysRevLett.107.101102, arXiv:1106.5512.

Heger, A., Fryer, C. L., Woosley, S. E., Langer, N., & Hartmann, D. H. (2003). How massive single stars end their life. *The Astrophysical Journal*, *591*, 288–300. https://doi.org/10.1086/375341, arXiv:astro-ph/0212469.

Hotokezaka, K., Kiuchi, K., Kyutoku, K., Muranushi, T., Sekiguchi, Y.-i., Shibata, M., & Taniguchi, K. (2013). Remnant massive neutron stars of binary neutron star mergers: Evolution process and gravitational waveform. *Physical Review D*, *88*, 044026. https://doi.org/10.1103/PhysRevD.88.044026, arXiv:1307.5888.

Ipser, J. R., & Lindblom, L. (1990). The oscillations of rapidly rotating Newtonian stellar models. *The Astrophysical Journal*, *355*, 226–240. https://doi.org/10.1086/168757.

Kastaun, W., & Galeazzi, F. (2015). Properties of hypermassive neutron stars formed in mergers of spinning binaries. *Physical Review D*, *91*, 064027. https://doi.org/10.1103/PhysRevD.91.064027, arXiv:1411.7975.

Kastaun, W., Willburger, B., & Kokkotas, K. D. (2010). Saturation amplitude of the f-mode instability. *Physical Review D*, *82*, 104036. https://doi.org/10.1103/PhysRevD.82.104036, arXiv:1006.3885.

Lee, W. H., & Ramirez-Ruiz, E. (2007). The progenitors of short gamma-ray bursts. *New Journal of Physics*, *9*, 17. https://doi.org/10.1088/1367-2630/9/1/017, arXiv:astro-ph/0701874.

Lignières, F., Rieutord, M., & Reese, D. (2006). Acoustic oscillations of rapidly rotating polytropic stars. I. Effects of the centrifugal distortion. *Astronomy & Astrophysics*, *455*, 607–620. https://doi.org/10.1051/0004-6361:20065015, arXiv:astro-ph/0604312.

Lockitch, K. H., & Friedman, J. L. (1999). Where are the R-modes of isentropic stars? *The Astrophysical Journal*, *521*, 764–788. https://doi.org/10.1086/307580, arXiv:gr-qc/9812019.

MacFadyen, A. I., Woosley, S. E., & Heger, A. (2001). Supernovae, jets, and collapsars. *The Astrophysical Journal*, *550*, 410–425. https://doi.org/10.1086/319698, arXiv:astro-ph/9910034.

Morsink, S. M. (2002). Nonlinear couplings between r-modes of rotating neutron stars. *The Astrophysical Journal*, *571*, 435–446. https://doi.org/10.1086/339858, arXiv:astro-ph/0202051.

Owen, B. J., Lindblom, L., Cutler, C., Schutz, B. F., Vecchio, A., & Andersson, N. (1998). Gravitational waves from hot young rapidly rotating neutron stars. *Physical Review D*, *58*, 084020. https://doi.org/10.1103/PhysRevD.58.084020, arXiv:gr-qc/9804044.

Passamonti, A., Gaertig, E., Kokkotas, K. D., & Doneva, D. (2013). Evolution of the f-mode instability in neutron stars and gravitational wave detectability. *Physical Review D*, *87*, 084010. https://doi.org/10.1103/PhysRevD.87.084010, arXiv:1209.5308.

Passamonti, A., Haskell, B., Andersson, N., Jones, D. I., & Hawke, I. (2009). Oscillations of rapidly rotating stratified neutron stars. *Monthly Notices of the Royal Astronomical Society*, *394*, 730–741. https://doi.org/10.1111/j.1365-2966.2009.14408.x, arXiv:0807.3457.

Passamonti, A. (2009). Time evolution of rapidly rotating stratified neutron stars. *Journal of Physics: Conference Series*, *189*, 012030. https://doi.org/10.1088/1742-6596/189/1/012030.

Pnigouras, P., & Kokkotas, K. D. (2016). Saturation of the f-mode instability in neutron stars. II. Applications and results. *Physics Review D*, *94*, 024053. https://doi.org/10.1103/PhysRevD.94.024053, arXiv:1607.03059.

Prakash, M., Lattimer, J. M., Pons, J. A., Steiner, A. W. ,& Reddy, S. (2001). Evolution of a neutron star from its birth to old age. In D. Blaschke, N. K. Glendenning, & A. Sedrakian (Eds.), *Physics of neutron star interiors* (vol. 578). Lecture notes in physics. Berlin: Springer. http://adsabs.harvard.edu/abs/2001LNP...578..364P, arXiv:astro-ph/0012136.

Press, W. H., Teukolsky, S. A., Vetterling, W. T., & Flannery, B. P. (1992). *Numerical recipes in FORTRAN 77* (vol. 1, 2nd ed.). Fortran numerical recipes. Cambridge: Cambridge University Press. http://adsabs.harvard.edu/abs/1992nrfa.book....P.

Press, W. H., Teukolsky, S. A., Vetterling, W. T., & Flannery, B. P. (1996). *Numerical recipes in FORTRAN 90* (vol. 2, 2nd ed.). Fortran numerical recipes. Cambridge: Cambridge University Press. http://adsabs.harvard.edu/abs/1992nrfa.book....P.

Ravi, V., & Lasky, P. D. (2014). The birth of black holes: Neutron star collapse times, gamma-ray bursts and fast radio bursts. *Monthly Notices of the Royal Astronomical Society*, *441*, 2433–2439. https://doi.org/10.1093/mnras/stu720, arXiv:1403.6327.

Reese, D., Lignières, F., & Rieutord, M. (2006). Acoustic oscillations of rapidly rotating polytropic stars. II. Effects of the Coriolis and centrifugal accelerations. *Astronomy & Astrophysics*, *455*, 621–637. https://doi.org/10.1051/0004-6361:20065269, arXiv:astro-ph/0605503.

Regimbau, T. (2011). The astrophysical gravitational wave stochastic background. *Research in Astronomy and Astrophysics*, *11*, 369–390. https://doi.org/10.1088/1674-4527/11/4/001, arXiv:1101.2762.

Rezzolla, L., & Kumar, P. (2015). A novel paradigm for short gamma-ray bursts with extended X-ray emission. *The Astrophysical Journal*, *802*, 95. https://doi.org/10.1088/0004-637X/802/2/95, arXiv:1410.8560.

Rowlinson, A. (2013). Studying the multi-wavelength signals from short GRBs. In *Proceedings of the 7th Huntsville gamma ray burst symposium*. Nashville, USA: eConf C1304143. http://adsabs.harvard.edu/abs/2013arXiv1308.1684R, arXiv:1308.1684.

Rowlinson, A., O'Brien, P. T., Metzger, B. D., Tanvir, N. R., & Levan, A. J. (2013). Signatures of magnetar central engines in short GRB light curves. *Monthly Notices of the Royal Astronomical Society, 430*, 1061–1087. https://doi.org/10.1093/mnras/sts683, arXiv:1301.0629.

Schenk, A. K., Arras, P., Flanagan, É. É., Teukolsky, S. A., & Wasserman, I. (2001). Nonlinear mode coupling in rotating stars and the r-mode instability in neutron stars. *Physical Review D, 65*, 024001. https://doi.org/10.1103/PhysRevD.65.024001, arXiv:gr-qc/0101092.

Stergioulas, N., & Friedman, J. L. (1998). Nonaxisymmetric neutral modes in rotating relativistic stars. *The Astrophysical Journal, 492*, 301–322. https://doi.org/10.1086/305030, arXiv:gr-qc/9705056.

Surace, M., Kokkotas, K. D., & Pnigouras, P. (2016). The stochastic background of gravitational waves due to the f-mode instability in neutron stars. *Astronomy & Astrophysics, 586*, A86. https://doi.org/10.1051/0004-6361/201527197, arXiv:1512.02502.

Unno, W., Osaki, Y., Ando, H., Saio, H., & Shibahashi, H. (1989). *Nonradial oscillations of stars* (2nd ed.). Tokyo: University of Tokyo Press. http://adsabs.harvard.edu/abs/1989nos.book....U.

Usov, V. V. (1992). Millisecond pulsars with extremely strong magnetic fields as a cosmological source of gamma-ray bursts. *Nature, 357*, 472–474. https://doi.org/10.1038/357472a0.

Wiringa, R. B., Fiks, V., & Fabrocini, A. (1988). Equation of state for dense nucleon matter. *Physical Review C, 38*, 1010–1037. https://doi.org/10.1103/PhysRevC.38.1010.

Woosley, S. E. (1993). Gamma-ray bursts from stellar mass accretion disks around black holes. *The Astrophysical Journal, 405*, 273–277. https://doi.org/10.1086/172359.

Yoshida, S., & Lee, U. (2000). Rotational modes of nonisentropic stars and the gravitational radiation-driven instability. *The Astrophysical Journal Supplement Series, 129*, 353–366. https://doi.org/10.1086/313410, arXiv:astro-ph/0002300.

Zhang, B., & Mészáros, P. (2001). Gamma-ray burst afterglow with continuous energy injection: Signature of a highly magnetized millisecond pulsar. *The Astrophysical Journal, 552*, L35–L38. https://doi.org/10.1086/320255, arXiv:astro-ph/0011133.

Zhu, X.-J., Fan, X.-L., & Zhu, Z.-H. (2011). Stochastic gravitational wave background from neutron star r-mode instability revisited. *The Astrophysical Journal, 729*, 59. https://doi.org/10.1088/0004-637X/729/1/59, arXiv:1102.2786.

Zink, B., Korobkin, O., Schnetter, E., & Stergioulas, N. (2010). Frequency band of the f-mode Chandrasekhar–Friedman–Schutz instability. *Physical Review D, 81*, 084055. https://doi.org/10.1103/PhysRevD.81.084055, arXiv:1003.0779.

Chapter 6
Final Remarks

Gravitational wave detection is about seeing
the biggest things that ever happen—the collisions, explosions,
and quakings of stars and black holes—by measuring
the smallest changes that have ever been measured.

Harry Collins, *Gravity's shadow* (2004)

6.1 Summary and Conclusions

Even though their existence has been known for a few decades now, there is still much uncertainty about neutron stars. Squeezing a star with the mass of the Sun down to the size of a large asteroid, the resulting object is bound to be fascinating! Nature's particle accelerators offer a unique opportunity to study matter and gravity at their very extremes, but the available observational data only gives us a glimpse of the physics involved, with crucial aspects of their structure eluding our current understanding.

The recent advent of gravitational-wave astronomy will hopefully shed some light on the —inaccessible by any other means— neutron star interior. Should gravitational radiation from individual sources be observable, much information about the equation of state of dense nuclear matter could be obtained. However, gravitational-wave asteroseismology will have to deal with very weak signals, generated by stellar oscillations. The fact that some of these oscillations are unstable to the emission of gravitational waves, due to the Chandrasekhar–Friedman–Schutz (CFS) mechanism, works to our advantage: the amplitude of the mode will grow, until such a point when nonlinear effects saturate the instability.

The saturation amplitude of an unstable mode determines the gravitational-wave strain of the generated signal and, thus, its detectability, but also affects the evolution of the neutron star through the instability region. If a fast-rotating newborn star enters the instability window during its cooling phase, it will traverse it at approximately

© Springer Nature Switzerland AG 2018

P. Pnigouras, *Saturation of the f-mode Instability in Neutron Stars*,
Springer Theses, https://doi.org/10.1007/978-3-319-98258-8_6

constant angular velocity, until thermal equilibrium is reached; then, at approximately constant temperature, the star will spin down, due to emission of gravitational radiation from the unstable mode, as well as magnetic braking, until it exits the window. The contribution of gravitational waves to the spin down phase and the time spent by the star inside the instability window are determined by the saturation amplitude.

In this work, we have studied the saturation of CFS-unstable (fundamental) f-modes, via low-order nonlinear mode coupling. We consider the quadratic-perturbation approximation, where resonantly coupled three-mode networks are formed throughout the star. The efficiency of the coupling among the three modes is determined by their coupling coefficient, mainly depending on their eigenfunctions, and by their detuning, which measures how close to resonance they are. The coupled triplet is prone to a parametric resonance instability, when the unstable (parent) mode crosses the so-called parametric instability threshold, at which point the two stable (daughter) modes start growing. The mode amplitudes approach a stable equilibrium and the parent mode saturates if certain conditions are satisfied, involving the modes' growth/damping rates and their detuning.

We have presented the first results about the saturation of the f-mode instability, due to nonlinear mode coupling, in neutron stars. Using Newtonian polytropes to describe both typical and supramassive neutron stars, we calculated all the couplings of the most unstable f-mode multipoles to other polar modes and obtained their saturation amplitude throughout their instability windows, by locating the triplet with the lowest stable parametric instability threshold.

Although it is usually treated as a constant (e.g., Passamonti et al. 2013), the saturation amplitude changes throughout the window, due to its temperature dependence and because different daughter pairs may set the lowest parametric instability threshold at different points. We found that the saturation amplitude is larger near the low- and high-temperature edges of the instability window (as high as $\approx 3 \times 10^{-4}$), and gradually decreases at intermediate temperatures (with values as low as $\sim 10^{-9}$; the definition used for the amplitude is $|Q| = \sqrt{E/Mc^2}$, where E is the mode energy, normalised to the rest mass energy of the star). These values are lower compared to previous work, where the saturation of unstable f-modes was studied either by nonlinear hydrodynamic simulations (Kastaun et al. 2010) or via large-amplitude viscous dissipation (Passamonti and Glampedakis 2012). Considering the highest value of the saturation amplitude obtained here, signals generated by the f-mode instability might be detectable even with Advanced LIGO, from sources in the Virgo cluster (≈ 20 Mpc; Passamonti et al. 2013; Doneva et al. 2015).

The perturbative nonlinear approach that we use is, in its core, simple and has many advantages. As long as the eigenfrequencies and eigenfunctions of the modes are provided, it allows us to easily identify the important couplings in the system and precisely track their effects on the modes' amplitude evolution. Furthermore, it helps us reveal and understand the richness of possible outcomes and offers a strong insight into the problem, letting us follow every parameter's contribution.

The calculation of the eigenfrequencies and eigenfunctions of the modes, however, can be a quite laborious task, with analytic solutions existing only in simple models (e.g., homogeneous star) and with no natural limit on the number of modes that should be considered —for instance, solar observations have shown very high p-mode multipole oscillations. In order to obtain as many modes as possible, the slow-rotation approximation was utilised, which is the origin of the major uncertainties in our results (correctness of models aside).

The Newtonian formalism provides an accurate qualitative description of the problem, at least for typical neutron stars. In principle, general relativity should change some key components of the setup (e.g., larger instability windows, shorter growth time scales for the parent), thus affecting the final results. Moreover, it is the only appropriate framework for modelling supramassive post-merger neutron star remnants, since our Newtonian calculation reflects only their rudimentary properties. Therefore, a relativistic nonlinear perturbation scheme needs to be developed in order to obtain conclusive results, especially considering that relativistic hydrodynamic simulations are still far from remaining stable during the secular time scales needed for the instability to grow.

Once the sensitivity of gravitational-wave detectors improves, gravitational radiation from neutron star oscillations will divulge information of key significance for astrophysics, nuclear physics, and gravitational physics. In the meantime, much work still has to be done, regarding the elimination of the major uncertainties and the refinement of the models, in order to reach confident conclusions. Thanks to the rapid development of technology, which enables us to measure "the smallest changes that have ever been measured",[1] and the advancement of theoretical and computational expertise, the future can only hold good omens for our deeper understanding of the Universe.

6.2   Epilogue: Notes on the Cosmic Staff

Two-and-a-half-thousand years after Pythagoras, we can positively identify the "Music of the spheres" as gravitational waves! Indeed, the ripples of spacetime represent, in a crude analogy, cosmic sounds, travelling via space before they reach our "ears", i.e., our detectors, which are now sensitive enough to listen to the whispers of the Universe. We can only imagine Albert Einstein performing the following passage on his violin:

---

[1] Harry Collins, *Gravity's shadow: The search for gravitational waves*, Introduction. Chicago: The University of Chicago Press (2004).

This is the actual gravitational-wave "sound" of a supramassive neutron star, evolving through the instability window of the quadrupole f-mode.[2] The music of the Cosmos is deceptively simple, and yet there is so much to learn, from every single note, about "the biggest things that ever happen".

Thus far, light has been our primary messenger from outer space. The dawn of gravitational-wave astronomy is upon us and finally allows the "dark side" of the Universe to take the floor; and after a lifetime of deafness, we probably have a lot of interesting things to hear!

References

Doneva, D. D., Kokkotas, K. D., & Pnigouras, P. (2015). Gravitational wave afterglow in binary neutron star mergers. *Physical Review D*, *92*, 104040. https://doi.org/10.1103/PhysRevD.92.104040, arXiv:1510.00673.

Kastaun, W., Willburger, B., & Kokkotas, K. D. (2010). Saturation amplitude of the f-mode instability. *Physical Review D*, *82*, 104036. https://doi.org/10.1103/PhysRevD.82.104036, arXiv:1006.3885.

Passamonti, A., & Glampedakis, K. (2012). Non-linear viscous damping and gravitational wave detectability of the f-mode instability in neutron stars. *Monthly Notices of the Royal Astronomical Society*, *422*, 3327–3338. https://doi.org/10.1111/j.1365-2966.2012.20849.x, arXiv:1112.3931.

Passamonti, A., Gaertig, E., Kokkotas, K. D., & Doneva, D. (2013). Evolution of the f-mode instability in neutron stars and gravitational wave detectability. *Physical Review D*, *87*, 084010. https://doi.org/10.1103/PhysRevD.87.084010, arXiv:1209.5308.

---

[2]Specifically, of a neutron star with baryon mass $M_b \approx 3\,M_\odot$, described by the WFF2 equation of state, used in Doneva et al. (2015). The star enters the instability window of the quadrupole $(l = m = 2)$ f-mode, rotating at the Kepler limit and emitting gravitational waves with a frequency of 810 Hz. After $\sim$ 10 min, thermal equilibrium is reached and the star starts descending the window. During this phase, which lasts $\sim$ 10 hrs, the gravitational-wave frequency continuously decreases to 360 Hz, until the star collapses to a black hole. The f-mode saturation amplitude is set, throughout the evolution, to $|Q| = 10^{-3}$.

Appendix A
Polytropic Stars

Following the definition of a polytropic process in thermodynamics, a star is said to be polytropic when its equation of state has the form

$$p = K\rho^{\Gamma}, \tag{A.1}$$

where K is the *polytropic constant* and Γ is the *polytropic exponent*, usually also written as

$$\Gamma = 1 + \frac{1}{n}, \tag{A.2}$$

where n is the *polytropic index*. To avoid confusion with the mode overtone n, defined in Sect. 2.4, the polytropic exponent Γ is mostly used in the main chapters, but the polytropic index n will be preferred throughout this appendix.

A.1 Nonrotating Polytropes

We write the equation of hydrostatic equilibrium for a nonrotating star [Eq. (2.1.5) with $\Omega = 0$] as

$$\frac{1}{r^2}\frac{d}{dr}\left(\frac{r^2}{\rho}\frac{dp}{dr}\right) = -4\pi G\rho,$$

where we used the fact that

$$\boldsymbol{g} = -\nabla\Phi = -\frac{GM_r}{r^2}\boldsymbol{e}_r,$$

$\boldsymbol{g}$ being the local gravitational acceleration, and replaced M_r from Eq. (2.3.42). Substituting Eqs. (A.1) and (A.2), we get

© Springer Nature Switzerland AG 2018

P. Pnigouras, *Saturation of the f-mode Instability in Neutron Stars*, Springer Theses, https://doi.org/10.1007/978-3-319-98258-8

$$\frac{1}{\xi^2} \frac{d}{d\xi} \left(\xi^2 \frac{d\theta}{d\xi} \right) = -\theta^n, \tag{A.1.1}$$

where the dimensionless variables θ and ξ are defined from the relations

$$\rho = \rho_c \theta^n \tag{A.1.2}$$

and

$$r = \left[\frac{(n+1)K}{4\pi G} \rho_c^{\frac{1}{n}-1} \right]^{1/2} \xi \tag{A.1.3}$$

respectively, ρ_c being the central density of the star. Equation (A.1.1) is the well-known *Lane-Emden equation*, for a polytrope with index n.[1]

At the centre of the star ($\xi \to 0$), the boundary conditions are (Shapiro and Teukolsky 1983, Sect. 3.3)

$$\theta(0) = 1 \tag{A.1.4}$$

and

$$\theta'(0) = 0, \tag{A.1.5}$$

where the prime denotes differentiation with respect to ξ. Using these, we can integrate Eq. (A.1.1) and show that $\theta(\xi)$, known as Emden's function, always has a finite root ξ_1 for $n < 5$. Hence, ξ_1 corresponds to the surface of the star, namely

$$R = \left[\frac{(n+1)K}{4\pi G} \rho_c^{\frac{1}{n}-1} \right]^{1/2} \xi_1,$$

whereas, using Eq. (2.3.42), the mass is found as

$$M = 4\pi \left[\frac{(n+1)K}{4\pi G} \right]^{3/2} \rho_c^{\frac{3-n}{2n}} \xi_1^2 \left| \theta'(\xi_1) \right|. \tag{A.1.6}$$

The Lane-Emden equation admits an analytic solution for the two extremes, namely, $n = 0$, which describes a homogeneous star with $\rho = \rho_c$ everywhere (see Sect. 2.5), and $n = 5$, known as the Roche model, for which the mass is concentrated towards the centre of a star with an infinite radius. Thus, density and pressure decrease faster towards the surface as the polytropic index increases, i.e., as the equation of state gets softer (see Sect. 1.3). An additional case with an analytic solution is the $n = 1$ polytrope, where Emden's function is given by

$$\theta(\xi) = \frac{\sin \xi}{\xi} \tag{A.1.7}$$

---

[1] The variable ξ used here should not be confused with the displacement vector $\boldsymbol{\xi}$ or its components, used throughout the main chapters. We did not choose a different variable, because of the popularity of the Lane-Emden formalism in this particular notation.

and its first root is $\xi_1 = \pi$, for which $\theta'(\xi_1) = -1/\pi$.

A.2 Rotating Polytropes

The Lane-Emden formalism was extended for rotationally and tidally distorted polytropes by Chandrasekhar (1933a, b, c, d). We will be concerned with the first case, for which the equation of hydrostatic equilibrium (2.1.5) takes the form

$$\frac{1}{r^2} \frac{\partial}{\partial r} \left(\frac{r^2}{\rho} \frac{\partial p}{\partial r} \right) + \frac{1}{r^2} \frac{\partial}{\partial x} \left(\frac{1 - x^2}{\rho} \frac{\partial p}{\partial x} \right) = -4\pi G \rho + 2\Omega^2,$$

where $x = \cos \theta$. Like in the case of nonrotating polytropes, we write it in the dimensionless form

$$\frac{1}{\xi^2} \frac{\partial}{\partial \xi} \left(\xi^2 \frac{\partial \Theta}{\partial \xi} \right) + \frac{1}{\xi^2} \frac{\partial}{\partial x} \left[\left(1 - x^2 \right) \frac{\partial \Theta}{\partial x} \right] = -\Theta^n + \tilde{\Omega}^2, \tag{A.2.1}$$

where ξ is given by Eq. (A.1.3), Θ is defined as

$$\rho = \rho_c \Theta^n, \tag{A.2.2}$$

and the angular velocity Ω is normalised as

$$\tilde{\Omega}^2 = \frac{\Omega^2}{2\pi G \rho_c}.$$

Expanding Θ in terms of $\tilde{\Omega}$ we get

$$\Theta = \theta + \Psi \tilde{\Omega}^2 + \mathcal{O} \left(\tilde{\Omega}^4 \right), \tag{A.2.3}$$

where θ is the nonrotating solution, obtained from Eq. (A.1.1). We can replace Eq. (A.2.3) in Eq. (A.2.1) to get an equation for the correction function Ψ, namely

$$\frac{1}{\xi^2} \frac{\partial}{\partial \xi} \left(\xi^2 \frac{\partial \Psi}{\partial \xi} \right) + \frac{1}{\xi^2} \frac{\partial}{\partial x} \left[\left(1 - x^2 \right) \frac{\partial \Psi}{\partial x} \right] = -n\theta^{n-1} \Psi + 1. \tag{A.2.4}$$

Then, we expand Ψ in terms of the Legendre polynomials P_i (e.g., see Abramowitz and Stegun 1972, Chap. 8] and, after a series of arguments and calculations (which can be found in Chandrasekhar 1933a), we obtain

$$\Psi = \psi_0(\xi) + A_2 \psi_2(\xi) P_2(x), \tag{A.2.5}$$

where the functions ψ_0 and ψ_2 satisfy the equations

$$\frac{1}{\xi^2}\frac{d}{d\xi}\left(\xi^2\frac{d\psi_0}{d\xi}\right) = -n\theta^{n-1}\psi_0 + 1 \qquad\qquad (A.2.6)$$

and

$$\frac{1}{\xi^2}\frac{d}{d\xi}\left(\xi^2\frac{d\psi_2}{d\xi}\right) = \left(-n\theta^{n-1} + \frac{6}{\xi^2}\right)\psi_2, \qquad\qquad (A.2.7)$$

and

$$A_2 = -\frac{5}{6}\frac{\xi_1^2}{3\psi_2(\xi_1) + \xi_1\psi_2'(\xi_1)}, \qquad\qquad (A.2.8)$$

with ξ_1 still denoting the first root of Emden's function $\theta(\xi)$.

Near the centre of the star ($\xi \to 0$), ψ_0 and ψ_2 behave like

$$\psi_0(\xi) = \frac{\xi^2}{6} + \mathcal{O}\left(\xi^4\right) \qquad\qquad (A.2.9)$$

and

$$\psi_2(\xi) = \xi^2 + \mathcal{O}\left(\xi^4\right). \qquad\qquad (A.2.10)$$

Thus, we can integrate Eqs. (A.2.6) and (A.2.7) to obtain the solution Θ for a rotationally distorted (to second order) polytrope. Then, the surface of the star is found from $\Theta(\xi_0) = 0$, or

$$\xi_0 = \xi_1 + \frac{\tilde{\Omega}^2}{|\theta'(\xi_1)|}\left[\psi_0(\xi_1) + A_2\psi_2(\xi_1)P_2(x)\right].$$

This shows that, to second order in rotation, the star expands (first correction term) and changes its shape to an oblate spheroid (second correction term). At the equator $P_2(0) = -1/2$ and at the poles $P_2(1) = 1$, so, using Eq. (A.2.8), the oblateness of the star f is found as

$$f = \frac{R_e - R_p}{R_e} = \frac{5}{4}\frac{\tilde{\Omega}^2}{|\theta'(\xi_1)|}\frac{\xi_1\psi_2(\xi_1)}{3\psi_2(\xi_1) + \xi_1\psi_2'(\xi_1)}, \qquad\qquad (A.2.11)$$

where R_e is the equatorial and R_p the polar radius. The mass of the star can also be obtained as

$$M = 4\pi\left[\frac{(n+1)K}{4\pi G}\right]^{3/2}\rho_c^{\frac{3-n}{2n}}\xi_1^2\left|\theta'(\xi_1)\right|\left[1 + \tilde{\Omega}^2\frac{\xi_1/3 - \psi_0'(\xi_1)}{|\theta'(\xi_1)|}\right]. \qquad (A.2.12)$$

Comparing Eqs. (A.1.6) and (A.2.12) we see that, if the rotating polytrope has the same central density ρ_c with its nonrotating counterpart, then its mass is larger by a factor of

$$\frac{M(\Omega)}{M(0)} = 1 + \tilde{\Omega}^2 \frac{\xi_1/3 - \psi_0'(\xi_1)}{|\theta'(\xi_1)|},$$
(A.2.13)

where $M(0)$ and $M(\Omega)$ are the masses of the nonrotating and the rotating star respectively. If instead we demand that the rotating polytrope have the same mass with the nonrotating one, then its central density changes by a factor of

$$\frac{\rho_c(\Omega)}{\rho_c(0)} = 1 - \tilde{\Omega}^2 \frac{2n}{3-n} \frac{\xi_1/3 - \psi_0'(\xi_1)}{|\theta'(\xi_1)|},$$
(A.2.14)

where $\rho_c(0)$ and $\rho_c(\Omega)$ are the central densities of the nonrotating and the rotating polytrope respectively. Hence, if $n < 3$ ($n > 3$) the central density of the rotating polytrope is lower (greater) than the central density of its nonrotating counterpart with the same mass (for the subtleties related to the $n = 3$ polytrope and the details of the calculations above, see Chandrasekhar 1933a).

Like with the Lane-Emden equation, there are analytic solutions for ψ_0 and ψ_2, for the $n = 0$, 1 and 5 polytropes (although the $n = 0$ and 5 cases need to be handled with caution; see Chandrasekhar 1933d). For $n = 1$, the solutions are given by

$$\psi_0(\xi) = 1 \quad \theta(\xi) = 1 \quad \frac{\sin \xi}{\xi}$$
(A.2.15)

and

$$\psi_2(\xi) = 15\sqrt{\frac{\pi}{2\xi}} J_{5/2}(\xi) = -15 \frac{3\xi \cos \xi + (\xi^2 - 3) \sin \xi}{\xi^3},$$
(A.2.16)

where J_i are the Bessel functions of the first kind (see, for example, Abramowitz and Stegun 1972, Chaps. 9 and 10]. For $\xi_1 = \pi$ the solutions above give $\psi_0(\xi_1) = 1$, $\psi_0'(\xi_1) = 1/\pi$, $\psi_2(\xi_1) = 45/\pi^2$, and $\psi_2'(\xi_1) = 15(1 - 9/\pi^2)/\pi$. Tables with numerical results on other polytropic indices can be found in Chandrasekhar (1933a) and Chandrasekhar and Lebovitz (1962).

References

1. Abramowitz, M., & Stegun, I. A. (1972). *Handbook of Mathematical Functions*. New York: Dover. http://adsabs.harvard.edu/abs/1972hmfw.book.....A.
2. Chandrasekhar, S. (1933a). The equilibrium of distorted polytropes. I. The rotational problem. *Monthly Notices of the Royal Astronomical Society, 93,* 390–406. http://adsabs.harvard.edu/abs/1933MNRAS..93..390C.
3. Chandrasekhar, S. (1933b). The equilibrium of distorted polytropes. II. The tidal problem. *Monthly Notices of the Royal Astronomical Society, 93,* 449. http://adsabs.harvard.edu/abs/1933MNRAS..93..449C.

4. Chandrasekhar, S. (1933c). The equilibrium of distorted polytropes. III. The double star problem. *Monthly Notices of the Royal Astronomical Society*, *93*, 462. http://adsabs.harvard.edu/abs/1933MNRAS..93..462C.
5. Chandrasekhar, S. (1933d). The equilibrium of distorted polytropes. IV. The rotational and the tidal distortions as functions of the density distribution. *Monthly Notices of the Royal Astronomical Society*, *93*, 539–574. http://adsabs.harvard.edu/abs/1933MNRAS..93..539C.
6. Chandrasekhar, S., & Lebovitz, N. R. (1962). On the oscillations and the stability of rotating gaseous masses. III. The distorted polytropes. *The Astrophysical Journal*, *136*, 1082. https://doi.org/10.1086/147459.
7. Shapiro, S. L., & Teukolsky, S. A. (1983). *Black holes, white dwarfs, and neutron stars: The physics of compact objects*. New York: Wiley. http://adsabs.harvard.edu/abs/1983bhwd.book.....S.

Appendix B
Polar Mode Rotational Corrections

B.1 First-Order Corrections

The first-order rotational corrections to the eigenfrequencies ω and eigenfunctions $\boldsymbol{\xi}$ of polar modes can be found from Eq. (2.6.9), namely

$$- \omega_\alpha^{(0)2} \boldsymbol{\xi}_\alpha^{(1)} + \mathcal{C}^{(0)} \left(\boldsymbol{\xi}_\alpha^{(1)} \right) - 2\omega_\alpha^{(0)} \omega_\alpha^{(1)} \boldsymbol{\xi}_\alpha^{(0)} + i \omega_\alpha^{(0)} \mathcal{B}^{(1)} \left(\boldsymbol{\xi}_\alpha^{(0)} \right) = \mathbf{0}. \qquad (B.1.1)$$

Following Unno et al. (1989, § 19), we will expand $\boldsymbol{\xi}_\alpha^{(0)}$ and $\boldsymbol{\xi}_\alpha^{(1)}$ in terms of the eigenfunctions $\boldsymbol{\xi}_\alpha$ of the nonrotating star, as

$$\boldsymbol{\xi}_\alpha^{(0)} = \sum_{m=-l}^{l} c_{\alpha m}^{(0)} \boldsymbol{\xi}_\alpha \qquad (B.1.2)$$

and

$$\boldsymbol{\xi}_\alpha^{(1)} = \sum_{\substack{\beta \\ \beta \neq \alpha}} c_{\alpha\beta}^{(1)} \boldsymbol{\xi}_\beta. \qquad (B.1.3)$$

The zeroth-order eigenfunction $\boldsymbol{\xi}_\alpha^{(0)}$ cannot be simply taken equal to $\boldsymbol{\xi}_\alpha$, because of the degeneracy of the eigenfrequencies with respect to m in the nonrotating limit (see Sect. 2.4). As a result, the $2l + 1$ eigenfunctions with the same degree l and overtone n correspond to the same eigenfrequency ω_α. So, we have to consider the zeroth-order eigenfunction as a linear combination of all the degenerate eigenfunctions with the same l and n, but different m (see, for instance, Mathews and Walker 1970, Chap. 10). In Eq. (B.1.2), mode $\boldsymbol{\xi}_\alpha$ is associated with the triplet (n, l, m), with the summation changing only m.

The first-order correction to the eigenfunction $\boldsymbol{\xi}_\alpha^{(1)}$ is taken to be the linear combination of all modes (including axial modes) of the nonrotating star (except for $\boldsymbol{\xi}_\alpha$

© Springer Nature Switzerland AG 2018
P. Pnigouras, *Saturation of the f-mode Instability in Neutron Stars*,
Springer Theses, https://doi.org/10.1007/978-3-319-98258-8

itself), where the correction coefficients $c_{\alpha\beta}^{(1)}$ are $\mathcal{O}\left(\Omega\right)$. In Eq. (B.1.3), the "quantum numbers" of the mode ξ_β are (n', l', m').

Replacing Eqs. (B.1.2) and (B.1.3) in Eq. (B.1.1) and taking the inner product, defined by Eq. (2.3.55), with ξ_γ [corresponding to (n'', l'', m'')], we get

$$\left(-\omega_\alpha^{(0)2} + \omega_\gamma^{(0)2}\right) c_{\alpha\gamma}^{(1)} I_\gamma - 2\omega_\alpha^{(0)}\omega_\alpha^{(1)} \delta_{n''n}\delta_{l''l} I_\alpha c_{\alpha m''}^{(0)} + i\omega_\alpha^{(0)} \sum_m c_{\alpha m}^{(0)} \langle \xi_\gamma, \mathcal{B}^{(1)}\left(\xi_\alpha\right)\rangle = 0,$$

(B.1.4)

with I_α defined in Eq. (2.3.55). For $(n'', l'') = (n, l)$ we obtain

$$\sum_m \mathcal{B}_{m''m}^{(1)} c_{\alpha m}^{(0)} = \omega_\alpha^{(1)} c_{\alpha m''}^{(0)},$$

(B.1.5)

where

$$\mathcal{B}_{m''m}^{(1)} = \frac{i}{2I_\alpha} \langle \xi_{m''}, \mathcal{B}^{(1)}\left(\xi_m\right)\rangle,$$

(B.1.6)

with ξ_m and $\xi_{m''}$ corresponding to mode ξ_α with (n, l, m) and (n, l, m'') respectively.

Equation (B.1.5) represents the m''-component of the matrix equation

$$B^{(1)} c_\alpha^{(0)} = \omega_\alpha^{(1)} c_\alpha^{(0)},$$

(B.1.7)

$B^{(1)}$ being the $(2l + 1) \times (2l + 1)$ matrix with components $\mathcal{B}_{m''m}^{(1)}$. The eigenvalues of this matrix give the first-order eigenfrequency corrections (one for each value of m) of the mode ξ_α, whereas its eigenvectors give the components of the zeroth-order eigenfunction $\xi_\alpha^{(0)}$.

If we now set $(n'', l'') = (n', l') \neq (n, l)$ in Eq. (B.1.4), we get for the first-order correction coefficients

$$c_{\alpha\beta}^{(1)} = \frac{i\omega_\alpha^{(0)}}{I_\beta\left(\omega_\alpha^{(0)2} - \omega_\beta^{(0)2}\right)} \sum_m c_{\alpha m}^{(0)} \langle \xi_\beta, \mathcal{B}^{(1)}\left(\xi_\alpha\right)\rangle.$$

(B.1.8)

We will proceed with the evaluation of the matrix $B^{(1)}$. Taking the angular velocity along the z axis ($\theta = 0$), namely

$$\mathbf{\Omega} = \left(\Omega\cos\theta, -\Omega\sin\theta, 0\right),$$

then, replacing the polar mode eigenfunction (2.3.19) in Eq. (B.1.6) and using the spherical harmonic orthogonality relation (2.3.14), it becomes

$$\mathcal{B}_{m''m}^{(1)} = \delta_{m''m} \frac{m\Omega}{I_\alpha} \int_0^R \left(2\xi_r\xi_h + \xi_h^2\right) \rho r^2 \mathrm{d}r.$$

The equation above shows that the matrix $B^{(1)}$ is diagonal, so its components are its eigenvalues and they are given by

$$\omega_{\alpha}^{(1)} = mC_1\Omega,$$ (B.1.9)

where, using Eq. (2.3.56),

$$C_1 = \frac{\displaystyle\int_0^R \left(2\xi_r\xi_h + \xi_h^2\right)\rho r^2 dr}{\displaystyle\int_0^R \left[\xi_r^2 + l(l+1)\xi_h^2\right]\rho r^2 dr}.$$ (B.1.10)

Then, for a certain eigenvalue (i.e., for a specific m), Eq. (B.1.7) can be written as

$$
\begin{pmatrix}
\mathcal{B}_{-l,-l}^{(1)} & 0 & \cdots & 0 & \cdots & 0 \\
0 & \mathcal{B}_{-l+1,-l+1}^{(1)} & \cdots & 0 & \cdots & 0 \\
\vdots & \vdots & \ddots & \vdots & \ddots & \vdots \\
0 & 0 & \cdots & \mathcal{B}_{m,m}^{(1)} & \cdots & 0 \\
\vdots & \vdots & \ddots & \vdots & \ddots & \vdots \\
0 & 0 & \cdots & 0 & \cdots & \mathcal{B}_{l,l}^{(1)}
\end{pmatrix}
\begin{pmatrix}
c_{\alpha,-l}^{(0)} \\
c_{\alpha,-l+1}^{(0)} \\
\vdots \\
c_{\alpha,m}^{(0)} \\
\vdots \\
c_{\alpha,l}^{(0)}
\end{pmatrix}
= \omega_{\alpha}^{(1)}
\begin{pmatrix}
c_{\alpha,-l}^{(0)} \\
c_{\alpha,-l+1}^{(0)} \\
\vdots \\
c_{\alpha,m}^{(0)} \\
\vdots \\
c_{\alpha,l}^{(0)}
\end{pmatrix}.
$$

So, the eigenvector $c_{\alpha}^{(0)}$ corresponding to the eigenvalue $\omega_{\alpha}^{(1)}$ is

$$
c_{\alpha}^{(0)} =
\begin{pmatrix}
0 \\
0 \\
\vdots \\
1 \\
\vdots \\
0
\end{pmatrix}
\begin{matrix}
\leftarrow \text{Position } -l \\
\leftarrow \text{Position } -l+1 \\
\vdots \\
\leftarrow \text{Position } m \\
\vdots \\
\leftarrow \text{Position } l
\end{matrix}
$$

Consequently, the zeroth-order eigenfunction (B.1.2) is given by a single spherical harmonic, namely

$$\boldsymbol{\xi}_{\alpha}^{(0)} = \boldsymbol{\xi}_{\alpha},$$ (B.1.11)

and Eq. (B.1.8) becomes

$$c_{\alpha\beta}^{(1)} = \frac{i\omega_{\alpha}^{(0)}}{I_{\beta}\left(\omega_{\alpha}^{(0)2} - \omega_{\beta}^{(0)2}\right)}\langle\boldsymbol{\xi}_{\beta},\mathcal{B}^{(1)}\left(\boldsymbol{\xi}_{\alpha}\right)\rangle.$$ (B.1.12)

Using these results, we can also derive the first-order inner product of two rotationally corrected eigenfunctions, $\boldsymbol{\xi}_{\alpha}(\Omega) = \boldsymbol{\xi}_{\alpha}^{(0)} + \boldsymbol{\xi}_{\alpha}^{(1)} + \mathcal{O}\left(\Omega^2\right)$, as (Schenk et al. 2001)

$$\langle \boldsymbol{\xi}_\alpha(\Omega), \boldsymbol{\xi}_\beta(\Omega) \rangle = \frac{i}{\omega_\alpha^{(0)} + \omega_\beta^{(0)}} \langle \boldsymbol{\xi}_\alpha, \boldsymbol{\mathcal{B}}^{(1)}(\boldsymbol{\xi}_\beta) \rangle + \mathcal{O}\left(\Omega^2\right), \qquad (B.1.13)$$

for $\alpha \neq \beta$, where we used the fact that operator $\mathcal{B}$ is anti-Hermitian[2] (Lynden-Bell and Ostriker 1967). From this relation one can see that the rotationally corrected eigenfunctions do not necessarily satisfy the orthogonality condition (2.3.55); Eq. (2.6.3) should be used instead as an orthogonality relation.

An Alternative Approach

From Eqs. (B.1.3) and (B.1.12) we see that the actual computation of the first-order corrections to the eigenfunctions can be cumbersome, since it is an expansion over all the modes of the nonrotating star. In practice, due to the form of $c_{\alpha\beta}^{(1)}$, only neighbouring modes with similar eigenfrequencies are considered, with the contribution from the rest of the modes being negligible.

However, there is an alternative way to obtain the first-order corrections to the eigenfunctions, first presented by Hansen et al. (1978) in the Cowling approximation (see Sect. 2.3.4) and then extended for the general case by Saio (1981). First, from Eqs. (B.1.3) and (2.3.53) we notice that axial modes do not contribute to the radial component $\xi_{\alpha,r}^{(1)}$ of $\boldsymbol{\xi}_\alpha^{(1)}$. From the radial component of Eq. (B.1.1) one can further show that

$$\int \xi_{\alpha,r}^{(1)} \xi_{\beta,r}^{(0)*} \rho \mathrm{d}^3 \boldsymbol{r} = 0,$$

for $\alpha \neq \beta$. Therefore, $\xi_r^{(1)}$ is proportional to a single spherical harmonic Y_l^m, namely

$$\xi_r^{(1)}(r, \theta, \phi) = \xi_r^{(1)}(r) Y_l^m(\theta, \phi) \qquad (B.1.14)$$

(we will omit the subscript α from now on, for simplicity). Then, using Eqs. (B.1.14) and (2.3.19), we take the θ and ϕ components of Eq. (B.1.1) to get

$$\xi_\theta^{(1)}(r, \theta, \phi) = \frac{1}{\omega^{(0)2} r} \left\{ \left[\chi^{(1)}(r) - \frac{2\omega^{(1)}}{\omega^{(0)}} \chi^{(0)}(r) \right] \frac{\partial}{\partial \theta} + \frac{2m\Omega}{\omega^{(0)}} \chi^{(0)}(r) \cot \theta \right\} Y_l^m(\theta, \phi) \qquad (B.1.15)$$

and

$$\xi_\phi^{(1)}(r, \theta, \phi) = \left\{ \frac{1}{\omega^{(0)2} r} \left[\chi^{(1)}(r) - \frac{2\omega^{(1)}}{\omega^{(0)}} \chi^{(0)}(r) \right] \frac{1}{\sin \theta} \frac{\partial}{\partial \phi} \right.$$
$$\left. + \frac{2i\Omega}{\omega^{(0)}} \left[\xi_r^{(0)}(r) \sin \theta + \frac{\chi^{(0)}(r)}{\omega^{(0)2} r} \cos \theta \frac{\partial}{\partial \theta} \right] \right\} Y_l^m(\theta, \phi), \qquad (B.1.16)$$

where the variable χ is defined as

---

[2] Namely, it satisfies the relation $\langle \boldsymbol{\xi}, \boldsymbol{\mathcal{B}} \cdot \boldsymbol{\xi}' \rangle = -\langle \boldsymbol{\mathcal{B}} \cdot \boldsymbol{\xi}, \boldsymbol{\xi}' \rangle$, for any $\boldsymbol{\xi}, \boldsymbol{\xi}'$ on the space of complex vector functions, with respect to the inner product $\langle \boldsymbol{\xi}, \boldsymbol{\xi}' \rangle = \int \boldsymbol{\xi} \cdot \boldsymbol{\xi}' \rho \mathrm{d}^3 \boldsymbol{r}$.

$$\chi(r, \theta, \phi) = \frac{\delta p}{\rho} + \delta\Phi. \tag{B.1.17}$$

It was shown in Sect. 2.3 that $\delta p^{(0)}$ and $\delta\Phi^{(0)}$ are proportional to Y_l^m. By replacing Eq. (B.1.3) in the perturbed continuity equation (2.2.4), the perturbed Poisson equation (2.2.6), and the perturbed equation of state (2.2.7), we see that the same applies to $\delta p^{(1)}$ and $\delta\Phi^{(1)}$, which means that χ can be expanded as

$$\chi(r, \theta, \phi) = \left[\chi^{(0)}(r) + \chi^{(1)}(r)\right] Y_l^m(\theta, \phi) + \mathcal{O}\left(\Omega^2\right).$$

It is obvious from Eq. (2.3.20) that $\chi^{(0)}(r) = \omega^{(0)2} r \xi_h^{(0)}(r)$. Comparing Eqs. (B.1.15) and (B.1.16) with Eq. (2.3.19) we see that the rotationally corrected eigenfunctions of polar modes do not follow the angular dependence of polar modes any more.

Given that $\omega^{(0)}$, $\xi_r^{(0)}$, and $\chi^{(0)}$ have been found from the integration of Eqs. (2.3.16)–(2.3.18) and their boundary conditions (2.3.24)–(2.3.27), whereas $\omega^{(1)}$ can be obtained from Eq. (B.1.9), it only remains to calculate $\xi_r^{(1)}$ and $\chi^{(1)}$. Applying a procedure similar to that followed for zeroth-order quantities in Sect. 2.3, we find a system of differential equations and boundary conditions for $\xi_r^{(1)}$, $\delta p^{(1)}$, and $\delta\Phi^{(1)}$. Using the dimensionless formulation presented in Sect. 2.3.3, they are written as

$$x\frac{dy_1^{(1)}}{dx} = \left(V_g - 3\right) y_1^{(1)} + \left[\frac{l(l+1)}{c_1\tilde{\omega}^{(0)2}} - V_g\right] y_2^{(1)} + V_g y_3^{(1)}$$
$$+ \frac{2m\Omega}{\omega^{(0)}} \left\{ y_1^{(0)} + \left[1 - \frac{\omega^{(1)}}{m\Omega}l(l+1)\right] \frac{y_2^{(0)}}{c_1\tilde{\omega}^{(0)2}} \right\}, \tag{B.1.18}$$

$$x\frac{dy_2^{(1)}}{dx} = \left(c_1\tilde{\omega}^{(0)2} - A^*\right) y_1^{(1)} + \left(A^* - U + 1\right) y_2^{(1)} - A^* y_3^{(1)}$$
$$+ \frac{2m\Omega}{\omega^{(0)}} \left(\frac{\omega^{(1)}}{m\Omega} c_1\tilde{\omega}^{(0)2} y_1^{(0)} - y_2^{(0)}\right), \tag{B.1.19}$$

$$x\frac{dy_3^{(1)}}{dx} = (1 - U) y_3^{(1)} + y_4^{(1)}, \tag{B.1.20}$$

$$x\frac{dy_4^{(1)}}{dx} = UA^* y_1^{(1)} + UV_g y_2^{(1)} + \left[l(l+1) - UV_g\right] y_3^{(1)} - U y_4^{(1)}, \tag{B.1.21}$$

whereas the boundary conditions at the centre $(x \to 0)$ and the surface $(x \to 1)$ are

$$l y_3^{(1)} - y_4^{(1)} = 0, \tag{B.1.22}$$

$$c_1\tilde{\omega}^{(0)2} y_1^{(1)} - l y_2^{(1)} + \frac{2m\Omega}{\omega^{(0)}} \left(\frac{\omega^{(1)}}{m\Omega} - \frac{1}{l}\right) c_1\tilde{\omega}^{(0)2} y_1^{(0)} = 0, \tag{B.1.23}$$

and

$$U y_1^{(1)} + (l+1) y_3^{(1)} + y_4^{(1)} = 0,$$ (B.1.24)

$$\left(1 - \frac{4 + c_1 \tilde{\omega}^{(0)2}}{V}\right) y_1^{(1)} + \left[\frac{l(l+1)}{c_1 \tilde{\omega}^{(0)2} V} - 1\right] y_2^{(1)} + \left(1 - \frac{l+1}{V}\right) y_3^{(1)}$$

$$+ \frac{2m\Omega}{\omega^{(0)} V}\left\{\left(1 - \frac{\omega^{(1)}}{m\Omega} c_1 \tilde{\omega}^{(0)2}\right) y_1^{(0)} + \left[1 - \frac{\omega^{(1)}}{m\Omega} l(l+1) + c_1 \tilde{\omega}^{(0)2}\right] \frac{y_2^{(0)}}{c_1 \tilde{\omega}^{(0)2}}\right\} = 0,$$
(B.1.25)

respectively. Now, Eqs. (B.1.18)–(B.1.25) can be solved as a boundary value problem for $y_1^{(1)}$, $y_2^{(1)}$, $y_3^{(1)}$, and $y_4^{(1)}$, from which $\xi_r^{(1)}$ and $\chi^{(1)}$ are obtained. It should be noted that the solutions are proportional to $m\Omega$, so they may be found for one value of m and then rescaled for the rest.

B.2   Second-Order Corrections

In order to obtain the second-order rotational corrections to the eigenfrequencies ω and eigenfunctions ξ of polar modes we need to use Eq. (2.6.10), namely

$$-\omega_\alpha^{(0)2} \xi_\alpha^{(2)} + \mathcal{C}^{(0)}\left(\xi_\alpha^{(2)}\right) - 2\omega_\alpha^{(0)} \omega_\alpha^{(1)} \xi_\alpha^{(1)} + i\omega_\alpha^{(0)} \mathcal{B}^{(1)}\left(\xi_\alpha^{(1)}\right)$$
$$-2\omega_\alpha^{(0)} \omega_\alpha^{(2)} \xi_\alpha^{(0)} - \omega_\alpha^{(1)2} \xi_\alpha^{(0)} + i\omega_\alpha^{(1)} \mathcal{B}^{(1)}\left(\xi_\alpha^{(0)}\right) + \mathcal{C}^{(2)}\left(\xi_\alpha^{(0)}\right) = 0.$$ (B.2.1)

The situation in this case gets much more complicated than with the calculation of $\mathcal{O}(\Omega)$ corrections, because, at second order in Ω, the equilibrium configuration is distorted by the centrifugal force [see Eq. (2.1.5)]. Hence, the various equilibrium quantities (i.e., density, pressure, and gravitational potential) do not depend only on the radial coordinate r, but also on the colatitude coordinate θ.

For the rigorous derivation of second-order corrections, the reader is referred to Saio (1981), whose basic steps we are going to reproduce in this section.[3] A similar formulation was also developed by Smeyers and Denis (1971), who applied it in a homogeneous, compressible star (see Sect. 2.5).

---

[3]In addition to second-order rotational effects on the polar modes of a star, Saio (1981) also considered the influence of tidal deformations from a companion star, whose orbital motion is parallel and synchronous to the rotation of the first star. These will be ignored in the current analysis.

B.2.1 Generic Formulation

In the presence of centrifugal acceleration, which is a second-order rotational effect, the star is deformed into an oblate spheroid. We define a new coordinate system (a, θ, ϕ), rotating with the star, where the radial coordinate r of spherical polar coordinates is replaced with a. Then, a distorted equipotential surface can be described as

$$r = a \left[1 + \varepsilon(a, \theta) \right], \tag{B.2.2}$$

with a remaining constant on each equipotential surface. If the central density of the nonrotating, spherical star, $\rho_{c,\,\text{sp}}$, changes due to rotation by $\Delta\rho_c \equiv \rho_c - \rho_{c,\,\text{sp}}$ (not to be confused with the Lagrangian perturbation Δ, which will not be used in this section), then the density on the rotating star can be expressed as

$$\rho \left[a(1 + \varepsilon) \right] = \rho_{\text{sp}}(a) \frac{\rho_c}{\rho_{c,\,\text{sp}}} = \rho_{\text{sp}}(a) \left(1 + \frac{\Delta\rho_c}{\rho_{c,\,\text{sp}}} \right), \tag{B.2.3}$$

where $\rho_{\text{sp}}(a)$ is the density on a sphere, with radius a, on the nonrotating star. Since density and pressure depend only on a, they can simply be written as

$$\rho(a) = \rho_{\text{sp}}(a) + \Delta\rho(a) \tag{B.2.4}$$

and likewise for the pressure.

In this notation, Eq. (B.2.1) becomes (omitting the mode index α from now on, for simplicity)

$$- \left(2\omega^{(2)} \omega^{(0)} + \omega^{(1)2} \right) \boldsymbol{\xi}^{(0)} - \omega^{(0)2} \left[\boldsymbol{\xi}^{(2)} + 2\varepsilon \boldsymbol{\xi}^{(0)} + a \xi_a^{(0)} \nabla\varepsilon + \boldsymbol{e}_a a \left(\boldsymbol{\xi}^{(0)} \cdot \nabla \right) \varepsilon \right]$$
$$- 2\omega^{(1)} \omega^{(0)} \boldsymbol{\xi}^{(1)} + \nabla\chi^{(2)} - \boldsymbol{e}_a \left(\frac{p\Gamma_1}{\rho} A \right)_{\text{sp}} \left[\nabla \cdot \boldsymbol{\xi}^{(2)} + \left(\boldsymbol{\xi}^{(0)} \cdot \nabla \right) \left(3\varepsilon + a \frac{\partial\varepsilon}{\partial a} \right) \right]$$
$$- i \left[\omega^{(0)} \boldsymbol{\mathcal{B}}^{(1)} \left(\boldsymbol{\xi}^{(1)} \right) + \omega^{(1)} \boldsymbol{\mathcal{B}}^{(1)} \left(\boldsymbol{\xi}^{(0)} \right) \right] - \Delta \left(\frac{p\Gamma_1}{\rho} A \right) \left(\nabla \cdot \boldsymbol{\xi}^{(0)} \right) \boldsymbol{e}_a = \mathbf{0}. \tag{B.2.5}$$

The displacement vector in the nonrotating limit, $\boldsymbol{\xi}^{(0)} = \left(\xi_a^{(0)}, \xi_\theta^{(0)}, \xi_\phi^{(0)} \right)$, admits the form

$$\boldsymbol{\xi}^{(0)}(a, \theta, \phi) = \left[\xi_a^{(0)}(a), \ \xi_h^{(0)}(a) \frac{\partial}{\partial\theta}, \ \xi_h^{(0)}(a) \frac{1}{\sin\theta} \frac{\partial}{\partial\phi} \right] Y_l^m(\theta, \phi), \tag{B.2.6}$$

which is the same as in Eq. (2.3.19), with r replaced by a (we are not going to separate the spherical harmonics yet though). Accordingly, $\boldsymbol{e}_a$ is the unit vector in the direction of increasing a. The nabla operator is defined as in spherical polar coordinates, but with a substituting r. Finally, the Schwarzschild discriminant A and the variable χ are given by Eqs. (2.3.8) and (B.1.17) respectively, with a replacing r.

Like for the case of first-order rotational corrections, we expand the second-order corrections as

$$\xi_\alpha^{(2)} = \sum_\beta c_{\alpha\beta}^{(2)} \xi_\beta^{(0)} \tag{B.2.7}$$

(we temporarily reintroduce mode indices here). With the help of the (expressed in the notation above) perturbed continuity equation (2.2.4), perturbed Poisson equation (2.2.6), and perturbed equation of state (2.2.7), we can also obtain an expansion for $\chi_\alpha^{(2)}$. Then, replacing $\xi_\alpha^{(2)}$ and $\chi_\alpha^{(2)}$ in Eq. (B.2.5) and taking the inner product, defined by Eq. (2.3.55), with $\xi_\alpha^{(0)}$, it becomes (omitting again the index α)

$$
\begin{aligned}
\left(2\omega^{(2)}\omega^{(0)} - \omega^{(1)2}\right) I^{(0)} &= -\omega^{(0)} \int \left[2\omega^{(1)}\xi^{(1)} + i\boldsymbol{B}^{(1)}\left(\xi^{(1)}\right)\right] \cdot \xi^{(0)*} \rho_{sp} d^3 r \\
&\quad -2\omega^{(0)2} \int \left[\varepsilon\xi^{(0)} \cdot \xi^{(0)*} + \mathrm{Re}\left(a\xi_a^{(0)}\xi^{(0)*} \cdot \nabla\varepsilon\right)\right] \rho_{sp} d^3 r \\
&\quad + \int \left(\frac{p\Gamma_1}{\rho}\right)_{sp} \left[\xi_a^{(0)*}\left(\frac{1}{\Gamma_1}\frac{d\ln p}{da}\right)_{sp} + \nabla \cdot \xi^{(0)*}\right]\left(\xi^{(0)} \cdot \nabla\right)\left(3\varepsilon + a\frac{\partial\varepsilon}{\partial a}\right) \rho_{sp} d^3 r \\
&\quad - \int \left[\xi_a^{(0)*}\left(\frac{d\ln\rho}{da}\right)_{sp} + \nabla \cdot \xi^{(0)*}\right]\Psi^{(2)} \rho_{sp} d^3 r - \int \xi^{(0)*} \cdot \boldsymbol{\mathcal{D}}^{(2)}\left(\xi^{(0)}\right) \rho_{sp} d^3 r.
\end{aligned}
\tag{B.2.8}
$$

Here, $I^{(0)}$ is defined by Eq. (2.3.55) as the inner product between zeroth-order eigenfunctions and Re denotes the real part of a complex quantity. $\Psi^{(2)}$ is part of the solution of the (corrected to second order in Ω) perturbed Poisson equation (2.2.6) and is given by

$$
\begin{aligned}
\Psi^{(2)} &= G \iiint \frac{\left\{\nabla \cdot \left[\rho_{sp}\xi^{(0)}\left(3\varepsilon + \frac{\partial\varepsilon}{\partial a}\right)\right]\right\}'}{|\boldsymbol{r} - \boldsymbol{r}'|} (a')^2 \sin\theta' da' d\theta' d\phi' \\
&\quad + G \iiint \left[\nabla \cdot \left(\rho_{sp}\xi^{(0)}\right)\right]' \left[\left(\varepsilon a\frac{\partial}{\partial a} + \varepsilon' a'\frac{\partial}{\partial a'}\right)\frac{1}{|\boldsymbol{r} - \boldsymbol{r}'|}\right] (a')^2 \sin\theta' da' d\theta' d\phi',
\end{aligned}
$$

where $\boldsymbol{r}$ is the position vector and the primed variables $\left(a', \theta', \phi'\right)$ are the integration variables, with the rest of the primed quantities being functions of these variables. Finally,

$$
\begin{aligned}
\boldsymbol{\mathcal{D}}^{(2)}\left(\xi^{(0)}\right) &= \nabla\left[\frac{1}{\rho_{sp}}\Delta\left(\frac{p\Gamma_1}{\rho}\right)\nabla \cdot \left(\rho_{sp}\xi^{(0)}\right) - \Delta\left(\frac{p\Gamma_1}{\rho}A\right)\xi_a^{(0)} - \frac{\Delta\rho_c}{\rho_{c,sp}}\chi^{(0)}\right] \\
&\quad + \Delta\left(\frac{p\Gamma_1}{\rho}A\right)\left(\nabla \cdot \xi^{(0)}\right)\boldsymbol{e}_a,
\end{aligned}
\tag{B.2.9}
$$

which contains all the terms associated with the difference operator Δ, defined above.

We can now calculate the second-order rotational corrections to the eigenfrequencies, $\omega^{(2)}$, for a given form of the deformation function ε. All the terms on the right-hand side of Eqs. (B.2.8) are related to the distortion of the star, induced by the centrifugal force, except the first term, which is due to the Coriolis force and requires the knowledge of the first-order rotational corrections to the eigenfunctions. For these we will use Eqs. (B.1.14)–(B.1.16) (with r replaced by a), which we substitute, together with Eq. (B.2.6), in the Coriolis term, to get

$$
-\omega^{(0)} \int \left[2\omega^{(1)} \boldsymbol{\xi}^{(1)} + i\boldsymbol{\mathcal{B}}^{(1)} \left(\boldsymbol{\xi}^{(1)} \right) \right] \cdot \boldsymbol{\xi}^{(0)*} \rho_{\mathrm{sp}} \mathrm{d}^3 r =
$$

$$
- 2\omega^{(1)}\omega^{(0)} \int_0^R \left[\xi_a^{(1)} \xi_a^{(0)} + \frac{l(l+1)}{\omega^{(0)4} a^2} \chi^{(0)} \chi^{(1)} \right] \rho_{\mathrm{sp}} a^2 \mathrm{d}a
$$

$$
+ 2m\Omega\omega^{(0)} \int_0^R \frac{1}{\omega^{(0)2} a} \left[\chi^{(1)} \left(\xi_a^{(0)} + \frac{\chi^{(0)}}{\omega^{(0)2} a} \right) + \chi^{(0)} \xi_a^{(1)} \right] \rho_{\mathrm{sp}} a^2 \mathrm{d}a
$$

$$
- 4m\Omega\omega^{(1)} \int_0^R \left(\frac{\chi^{(0)}}{\omega^{(0)2} a} \right)^2 \rho_{\mathrm{sp}} a^2 \mathrm{d}a - 4\omega^{(1)2} \int_0^R \xi_a^{(0)2} \rho_{\mathrm{sp}} a^2 \mathrm{d}a
$$

$$
+ 4\Omega^2 \int_0^R \left\{ \frac{2}{3} \left(1 - \mathcal{I}_{l,m} \right) \xi_a^{(0)2} - 2\mathcal{I}_{l,m} \xi_a^{(0)} \frac{\chi^{(0)}}{\omega^{(0)2} a} \right.
$$

$$
\left. + \left(\frac{\chi^{(0)}}{\omega^{(0)2} a} \right)^2 \left[\frac{l(l+1)}{3} \left(2\mathcal{I}_{l,m} + 1 \right) - 2\mathcal{I}_{l,m} \right] \right\} \rho_{\mathrm{sp}} a^2 \mathrm{d}a, \quad (\mathrm{B.2.10})
$$

where

$$
\mathcal{I}_{l,m} = \iint Y_l^m(\theta, \phi) Y_l^{m*}(\theta, \phi) P_2(\cos \theta) \sin \theta \mathrm{d}\theta \mathrm{d}\phi
$$

$$
= \frac{3}{2(2l+1)} \left[\frac{(l+1)^2 - m^2}{2l+3} + \frac{l^2 - m^2}{2l-1} \right] - \frac{1}{2}, \quad (\mathrm{B.2.11})
$$

with P_l denoting the Legendre polynomials (see, for example, Abramowitz and Stegun 1972, Chap. 8). Here, $\xi_a^{(0)}$, $\xi_a^{(1)}$, $\chi^{(0)}$, and $\chi^{(1)}$ depend only on a; as discussed earlier, zeroth-order quantities are found from Eqs. (2.3.16)–(2.3.18) and (2.3.24)–(2.3.27) and first-order ones from Eqs. (B.1.18)–(B.1.25). Finally, $\omega^{(1)}$ is given by Eq. (B.1.9).

B.2.2 Application to Polytropes

We are going to apply the generic formalism presented above in polytropes (see Appendix A). In order to simplify the process, we will assume that central density and pressure changes, induced by rotation, have a negligible effect on the calculation of second-order eigenfrequency corrections, i.e.,

$$\Delta\rho_c = \Delta p_c = 0.$$

This choice makes the rotating star have a larger mass than its nonrotating counterpart [see Appendix A.2 and specifically Eq. (A.2.13); cf. also Christensen-Dalsgaard and Thompson 1999]. Based on Eqs. (B.2.3) and (B.2.4), this implies

$$\rho(a) = \rho_{sp}(a),$$

which means that the density on an equipotential surface, corresponding to a, on the rotating star coincides with the density on a sphere, with radius a, on the nonrotating star. Then, from Eq. (B.2.9) we see that

$$\mathcal{D}^{(2)}\left(\boldsymbol{\xi}^{(0)}\right) = \mathbf{0}.$$

In order to obtain an expression for the deformation function ε, we need to use the results obtained in Appendix A.2. To avoid conflict between the Lane-Emden variable ξ and the displacement vector $\boldsymbol{\xi}$, we will replace the former with ζ. In our current notation, we may represent ζ as

$$\zeta = \bar{\zeta}(1 + \varepsilon), \tag{B.2.12}$$

where $\bar{\zeta}$ is the Lane-Emden analogue of a, namely, from Eqs. (A.1.3), (B.2.2) and (B.2.12),

$$a = \left[\frac{(n+1)K}{4\pi G}\rho_c^{\frac{1}{n}-1}\right]^{1/2}\bar{\zeta}.$$

Then, according to Eq. (B.2.3), the solution $\theta(\zeta)$ of the Lane-Emden equation (A.1.1) is related to the solution $\Theta(\zeta)$ of the equation for rotating polytropes (A.2.1) by

$$\theta(\bar{\zeta}) = \Theta\left[\bar{\zeta}(1 + \varepsilon)\right],$$

from which we get

$$\varepsilon(a, \theta) = \frac{1}{3}\frac{\Omega^2}{GM/R^3}\left(\frac{a}{R}\right)^3\frac{M}{M_r}\left[\alpha(a) - \beta(a)P_2(\cos\theta)\right], \tag{B.2.13}$$

where

$$\alpha(a) = \frac{6\psi_0(\bar{\zeta})}{\bar{\zeta}^2}$$

and

$$\beta(a) = -\frac{6A_2\psi_2(\bar{\zeta})}{\bar{\zeta}^2},$$

with Chandrasekhar's functions, ψ_0 and ψ_2, obtained from Eqs. (A.2.6) and (A.2.7), and A_2 given by Eq. (A.2.8). Also, M_r is defined in Eq. (2.3.42), with r replaced by a, whereas M and R denote the mass and radius of the (nonrotating) star respectively.

We can now substitute Eq. (B.2.13) in Eq. (B.2.8), to get (omitting the subscript sp)

$$\frac{\omega^{(2)}}{\omega^{(0)}}I^{(0)} = \frac{1}{2}\left(\frac{\omega^{(1)}}{\omega^{(0)}}\right)^2 I^{(0)} - \frac{1}{2\omega^{(0)}}\int\left[2\omega^{(1)}\boldsymbol{\xi}^{(1)} + i\boldsymbol{B}^{(1)}\left(\boldsymbol{\xi}^{(1)}\right)\right]\cdot\boldsymbol{\xi}^{(0)*}\rho d^3\boldsymbol{r}$$

$$- S\int_0^R\left\{\left[(4-U)\xi_a^{(0)2} + l(l+1)\left(\frac{\chi^{(0)}}{\omega^{(0)2}a}\right)^2\right]\alpha + \xi_a^{(0)2}\frac{d\alpha}{d\ln a}\right\}wda$$

$$+ DI_{l,m}\int_0^R\left\{\left[(4-U)\beta + \frac{d\beta}{d\ln a}\right]\xi_a^{(0)2} + 3\beta\xi_a^{(0)}\frac{\chi^{(0)}}{\omega^{(0)2}a}\right.$$

$$\left. + [l(l+1)-3]\left(\frac{\chi^{(0)}}{\omega^{(0)2}a}\right)^2\beta\right\}wda$$

$$\frac{S}{2\omega^{(0)2}}\int_0^R\left(\chi^{(0)}\quad\delta\Phi^{(0)}\right)\xi_a^{(0)}f(\alpha)\frac{w}{a}da$$

$$+ \frac{DI_{l,m}}{2\omega^{(0)2}}\int_0^R\left(\chi^{(0)} - \delta\Phi^{(0)}\right)\left\{\xi_a^{(0)}f(\beta) + 3\left[(6-U)\beta + \frac{d\beta}{d\ln a}\right]\frac{\chi^{(0)}}{\omega^{(0)2}a}\right\}\frac{w}{a}da$$

$$- \frac{4\pi G}{2\omega^{(0)2}(2l+1)}\int_0^R\left[\xi_a^{(0)}A - \frac{\rho}{\Gamma_1 p}\left(\chi^{(0)} - \delta\Phi^{(0)}\right)\right]$$

$$\times\left\{\frac{1}{a^{l+1}}\int_0^a (a')^{l+2}\left[\mathbb{S}(\alpha)S - \mathbb{D}(\beta)DI_{l,m}\right]da'\right.$$

$$\left. + a^l\int_a^R\frac{1}{(a')^{l-1}}\left[\mathbb{S}(\alpha)S - \mathbb{D}(\beta)DI_{l,m}\right]da'\right\}\rho a^2 da, \qquad (B.2.14)$$

where[4]

$$w = c_1\rho a^2,$$

$$S = D = \frac{1}{3}\frac{\Omega^2}{GM/R^3},$$

[4]In the original work of Saio (1981), the parameter D also includes the effects of tidal deformations from a companion star and is not the same as parameter S.

and

$$f \begin{pmatrix} \alpha \\ \beta \end{pmatrix} = \left[2 (3 - U)^2 + (A^* + V_g) U + (9 - 2U) \frac{d}{d \ln a} + \frac{d^2}{d(\ln a)^2} \right] \begin{pmatrix} \alpha \\ \beta \end{pmatrix},$$

$$\mathbb{S}(\alpha) = \left[\left[c_1 \rho \left\{ \left[\xi_a^{(0)} A - \frac{\rho}{\Gamma_1 p} \left(\chi^{(0)} - \delta \Phi^{(0)} \right) \right] \left[(6 - U) \alpha + \frac{d\alpha}{d \ln a} \right] + \frac{\xi_a^{(0)}}{a} f(\alpha) \right\} \right] \right]_{a=a'},$$

$$\mathbb{D}(\beta) = \mathbb{S}(\beta) + \left\{ \frac{3 c_1 \rho}{a} \frac{\chi^{(0)}}{\omega^{(0)2} a} \left[(6 - U) \beta + \frac{d\beta}{d \ln a} \right] \right\}_{a=a'}.$$

The variables c_1, U, A^*, and V_g are defined in Sect. 2.3.3 (with a replacing r), $\mathcal{I}_{l,m}$ is given by Eq. (B.2.11), and the second term on the right-hand side of Eq. (B.2.14) was evaluated in Eq. (B.2.10).

We may now write the second-order eigenfrequency correction in the form

$$\omega^{(2)} = C_2 \omega^{(0)} \left(\frac{\Omega}{\omega^{(0)}} \right)^2. \tag{B.2.15}$$

The parameter C_2 can be decomposed as

$$C_2 = X_1 + m^2 Y_1 + X_2 + m^2 Y_2 + Z, \tag{B.2.16}$$

where the term $X_1 + m^2 Y_1$ corresponds to the effects of the Coriolis force, included in the first two terms on the right-hand side of Eq. (B.2.14), whereas $X_2 + m^2 Y_2$ and Z are due to the deformation of the star and comprise terms proportional to D and S respectively.

References

1. Abramowitz, M., & Stegun, I. A. (1972). *Handbook of Mathematical Functions*. New York: Dover. http://adsabs.harvard.edu/abs/1972hmfw.book.....A.
2. Christensen-Dalsgaard, J., & Thompson, M. J. (1999). A note on Saio's estimate of second-order effects of rotation on stellar oscillation frequencies. *Astronomy & Astrophysics, 350*, 852–854. http://adsabs.harvard.edu/abs/1999A%26A...350..852C.
3. Hansen, C. J., Cox, J. P., & Carroll, B. W. (1978). The quasi-adiabatic analysis of nonradial modes of stellar oscillation in the presence of slow rotation. *The Astrophysical Journal, 226*, 210–221. https://doi.org/10.1086/156600.
4. Lynden-Bell, D., & Ostriker, J. P. (1967). On the stability of differentially rotating bodies. *Monthly Notices of the Royal Astronomical Society, 136*, 293. http://adsabs.harvard.edu/abs/1967MNRAS.136..293L.
5. Mathews, J., & Walker, R. L. (1970). *Mathematical methods of physics* (2nd ed.). California: The Benjamin/Cummings Publishing Company. http://adsabs.harvard.edu/abs/1970mmp..book.....M.
6. Saio, H. (1981). Rotational and tidal perturbations of nonradial oscillations in a polytropic star. *The Astrophysical Journal, 244*, 299–315. https://doi.org/10.1086/158708.
7. Schenk, A. K., Arras, P., Flanagan, É. É., Teukolsky, S. A., & Wasserman, I. (2001). Nonlinear mode coupling in rotating stars and the r-mode instability in neutron stars. *Physical Review D, 65*, 024001. https://doi.org/10.1103/PhysRevD.65.024001, arXiv:gr-qc/0101092.

8. Smeyers, P., & Denis, J. (1971). Second order rotational perturbation of non-radial oscillations of a star. *Astronomy & Astrophysics*, *14*, 311. http://adsabs.harvard.edu/abs/1971A%26A....14..311S.

9. Unno, W., Osaki, Y., Ando, H., Saio, H., & Shibahashi, H. (1989). *Nonradial oscillations of stars* (2nd ed.). Tokyo: University of Tokyo Press. http://adsabs.harvard.edu/abs/1989nos..book.....U.

Appendix C
Polar Mode Growth/Damping Rates

We will evaluate the polar mode growth/damping rate γ, including contributions from gravitational waves, shear viscosity, and bulk viscosity, namely

$$\gamma = \gamma_{\text{GW}} + \gamma_{\text{SV}} + \gamma_{\text{BV}},$$

or, using Eq. (3.6.7),

$$\gamma = \frac{1}{2E} \left[\left(\frac{\mathrm{d}E}{\mathrm{d}t} \right)_{\text{GW}} + \left(\frac{\mathrm{d}E}{\mathrm{d}t} \right)_{\text{SV}} + \left(\frac{\mathrm{d}E}{\mathrm{d}t} \right)_{\text{BV}} \right],$$

where the mode energy E is given by Eq. (2.6.4) (evaluated at unit amplitude) and its rate of change due to gravitational waves, shear viscosity, and bulk viscosity is defined in Eqs. (3.5.1), (3.6.1), (3.6.4), respectively. We will consider mode eigenfunctions as obtained in the nonrotating limit, due to their simple spherical harmonic dependence, which makes the various integrals analytically tractable.

C.1 Gravitational Waves

From Eq. (3.5.1) we have

$$\left(\frac{\mathrm{d}E}{\mathrm{d}t} \right)_{\text{GW}} = -\sum_{l_{\min}}^{\infty} N_l \, \omega \, (\omega - m\Omega)^{2l+1} \left(|\delta D_l^m|^2 + |\delta J_l^m|^2 \right), \qquad \text{(C.1.1)}$$

with the constant N_l given by Eq. (3.5.3), and the mass and current multipoles, δD_l^m and δJ_l^m, defined in Eqs. (3.5.4) and (3.5.5) respectively. In the nonrotating limit, polar mode eigenfunctions are given by Eq. (2.3.19), which means that their current multipoles are zero, namely

© Springer Nature Switzerland AG 2018
P. Pnigouras, *Saturation of the f-mode Instability in Neutron Stars*,
Springer Theses, https://doi.org/10.1007/978-3-319-98258-8

$$\delta J_l^m = \frac{2i\omega}{c(l+1)} \int \left(\xi_r Y_l^m \boldsymbol{e}_r + \xi_h r \nabla Y_l^m\right) \cdot \left(r\boldsymbol{e}_r \times \nabla Y_l^{*m}\right) \rho r^l \mathrm{d}^3 \boldsymbol{r} = 0, \quad \text{(C.1.2)}$$

where we used the fact that $\delta \boldsymbol{v} = i\omega \boldsymbol{\xi}$ and $\boldsymbol{v} = \boldsymbol{0}$. Mass multipoles are accordingly written as

$$\delta D_l^m = \int_0^R \delta \rho(r) r^{l+2} \mathrm{d}r, \quad \text{(C.1.3)}$$

where the orthogonality of spherical harmonics (2.3.14) was used. The (Eulerian) density perturbation can be written as a function of $\boldsymbol{\xi}$ by using the perturbed continuity equation (2.2.4) as

$$\delta \rho(r, \theta, \phi) = -\rho \nabla \cdot \boldsymbol{\xi} - \xi_r(r) Y_l^m(\theta, \phi) \frac{\mathrm{d}\rho}{\mathrm{d}r},$$

with the divergence of the displacement vector given by

$$\nabla \cdot \boldsymbol{\xi} = \left[\frac{\mathrm{d}\xi_r}{\mathrm{d}r} + \frac{2}{r}\xi_r - l(l+1)\frac{\xi_h}{r}\right] Y_l^m(\theta, \phi), \quad \text{(C.1.4)}$$

where Eq. (2.3.13) was used. Alternatively, using Eq. (2.3.7), $\delta\rho$ is expressed in terms of the dimensionless variables defined in Sect. 2.3.3 as

$$\delta \rho(r) = \rho \left[V_g(y_2 - y_3) + A^* y_1\right].$$

Note that, since in the nonrotating star there can be no instability, the eigenfrequency ω in Eq. (C.1.1) cannot be evaluated in the nonrotating limit.

C.2   Shear Viscosity

According to Eq. (3.6.1),

$$\left(\frac{\mathrm{d}E}{\mathrm{d}t}\right)_{\mathrm{SV}} = -\int 2\eta \, \delta\sigma^{ab} \delta\sigma_{ab}^* \mathrm{d}^3 \boldsymbol{r}, \quad \text{(C.2.1)}$$

with the shear tensor $\delta\sigma^{ab}$ defined in Eq. (3.6.2) and the shear viscosity coefficient η given by Eq. (3.6.3). An expression for Eq. (C.2.1) in terms of the components of the polar mode eigenfunction (2.3.19) can be found in Cutler et al. (1990), namely[5]

---

[5]Cutler et al. (1990) give this expression in terms of the covariant components of the displacement vector, which, in their notation, are

$$\xi_a = \left[\frac{W(r)}{r} Y_l^m \nabla_a r - V(r)\nabla_a Y_l^m\right] r^l e^{i\omega t}.$$

$$\left(\frac{dE}{dt}\right)_{\mathrm{SV}} = -2\omega^2 \frac{\tilde{\eta}}{T^2} \int_0^R \left\{ \frac{3}{2}\alpha_1^2 + 2l(l+1)\alpha_2^2 + l(l+1)\left[\frac{1}{2}l(l+1)-1\right]\xi_h^2 \right\} \rho^{9/4} dr,$$

$$\text{(C.2.2)}$$

where

$$\alpha_1 = \frac{r^2}{3}\left\{ \frac{2}{r}\left[\frac{d\xi_r}{dr} - \frac{\xi_r}{r}\right] + l(l+1)\frac{\xi_h}{r^2} \right\}$$

and

$$\alpha_2 = -\frac{r}{2}\left[\frac{d\xi_h}{dr} - \frac{\xi_h}{r} + \frac{\xi_r}{r}\right].$$

Moreover, the eigenfrequency ω is evaluated in the nonrotating limit and the viscosity coefficient η is rescaled, for convenience, as

$$\tilde{\eta} = \frac{\eta}{\rho^{9/4}T^{-2}} = 347 \text{ g}^{-5/4} \text{ cm}^{23/4} \text{ s}^{-1} \text{ K}^2.$$

Using the dimensionless variables of Sect. 2.3.3, Eq. (C.2.2) becomes

$$\left(\frac{dE}{dt}\right)_{\mathrm{SV}} = -2\omega^2 \frac{\tilde{\eta}}{T^2} \int_0^R \left\{ \frac{3}{2}\alpha_1^2 + 2l(l+1)\alpha_2^2 + l(l+1)\left[\frac{1}{2}l(l+1)-1\right]\left(\frac{ry_2}{c_1\tilde{\omega}^2}\right)^2 \right\} \rho^{9/4} dr,$$

with

$$\alpha_1 = \frac{2}{3}r^2\frac{dy_1}{dr} + \frac{1}{3}l(l+1)\frac{ry_2}{c_1\tilde{\omega}^2}$$

and

Since we are working with spherical coordinates, these components do not coincide with the physical components $\xi_{(a)}$, i.e., the components expressed in the (e_r, e_θ, e_ϕ) basis. To get the latter, we have to use the transformation $\xi^a = \xi_{(a)}/\sqrt{g_{aa}}$, or $\xi_a = g_{ab}\xi^b = g_{ab}\xi_{(b)}/\sqrt{g_{bb}}$, where the metric tensor of flat space is expressed in spherical coordinates as

$$g_{ab} = \begin{pmatrix} 1 & 0 & 0 \\ 0 & r^2 & 0 \\ 0 & 0 & r^2\sin^2\theta \end{pmatrix}.$$

Then, in our notation,

$$\xi = \left(\frac{W}{r}r^l, -\frac{V}{r}r^l\frac{\partial}{\partial\theta}, -\frac{V}{r}r^l\frac{1}{\sin\theta}\frac{\partial}{\partial\phi}\right) Y_l^m e^{i\omega t}.$$

Comparing to Eq. (2.3.19), this means that $\xi_r = Wr^{l-1}$ and $\xi_h = -Vr^{l-1}$, from which Eq. (C.2.2) can be obtained. Furthermore, our parameters α_1 and α_2 differ from the ones defined in Cutler et al. (1990) by a factor r^{l-1}.

$$\alpha_2 = -\frac{r}{2}\left\{\frac{1}{c_1\tilde{\omega}^2}\left[r\frac{dy_2}{dr} - y_2(3-U)\right] + y_1\right\}.$$

C.3   Bulk Viscosity

The energy rate due to bulk viscosity is given by Eq. (3.6.4), i.e.,

$$\left(\frac{dE}{dt}\right)_{BV} = -\int \zeta\delta\sigma\delta\sigma^*d^3r, \tag{C.3.1}$$

with the expansion scalar $\delta\sigma$ defined in Eq. (3.6.5) and the bulk viscosity coefficient ζ given by Eq. (3.6.6). Using $\delta v = i\omega\xi$, we get

$$\delta\sigma = i\omega\nabla\cdot\xi,$$

where the divergence of the displacement vector is given by Eq. (C.1.4). We further rescale the bulk viscosity coefficient ζ as

$$\tilde{\zeta} = \frac{\zeta}{\rho^2\omega^{-2}T^6} = 6\times10^{-59}\ \text{g}^{-1}\,\text{cm}^5\,\text{s}^{-3}\,\text{K}^{-6},$$

to get

$$\left(\frac{dE}{dt}\right)_{BV} = -\tilde{\zeta}T^6\int_0^R\left[\frac{d\xi_r}{dr} + \frac{2}{r}\xi_r - l(l+1)\frac{\xi_h}{r}\right]^2\rho^2r^2dr, \tag{C.3.2}$$

where the orthogonality of spherical harmonics (2.3.14) was used. Alternatively, we can use the perturbed continuity equation (2.2.4) to get

$$\nabla\cdot\xi = -\frac{1}{\rho}\left[\delta\rho(r)Y_l^m + \xi_r(r)Y_l^m\frac{d\rho}{dr}\right],$$

which, with the help of Eqs. (2.3.7) and (2.3.8), can be expressed in terms of the dimensionless variables of Sect. 2.3.3 as

$$\nabla\cdot\xi = V_g(y_1 - y_2 + y_3)Y_l^m.$$

C.4   Mode Energy

The mode energy E is given by Eq. (2.6.4) (evaluated at unit amplitude), namely

$$E = \omega \left[2\omega\langle \boldsymbol{\xi}, \boldsymbol{\xi} \rangle - \langle \boldsymbol{\xi}, i\boldsymbol{B}(\boldsymbol{\xi}) \rangle \right]. \tag{C.4.1}$$

Since we consider the eigenfunctions in the nonrotating limit, $\langle \boldsymbol{\xi}, \boldsymbol{\xi} \rangle$ is the moment of inertia I of the perturbation, given by Eq. (2.3.56). Also, in Appendix B.1 we proved that

$$\omega^{(1)} = \frac{1}{2I} \langle \boldsymbol{\xi}, i\boldsymbol{B}(\boldsymbol{\xi}) \rangle,$$

where $\omega^{(1)}$ is the first-order rotational correction to the mode eigenfrequency. Thus, Eq. (C.4.1) is written as

$$E = 2I\omega \left(\omega - \omega^{(1)} \right). \tag{C.4.2}$$

In Eq. (C.4.2) the eigenfrequency ω has to include rotational corrections. If the eigenfrequency is also evaluated in the nonrotating limit, the mode energy is simply given by Eq. (2.3.58) (evaluated at unit amplitude).

Reference

1. Cutler, C., Lindblom, L., & Splinter, R. J. (1990). Damping times for neutron star oscillations. *The Astrophysical Journal*, *363*, 603–611. https://doi.org/10.1086/169370.

Appendix D
Equations of Motion

D.1 Equation of Motion for Quadratic Perturbations

The equation of motion for quadratic perturbations, in terms of the velocity $v \equiv \delta v$, is easily derived by differentiating the perturbed Euler equation (4.1.4) with respect to time. Then,

$$\ddot{v} + \mathcal{B}(\dot{v}) + \mathcal{C}(v) + \mathcal{N} = 0, \qquad (D.1.1)$$

where

$$\mathcal{B}(v) = 2\mathbf{\Omega} \times v, \qquad (D.1.2)$$

$$\mathcal{C}(v) = \frac{1}{\rho}\nabla\left(\frac{\partial\delta_1 p}{\partial t}\right) - \frac{\nabla p}{\rho^2}\frac{\partial\delta_1\rho}{\partial t} + \nabla\left(\frac{\partial\delta_1\Phi}{\partial t}\right), \qquad (D.1.3)$$

and

$$\mathcal{N} = \frac{\partial}{\partial t}\left[(v\cdot\nabla)v + \frac{\nabla\delta_2 p}{\rho} + \delta_1\left(\frac{1}{\rho}\right)\nabla\delta_1 p + \delta_2\left(\frac{1}{\rho}\right)\nabla p + \nabla\delta_2\Phi\right]. \quad (D.1.4)$$

We will attempt to derive expressions for the perturbations in terms of the velocity and first-order terms. From the perturbed continuity equation (4.1.3), distinguishing first- and second-order terms, we get

$$\frac{\partial\delta_1\rho}{\partial t} = -\rho\nabla\cdot v - v\cdot\nabla\rho \qquad (D.1.5)$$

and

$$\frac{\partial\delta_2\rho}{\partial t} = -\delta_1\rho\nabla\cdot v - v\cdot\nabla\delta_1\rho. \qquad (D.1.6)$$

Accordingly, the perturbed Poisson equation (4.1.5) gives

© Springer Nature Switzerland AG 2018
P. Pnigouras, *Saturation of the f-mode Instability in Neutron Stars*,
Springer Theses, https://doi.org/10.1007/978-3-319-98258-8

$$\nabla^2 \delta_1 \Phi = 4\pi G \delta_1 \rho$$

and

$$\nabla^2 \delta_2 \Phi = 4\pi G \delta_2 \rho,$$

whose solutions are

$$\delta_1 \Phi = -G \int \frac{\delta_1 \rho(\boldsymbol{r}')}{|\boldsymbol{r} - \boldsymbol{r}'|} d^3 \boldsymbol{r}'$$

and

$$\delta_2 \Phi = -G \int \frac{\delta_2 \rho(\boldsymbol{r}')}{|\boldsymbol{r} - \boldsymbol{r}'|} d^3 \boldsymbol{r}',$$

or, differentiating with respect to time and using Eqs. (D.1.5) and (D.1.6),

$$\frac{\partial \delta_1 \Phi}{\partial t} = G \int \frac{\nabla' \cdot (\rho \boldsymbol{v})}{|\boldsymbol{r} - \boldsymbol{r}'|} d^3 \boldsymbol{r}' \qquad (D.1.7)$$

and

$$\frac{\partial \delta_2 \Phi}{\partial t} = G \int \frac{\nabla' \cdot (\delta_1 \rho \boldsymbol{v})}{|\boldsymbol{r} - \boldsymbol{r}'|} d^3 \boldsymbol{r}', \qquad (D.1.8)$$

where ∇' denotes differentiation with respect to $\boldsymbol{r}'$.

In order to derive the perturbed equation of state (4.1.6), we applied a Lagrangian perturbation to the equation of state (2.1.4), which, to quadratic order, gives

$$\Delta p = \left(\frac{\partial p}{\partial \rho} \right)_\mu \Delta \rho + \frac{1}{2} \left(\frac{\partial^2 p}{\partial \rho^2} \right)_\mu (\Delta_1 \rho)^2,$$

considering frozen or adiabatic perturbations ($\Delta \mu \approx 0$; see Sect. 2.2). After some manipulation, we get Eq. (4.1.6), which, making use of Eq. (4.1.2), can be decomposed into first- and second-order Eulerian perturbations, namely

$$\frac{\partial \delta_1 p}{\partial t} = -\boldsymbol{v} \cdot \nabla p - p \Gamma_1 \nabla \cdot \boldsymbol{v} \qquad (D.1.9)$$

and

$$\frac{\partial \delta_2 p}{\partial t} = \nabla \cdot \boldsymbol{v} \left\{ \boldsymbol{\xi} \cdot \nabla(p \Gamma_1) + p \Gamma_1 \left[\Gamma_1 + \left(\frac{\partial \ln \Gamma_1}{\partial \ln \rho} \right)_\mu \right] \nabla \cdot \boldsymbol{\xi} \right\} - \boldsymbol{v} \cdot \nabla \delta_1 p.$$

$$(D.1.10)$$

Finally, an expression for the Eulerian perturbation of $1/\rho$ can be obtained as

$$
\frac{1}{\rho + \delta\rho} = \frac{1}{\rho\left(1 + \frac{\delta\rho}{\rho}\right)} = \frac{1}{\rho}\left[1 - \frac{\delta\rho}{\rho} + \left(\frac{\delta_1\rho}{\rho}\right)^2 + \mathcal{O}\left(\xi^3\right)\right],
$$

which implies

$$
\delta_1\left(\frac{1}{\rho}\right) = -\frac{\delta_1\rho}{\rho^2} \tag{D.1.11}
$$

and

$$
\delta_2\left(\frac{1}{\rho}\right) = -\frac{\delta_2\rho}{\rho^2} + \frac{(\delta_1\rho)^2}{\rho^3}. \tag{D.1.12}
$$

Using the equations above, Eq. (D.1.4) can be reduced to Eq. (4.1.10), i.e.,

$$
\mathcal{N} = \frac{\partial}{\partial t}\left[\nabla\left(\frac{\boldsymbol{v}\cdot\boldsymbol{v}}{2}\right) - \boldsymbol{v}\times(\nabla\times\boldsymbol{v}) - \frac{\delta_1\rho}{\rho^2}\nabla\delta_1 p + \frac{(\delta_1\rho)^2}{2\rho^3}\nabla p\right]
$$
$$
+ \frac{1}{\rho}\nabla\left[\left[\nabla\cdot\boldsymbol{v}\left\{\boldsymbol{\xi}\cdot\nabla(p\Gamma_1) + p\Gamma_1\left[\Gamma_1 + \left(\frac{\partial\ln\Gamma_1}{\partial\ln\rho}\right)_\mu\right]\nabla\cdot\boldsymbol{\xi}\right\} - \boldsymbol{v}\cdot\nabla\delta_1 p\right]\right]
$$
$$
+ \frac{\nabla p}{\rho}(\boldsymbol{v}\cdot\nabla)\left(\frac{\delta_1\rho}{\rho}\right) + G\nabla\left[\int\frac{\nabla'\cdot(\delta_1\rho\boldsymbol{v})}{|\boldsymbol{r}-\boldsymbol{r}'|}\mathrm{d}^3 r'\right], \tag{D.1.13}
$$

where the quadratic velocity term was rewritten using the vectorial identity

$$
\nabla(\boldsymbol{A}\cdot\boldsymbol{B}) = (\boldsymbol{A}\cdot\nabla)\boldsymbol{B} + (\boldsymbol{B}\cdot\nabla)\boldsymbol{A} + \boldsymbol{A}\times(\nabla\times\boldsymbol{B}) + \boldsymbol{B}\times(\nabla\times\boldsymbol{A}).
$$

D.2 Amplitude Equation of Motion

In order to derive the equation of motion for the amplitude (4.1.13) we have to replace the velocity expansion (4.1.11) in the equation of motion for quadratic perturbations (4.1.7) [or (D.1.1)].

Starting from the linear terms and using the eigenvalue equation for the perturbations (2.2.12), we get

$$
\ddot{\boldsymbol{v}} + \boldsymbol{\mathcal{B}}(\dot{\boldsymbol{v}}) + \boldsymbol{\mathcal{C}}(\boldsymbol{v}) = \sum_\alpha\left[\left(i\omega_\alpha\ddot{Q}_\alpha - \omega_\alpha^2\dot{Q}_\alpha\right)\boldsymbol{\xi}_\alpha e^{i\omega_\alpha t} + \left(-i\omega_\alpha\ddot{Q}_\alpha^* - \omega_\alpha^2\dot{Q}_\alpha^*\right)\boldsymbol{\xi}_\alpha^* e^{-i\omega_\alpha t}\right]
$$
$$
- \sum_\alpha\left[\dot{Q}_\alpha\boldsymbol{\mathcal{C}}(\boldsymbol{\xi}_\alpha)e^{i\omega_\alpha t} + \dot{Q}_\alpha^*\boldsymbol{\mathcal{C}}(\boldsymbol{\xi}_\alpha^*)e^{-i\omega_\alpha t}\right]. \tag{D.2.1}
$$

Now, from Eq. (2.6.2) it is implied that

$$\sum_\alpha \left(\dot{Q}_\alpha \boldsymbol{\xi}_\alpha e^{i\omega_\alpha t} + \dot{Q}_\alpha^* \boldsymbol{\xi}_\alpha^* e^{-i\omega_\alpha t} \right) = \mathbf{0}. \tag{D.2.2}$$

Hence, since $\mathcal{C}$ is a linear operator, the second sum in Eq. (D.2.1) vanishes. In order to isolate a single Q_α, we need to use the mode orthogonality condition (2.6.3). We add and subtract terms containing the eigenfrequency $\omega_\beta \neq \omega_\alpha$, to obtain

$$\ddot{\boldsymbol{v}} + \mathcal{B}(\dot{\boldsymbol{v}}) + \mathcal{C}(\boldsymbol{v}) = i \sum_\alpha \left\{ \left[(\omega_\beta + \omega_\alpha) \boldsymbol{\xi}_\alpha - i \mathcal{B}(\boldsymbol{\xi}_\alpha) \right] \left(\ddot{Q}_\alpha + i \omega_\alpha \dot{Q}_\alpha \right) e^{i\omega_\alpha t} \right.$$
$$+ \left[(\omega_\beta - \omega_\alpha) \boldsymbol{\xi}_\alpha^* - i \mathcal{B}(\boldsymbol{\xi}_\alpha^*) \right] \left(\ddot{Q}_\alpha^* - i \omega_\alpha \dot{Q}_\alpha^* \right) e^{-i\omega_\alpha t} \Big\}$$
$$- i \sum_\alpha \left\{ \left[\omega_\beta \boldsymbol{\xi}_\alpha - i \mathcal{B}(\boldsymbol{\xi}_\alpha) \right] \left(\ddot{Q}_\alpha + i \omega_\alpha \dot{Q}_\alpha \right) e^{i\omega_\alpha t} \right.$$
$$+ \left[\omega_\beta \boldsymbol{\xi}_\alpha^* - i \mathcal{B}(\boldsymbol{\xi}_\alpha^*) \right] \left(\ddot{Q}_\alpha^* - i \omega_\alpha \dot{Q}_\alpha^* \right) e^{-i\omega_\alpha t} \Big\}. \tag{D.2.3}$$

Differentiating Eq. (D.2.2) with respect to time, we get

$$\sum_\alpha \left(\ddot{Q}_\alpha \boldsymbol{\xi}_\alpha e^{i\omega_\alpha t} + i \omega_\alpha \dot{Q}_\alpha \boldsymbol{\xi}_\alpha e^{i\omega_\alpha t} + \ddot{Q}_\alpha^* \boldsymbol{\xi}_\alpha^* e^{-i\omega_\alpha t} - i \omega_\alpha \dot{Q}_\alpha^* \boldsymbol{\xi}_\alpha^* e^{-i\omega_\alpha t} \right) = \mathbf{0}, \tag{D.2.4}$$

which implies that the second sum in Eq. (D.2.3) vanishes.

Taking the inner product of the remaining terms with $\boldsymbol{\xi}_\beta$, then, based on the orthogonality relation (2.6.3), all terms for which $\alpha \neq \beta$ vanish. Hence,

$$i b_\alpha \left(\ddot{Q}_\alpha + i \omega_\alpha \dot{Q}_\alpha \right) e^{i\omega_\alpha t} = -\langle \boldsymbol{\xi}_\alpha, \mathcal{N} \rangle,$$

or

$$\ddot{Q}_\alpha + i \omega_\alpha \dot{Q}_\alpha = \frac{i}{b_\alpha} \langle \boldsymbol{\xi}_\alpha, \mathcal{N} \rangle e^{-i\omega_\alpha t}. \tag{D.2.5}$$

We shall further assume that the amplitude Q_α changes on a time scale much larger than the mode oscillation period (Dziembowski 1982; see Sect. 4.2.1), i.e.,

$$|\dot{Q}_\alpha| \ll \omega_\alpha |Q_\alpha|, \tag{D.2.6}$$

so second-order derivatives of Q can be ignored. Then, Eq. (D.2.5) takes the form of Eq. (4.1.12), namely

$$\dot{Q}_\alpha = \frac{1}{\omega_\alpha b_\alpha} \langle \boldsymbol{\xi}_\alpha, \mathcal{N} \rangle e^{-i\omega_\alpha t}. \tag{D.2.7}$$

We will now proceed with replacing the velocity expansion (4.1.11) into the quadratic terms $\mathcal{N}$, for which an expression was derived in Appendix D.1. For instance, the first term of Eq. (4.1.10) [or (D.1.13)] becomes

$$\frac{\partial}{\partial t}\left[\nabla\left(\frac{\boldsymbol{v}\cdot\boldsymbol{v}}{2}\right)\right] = -\sum_\beta\sum_\gamma\omega_\beta\omega_\gamma\Big[Q_\beta\left(\dot{Q}_\gamma + i\omega_\gamma Q_\gamma\right)\nabla\left(\boldsymbol{\xi}_\beta\cdot\boldsymbol{\xi}_\gamma\right)e^{i(\omega_\beta+\omega_\gamma)t}$$

$$-Q_\beta\left(\dot{Q}_\gamma^* - i\omega_\gamma Q_\gamma^*\right)\nabla\left(\boldsymbol{\xi}_\beta\cdot\boldsymbol{\xi}_\gamma^*\right)e^{i(\omega_\beta-\omega_\gamma)t}$$

$$-Q_\beta^*\left(\dot{Q}_\gamma + i\omega_\gamma Q_\gamma\right)\nabla\left(\boldsymbol{\xi}_\beta^*\cdot\boldsymbol{\xi}_\gamma\right)e^{i(-\omega_\beta+\omega_\gamma)t}$$

$$+Q_\beta^*\left(\dot{Q}_\gamma^* - i\omega_\gamma Q_\gamma^*\right)\nabla\left(\boldsymbol{\xi}_\beta^*\cdot\boldsymbol{\xi}_\gamma^*\right)e^{-i(\omega_\beta+\omega_\gamma)t}\Big].$$

Due to Eq. (D.2.6), the terms involving derivatives of Q will be ignored. The rest of the quadratic terms can be expanded accordingly, using the equations of Appendix D.1. Then, taking the inner product of $\mathcal{N}$ with $\boldsymbol{\xi}_\alpha$, we obtain Eq. (4.1.13), i.e.,

$$\dot{Q}_\alpha(t) = \frac{i}{b_\alpha}\sum_\beta\sum_\gamma\Big[F_{\alpha\beta\gamma}Q_\beta Q_\gamma e^{i(-\omega_\alpha+\omega_\beta+\omega_\gamma)t} + F_{\alpha\bar{\beta}\gamma}Q_\beta^* Q_\gamma e^{i(-\omega_\alpha-\omega_\beta+\omega_\gamma)t}$$

$$+F_{\alpha\beta\bar{\gamma}}Q_\beta Q_\gamma^* e^{i(-\omega_\alpha+\omega_\beta-\omega_\gamma)t} + F_{\alpha\bar{\beta}\bar{\gamma}}Q_\beta^* Q_\gamma^* e^{i(-\omega_\alpha-\omega_\beta-\omega_\gamma)t}\Big],$$

$$(D.2.8)$$

where F is the *coupling coefficient*, generally given by

$$F_{\alpha\beta\gamma} = \frac{1}{i\omega_\alpha}\langle\boldsymbol{\xi}_\alpha, \mathcal{N}(\boldsymbol{\xi}_\beta, \boldsymbol{\xi}_\gamma)\rangle. \qquad (D.2.9)$$

A bar over an index means that the corresponding mode eigenfunction in $\mathcal{N}$ has to be complex conjugated and its eigenfrequency sign reversed. The explicit form of the coupling coefficient is

$$F_{\alpha\beta\gamma} = \frac{1}{\omega_\alpha}\left(\omega_\beta S_{\alpha\beta\gamma} + \omega_\gamma S_{\alpha\gamma\beta}\right), \qquad (D.2.10)$$

where

$$S_{\alpha\beta\gamma} = \int\Big[\Big[\rho\omega_\beta\omega_\gamma[-\nabla\left(\boldsymbol{\xi}_\beta\cdot\boldsymbol{\xi}_\gamma\right) + \boldsymbol{\xi}_\beta\times\left(\nabla\times\boldsymbol{\xi}_\gamma\right) + \boldsymbol{\xi}_\gamma\times\left(\nabla\times\boldsymbol{\xi}_\beta\right)]$$

$$-\frac{1}{\rho}[\nabla\cdot\left(\rho\boldsymbol{\xi}_\beta\right)\nabla\left(\boldsymbol{\xi}_\gamma\cdot\nabla p + p\Gamma_1\nabla\cdot\boldsymbol{\xi}_\gamma\right) +$$

$$\nabla\cdot\left(\rho\boldsymbol{\xi}_\gamma\right)\nabla\left(\boldsymbol{\xi}_\beta\cdot\nabla p + p\Gamma_1\nabla\cdot\boldsymbol{\xi}_\beta\right)]$$

$$+\nabla\cdot\left(\rho\boldsymbol{\xi}_\beta\right)\nabla\cdot\left(\rho\boldsymbol{\xi}_\gamma\right)\frac{\nabla p}{\rho^2} - \left\{\boldsymbol{\xi}_\beta\cdot\nabla\left[\frac{\nabla\cdot\left(\rho\boldsymbol{\xi}_\gamma\right)}{\rho}\right]\right\}\nabla p$$

$$-G\rho\nabla\left\{\int\frac{\nabla'\cdot[\boldsymbol{\xi}_\beta\nabla\cdot\left(\rho\boldsymbol{\xi}_\gamma\right)]}{|\boldsymbol{r}-\boldsymbol{r}'|}d^3r'\right\}$$

$$+\nabla\left\{\boldsymbol{\xi}_\beta\cdot\nabla\left(\boldsymbol{\xi}_\gamma\cdot\nabla p + p\Gamma_1\nabla\cdot\boldsymbol{\xi}_\gamma\right) + \left(\nabla\cdot\boldsymbol{\xi}_\beta\right)\boldsymbol{\xi}_\gamma\cdot\nabla\left(p\Gamma_1\right)\right.$$

$$+ p\Gamma_1 \left[\Gamma_1 + \left(\frac{\partial \ln \Gamma_1}{\partial \ln \rho} \right)_{\mu} \right] (\nabla \cdot \boldsymbol{\xi}_\beta) (\nabla \cdot \boldsymbol{\xi}_\gamma) \Bigg\} \Bigg] \cdot \boldsymbol{\xi}_\alpha^* \mathrm{d}^3 r. \quad \text{(D.2.11)}$$

Reference

1. Dziembowski, W. (1982). Nonlinear mode coupling in oscillating stars. I. Second order theory of the coherent mode coupling. *Acta Astronomica*, *32*, 147–171. http://adsabs.harvard.edu/abs/1982AcA....32..147D.

Appendix E
Polar Mode Coupling Coefficient

Following Dziembowski (1982), we are going to derive an expression for the coupling coefficient (D.2.10) in the nonrotating limit, assuming that the resonant coupled triplet (see Sect. 4.2.1) consists of polar modes, i.e., modes whose eigenfunctions are given by Eq. (2.3.19). This will simplify the calculation greatly, due to the simple spherical harmonic dependence of the eigenfunctions. In this sense, we will find an expression for the *zeroth-order component* (with respect to rotation) of the coupling coefficient.

For the sake of generality though, we will keep assuming that the eigenfrequencies which appear in the formula for the coupling coefficient (D.2.10) have been obtained for the case of a rotating star. This way, we take into account rotational corrections in the zeroth-order component of the coupling coefficient only through the eigenfrequencies. Following the notation of Sect. 2.6.2, we will denote the eigenfrequency in the nonrotating limit, whenever needed, as $\omega^{(0)}$.

E.1 Parametrised Form

For notational convenience, we define the following variables:

$$\mu_\alpha = -\boldsymbol{\xi}_\alpha \cdot \nabla p - p\Gamma_1 \nabla \cdot \boldsymbol{\xi}_\alpha, \tag{E.1.1}$$

$$\lambda_\alpha = \frac{1}{\rho} \nabla \cdot \left(\rho \boldsymbol{\xi}_\alpha \right), \tag{E.1.2}$$

$$w_\alpha = G \int \frac{\rho \lambda_\alpha}{|\boldsymbol{r} - \boldsymbol{r}'|} \mathrm{d}^3 r', \tag{E.1.3}$$

and

$$\chi = \Gamma_1 + \left(\frac{\partial \ln \Gamma_1}{\partial \ln \rho} \right)_\mu. \tag{E.1.4}$$

© Springer Nature Switzerland AG 2018
P. Pnigouras, *Saturation of the f-mode Instability in Neutron Stars*,
Springer Theses, https://doi.org/10.1007/978-3-319-98258-8

Then, the eigenvalue equation for the perturbations in the nonrotating limit, given by Eq. (2.3.54) [or (2.6.8)], can be expressed as

$$\omega_\alpha^{(0)2}\boldsymbol{\xi}_\alpha = \mathcal{C}(\boldsymbol{\xi}_\alpha) = \frac{\nabla\mu_\alpha}{\rho} + \frac{\nabla p}{\rho}\lambda_\alpha + \nabla w_\alpha, \tag{E.1.5}$$

where we used Eq. (2.2.11) and various first-order relations from Appendix D.1. The coupling coefficient $F_{\alpha\beta\gamma}$, defined by Eq. (D.2.10), is written as

$$\omega_\alpha F_{\alpha\beta\gamma} = \omega_\beta S_{\alpha\beta\gamma} + \omega_\gamma S_{\alpha\gamma\beta}, \tag{E.1.6}$$

where

$$S_{\alpha\beta\gamma} = \int \Bigg\{ \underbrace{-\rho\omega_\beta\omega_\gamma\nabla\left(\boldsymbol{\xi}_\beta\cdot\boldsymbol{\xi}_\gamma\right)}_{T_1} \underbrace{+\rho\omega_\beta\omega_\gamma\boldsymbol{\xi}_\beta\times\left(\nabla\times\boldsymbol{\xi}_\gamma\right)+\rho\omega_\beta\omega_\gamma\boldsymbol{\xi}_\gamma\times\left(\nabla\times\boldsymbol{\xi}_\beta\right)}_{T_2}$$

$$\underbrace{+\lambda_\beta\nabla\mu_\gamma+\lambda_\gamma\nabla\mu_\beta+\lambda_\beta\lambda_\gamma\nabla p}_{T_3} \underbrace{-\left(\boldsymbol{\xi}_\beta\cdot\nabla\lambda_\gamma\right)\nabla p}_{T_4} \underbrace{-G\rho\nabla\left[\int\frac{\nabla'\cdot\left(\rho\boldsymbol{\xi}_\beta\lambda_\gamma\right)}{|\boldsymbol{r}-\boldsymbol{r}'|}\mathrm{d}^3r'\right]}_{T_5}$$

$$\underbrace{-\nabla\left(\boldsymbol{\xi}_\beta\cdot\nabla\mu_\gamma\right)}_{T_6} \underbrace{+\nabla\left[\left(\nabla\cdot\boldsymbol{\xi}_\beta\right)\boldsymbol{\xi}_\gamma\cdot\nabla\left(p\Gamma_1\right)\right]}_{T_7} \underbrace{+\nabla\left[p\Gamma_1\chi\left(\nabla\cdot\boldsymbol{\xi}_\beta\right)\left(\nabla\cdot\boldsymbol{\xi}_\gamma\right)\right]}_{T_8} \Bigg\}\cdot\boldsymbol{\xi}_\alpha^*\mathrm{d}^3r. \tag{E.1.7}$$

We are going to study each term of $\omega_\alpha F_{\alpha\beta\gamma}$ separately (including the corresponding terms from $S_{\alpha\gamma\beta}$). Also, in the following calculation, we will neglect the eigenfrequency detuning $\Delta\omega$, i.e., we shall consider $\omega_\alpha \approx \omega_\beta + \omega_\gamma$.

Term 1

$$T_1 = -(\omega_\beta+\omega_\gamma)\int\rho\omega_\beta\omega_\gamma\nabla\left(\boldsymbol{\xi}_\beta\cdot\boldsymbol{\xi}_\gamma\right)\cdot\boldsymbol{\xi}_\alpha^*\mathrm{d}^3r = -\omega_\alpha\int\rho\omega_\beta\omega_\gamma\nabla\left(\boldsymbol{\xi}_\beta\cdot\boldsymbol{\xi}_\gamma\right)\cdot\boldsymbol{\xi}_\alpha^*\mathrm{d}^3r.$$

Using Gauss's theorem, we get

$$T_1 = -\omega_\alpha\omega_\beta\omega_\gamma\left(\oint\boldsymbol{\xi}_\beta\cdot\boldsymbol{\xi}_\gamma\rho\boldsymbol{\xi}_\alpha^*\cdot\mathrm{d}\boldsymbol{S} - \int\boldsymbol{\xi}_\beta\cdot\boldsymbol{\xi}_\gamma\lambda_\alpha^*\rho\mathrm{d}^3r\right),$$

where $\mathrm{d}\boldsymbol{S}$ is the differential normal area vector at the stellar surface. Since the density ρ vanishes at the surface, the first term is neglected (for similar reasons, all surface integrals will be neglected from now on). So, finally,

$$T_1 = \omega_\alpha\omega_\beta\omega_\gamma K_{\alpha\beta\gamma}, \tag{E.1.8}$$

where

$$K_{\alpha\beta\gamma} = \int\boldsymbol{\xi}_\beta\cdot\boldsymbol{\xi}_\gamma\lambda_\alpha^*\rho\mathrm{d}^3r. \tag{E.1.9}$$

Term 2

$$T_2 = (\omega_\beta + \omega_\gamma) \int \rho \omega_\beta \omega_\gamma \left[\boldsymbol{\xi}_\beta \times (\nabla \times \boldsymbol{\xi}_\gamma) + \boldsymbol{\xi}_\gamma \times (\nabla \times \boldsymbol{\xi}_\beta) \right] \cdot \boldsymbol{\xi}_\alpha^* \mathrm{d}^3 r,$$

or

$$T_2 = \omega_\alpha \omega_\beta \omega_\gamma \left(N_{\alpha\beta\gamma} + N_{\alpha\gamma\beta} \right), \tag{E.1.10}$$

where

$$N_{\alpha\beta\gamma} = \int \rho \left[\boldsymbol{\xi}_\beta \times (\nabla \times \boldsymbol{\xi}_\gamma) \right] \cdot \boldsymbol{\xi}_\alpha^* \mathrm{d}^3 r. \tag{E.1.11}$$

Term 3

$$T_3 = (\omega_\beta + \omega_\gamma) \int \left(\lambda_\beta \nabla \mu_\gamma + \lambda_\gamma \nabla \mu_\beta + \lambda_\beta \lambda_\gamma \nabla p \right) \cdot \boldsymbol{\xi}_\alpha^* \mathrm{d}^3 r,$$

or

$$T_3 = \omega_\alpha J_{\alpha\beta\gamma}, \tag{E.1.12}$$

where

$$J_{\alpha\beta\gamma} = \int \left(\lambda_\beta \nabla \mu_\gamma + \lambda_\gamma \nabla \mu_\beta + \lambda_\beta \lambda_\gamma \nabla p \right) \cdot \boldsymbol{\xi}_\alpha^* \mathrm{d}^3 r. \tag{E.1.13}$$

Term 4

$$T_4 = -\omega_\beta \int (\boldsymbol{\xi}_\beta \cdot \nabla \lambda_\gamma) \nabla p \cdot \boldsymbol{\xi}_\alpha^* \mathrm{d}^3 r - \omega_\gamma \int (\boldsymbol{\xi}_\gamma \cdot \nabla \lambda_\beta) \nabla p \cdot \boldsymbol{\xi}_\alpha^* \mathrm{d}^3 r. \tag{E.1.14}$$

Term 5

$$T_5 = -\omega_\beta \underbrace{\int G\rho \nabla \left[\int \frac{\nabla' \cdot (\rho \boldsymbol{\xi}_\beta \lambda_\gamma)}{|r - r'|} \mathrm{d}^3 r' \right] \cdot \boldsymbol{\xi}_\alpha^* \mathrm{d}^3 r}_{T_{5a}} - \omega_\gamma \underbrace{\int G\rho \nabla \left[\int \frac{\nabla' \cdot (\rho \boldsymbol{\xi}_\gamma \lambda_\beta)}{|r - r'|} \mathrm{d}^3 r' \right] \cdot \boldsymbol{\xi}_\alpha^* \mathrm{d}^3 r}_{T_{5b}}.$$

Using Gauss's theorem, we get

$$T_{5a} = G \oint \left[\int \frac{\nabla' \cdot (\rho \boldsymbol{\xi}_\beta \lambda_\gamma)}{|r - r'|} \mathrm{d}^3 r' \right] \rho \boldsymbol{\xi}_\alpha^* \cdot \mathrm{d}S - G \iint \frac{\nabla' \cdot (\rho \boldsymbol{\xi}_\beta \lambda_\gamma)}{|r - r'|} \rho \lambda_\alpha^* \mathrm{d}^3 r' \mathrm{d}^3 r.$$

Neglecting the first term and taking into account the symmetry of $1/|r - r'|$, we obtain

$$T_{5a} = -\int \nabla \cdot (\rho \boldsymbol{\xi}_\beta \lambda_\gamma) \left[\int \frac{G\rho \lambda_\alpha^*}{|\boldsymbol{r}-\boldsymbol{r}'|} d^3 r' \right] d^3 r = -\int \nabla \cdot (\rho \boldsymbol{\xi}_\beta \lambda_\gamma) \, w_\alpha^* d^3 r.$$

Using Gauss's theorem one more time, T_{5a} gives

$$T_{5a} = -\oint \rho \lambda_\gamma w_\alpha^* \boldsymbol{\xi}_\beta \cdot d\boldsymbol{S} + \int \rho \lambda_\gamma \boldsymbol{\xi}_\beta \cdot \nabla w_\alpha^* d^3 r.$$

Neglecting the surface integral and using Eq. (E.1.5) to eliminate ∇w_α^* from the remaining term, the equation above becomes

$$T_{5a} = \int \omega_\alpha^{(0)2} \rho \lambda_\gamma \boldsymbol{\xi}_\beta \cdot \boldsymbol{\xi}_\alpha^* d^3 r - \int \lambda_\gamma \boldsymbol{\xi}_\beta \cdot \nabla \mu_\alpha^* d^3 r - \int \lambda_\alpha^* \lambda_\gamma \boldsymbol{\xi}_\beta \cdot \nabla p d^3 r.$$

The corresponding expression for T_{5b} is the same, with the indices β and γ interchanged. So, finally, T_5 can be written as

$$T_5 = \underbrace{-\omega_\alpha^{(0)2} \left(\omega_\beta K_{\gamma\alpha\beta} + \omega_\gamma K_{\beta\alpha\gamma} \right)}_{T_{5.1}}$$

$$\underbrace{+\omega_\beta \int \lambda_\gamma \boldsymbol{\xi}_\beta \cdot \nabla \mu_\alpha^* d^3 r + \omega_\gamma \int \lambda_\beta \boldsymbol{\xi}_\gamma \cdot \nabla \mu_\alpha^* d^3 r}_{T_{5.2}}$$

$$\underbrace{+\omega_\beta \int \lambda_\alpha^* \lambda_\gamma \boldsymbol{\xi}_\beta \cdot \nabla p d^3 r + \omega_\gamma \int \lambda_\alpha^* \lambda_\beta \boldsymbol{\xi}_\gamma \cdot \nabla p d^3 r}_{T_{5.3}}, \qquad (E.1.15)$$

with $K_{\alpha\beta\gamma}$ defined in Eq. (E.1.9).

Term 6

$$T_6 = -\omega_\beta \int \nabla \left[\boldsymbol{\xi}_\beta \cdot \nabla \mu_\gamma \right] \cdot \boldsymbol{\xi}_\alpha^* d^3 r - \omega_\gamma \int \nabla \left[\boldsymbol{\xi}_\gamma \cdot \nabla \mu_\beta \right] \cdot \boldsymbol{\xi}_\alpha^* d^3 r.$$

Applying Gauss's theorem, the first integral becomes

$$\int \nabla \left[\boldsymbol{\xi}_\beta \cdot \nabla \mu_\gamma \right] \cdot \boldsymbol{\xi}_\alpha^* d^3 r = \oint \boldsymbol{\xi}_\beta \cdot \nabla \mu_\gamma \boldsymbol{\xi}_\alpha^* \cdot d\boldsymbol{S} - \int \boldsymbol{\xi}_\beta \cdot \nabla \mu_\gamma \nabla \cdot \boldsymbol{\xi}_\alpha^* d^3 r.$$

Neglecting the surface integral, the remaining terms give

$$T_6 = \omega_\beta \int \boldsymbol{\xi}_\beta \cdot \nabla \mu_\gamma \nabla \cdot \boldsymbol{\xi}_\alpha^* d^3 r + \omega_\gamma \int \boldsymbol{\xi}_\gamma \cdot \nabla \mu_\beta \nabla \cdot \boldsymbol{\xi}_\alpha^* d^3 r. \qquad (E.1.16)$$

Term 7

$$T_7 = \omega_\beta \int \nabla \left[(\nabla \cdot \boldsymbol{\xi}_\beta) \, \boldsymbol{\xi}_\gamma \cdot \nabla \, (p\Gamma_1) \right] \cdot \boldsymbol{\xi}_\alpha^* d^3 r$$

$$+ \omega_\gamma \int \nabla \left[(\nabla \cdot \boldsymbol{\xi}_\gamma) \, \boldsymbol{\xi}_\beta \cdot \nabla \, (p\Gamma_1) \right] \cdot \boldsymbol{\xi}_\alpha^* d^3 r.$$

Using Gauss's theorem, the first term becomes

$$\int \nabla \left[(\nabla \cdot \boldsymbol{\xi}_\beta) \, \boldsymbol{\xi}_\gamma \cdot \nabla \, (p\Gamma_1) \right] \cdot \boldsymbol{\xi}_\alpha^* d^3 r = \oint (\nabla \cdot \boldsymbol{\xi}_\beta) \, \boldsymbol{\xi}_\gamma \cdot \nabla \, (p\Gamma_1) \, \boldsymbol{\xi}_\alpha^* \cdot d\boldsymbol{S}$$

$$- \int (\nabla \cdot \boldsymbol{\xi}_\beta) \, \boldsymbol{\xi}_\gamma \cdot \nabla \, (p\Gamma_1) \, \nabla \cdot \boldsymbol{\xi}_\alpha^* d^3 r.$$

With the surface integral neglected, we get

$$T_7 = -\omega_\beta \int (\nabla \cdot \boldsymbol{\xi}_\beta) \, \boldsymbol{\xi}_\gamma \cdot \nabla \, (p\Gamma_1) \, \nabla \cdot \boldsymbol{\xi}_\alpha^* d^3 r$$

$$-\omega_\gamma \int (\nabla \cdot \boldsymbol{\xi}_\gamma) \, \boldsymbol{\xi}_\beta \cdot \nabla \, (p\Gamma_1) \, \nabla \cdot \boldsymbol{\xi}_\alpha^* d^3 r. \qquad (E.1.17)$$

Term 8

$$T_8 = (\omega_\beta + \omega_\gamma) \int \nabla \left[p\Gamma_1 \chi \left(\nabla \cdot \boldsymbol{\xi}_\beta \right) \left(\nabla \cdot \boldsymbol{\xi}_\gamma \right) \right] \cdot \boldsymbol{\xi}_\alpha^* d^3 r.$$

Applying Gauss's theorem and neglecting the surface term, we obtain

$$T_8 = \omega_\alpha M_{\alpha\beta\gamma}, \qquad (E.1.18)$$

where

$$M_{\alpha\beta\gamma} = -\int p\Gamma_1 \chi \left(\nabla \cdot \boldsymbol{\xi}_\alpha^* \right) \left(\nabla \cdot \boldsymbol{\xi}_\beta \right) \left(\nabla \cdot \boldsymbol{\xi}_\gamma \right) d^3 r. \qquad (E.1.19)$$

Terms 4, 5.2, 5.3, 6 and 7 are combined to give

$$T' = \omega_\beta L_{\beta\alpha\gamma} + \omega_\gamma L_{\gamma\alpha\beta}, \qquad (E.1.20)$$

where

$$L_{\beta\alpha\gamma} = \int \left[- \boldsymbol{\xi}_\alpha^* \cdot \nabla p \boldsymbol{\xi}_\beta \cdot \nabla \lambda_\gamma + \boldsymbol{\xi}_\beta \cdot \nabla \mu_\alpha^* \lambda_\gamma + \lambda_\alpha^* \lambda_\gamma \boldsymbol{\xi}_\beta \cdot \nabla p \right.$$

$$\left. + \left(\nabla \cdot \boldsymbol{\xi}_\alpha^* \right) \boldsymbol{\xi}_\beta \cdot \nabla \mu_\gamma - \left(\nabla \cdot \boldsymbol{\xi}_\alpha^* \right) \left(\nabla \cdot \boldsymbol{\xi}_\beta \right) \boldsymbol{\xi}_\gamma \cdot \nabla (p\Gamma_1) \right] d^3 r. \quad (E.1.21)$$

Combining terms 1, 2, 3, 5.1, 8, and T' above, we finally get the more compact expression

$$\omega_\alpha F_{\alpha\beta\gamma} = \omega_\alpha \omega_\beta \omega_\gamma K_{\alpha\beta\gamma} - \omega_\alpha^{(0)2} \omega_\beta K_{\gamma\alpha\beta} - \omega_\alpha^{(0)2} \omega_\gamma K_{\beta\alpha\gamma} + \omega_\alpha \omega_\beta \omega_\gamma \left(N_{\alpha\beta\gamma} + N_{\alpha\gamma\beta} \right)$$
$$+ \omega_\alpha M_{\alpha\beta\gamma} + \omega_\alpha J_{\alpha\beta\gamma} + \omega_\beta L_{\beta\alpha\gamma} + \omega_\gamma L_{\gamma\alpha\beta}. \qquad (E.1.22)$$

E.2 Angular Part

Henceforth, we are going to use the dimensionless formulation presented in Sect. 2.3.3. The polar mode eigenfunction (2.3.19) is written as

$$\boldsymbol{\xi}_\alpha = r y_{1,\alpha} Y_\alpha \boldsymbol{e}_r + \frac{r y_{2,\alpha}}{c_1 \tilde{\omega}_\alpha^{(0)2}} r \nabla_\perp Y_\alpha, \qquad (E.2.1)$$

where

$$Y_\alpha \equiv Y_{l_\alpha}^{m_\alpha} \qquad (E.2.2)$$

and $\nabla_\perp$ is the horizontal component of the gradient operator, defined by Eq. (2.3.3). Furthermore, in this notation,

$$w_\alpha = g r y_{3,\alpha} Y_\alpha \qquad (E.2.3)$$

and

$$\frac{\partial w_\alpha}{\partial r} = g y_{4,\alpha} Y_\alpha. \qquad (E.2.4)$$

Equations (E.2.1), (E.2.3) and (E.2.4) will be used throughout the following calculations without reference.

The horizontal component of Eq. (E.1.5) gives

$$\mu_\alpha = g \rho r z_\alpha Y_\alpha, \qquad (E.2.5)$$

with

$$z_\alpha = y_{2,\alpha} - y_{3,\alpha}. \qquad (E.2.6)$$

Using Eq. (E.2.5) in Eq. (E.1.1), we also get

$$\nabla \cdot \boldsymbol{\xi}_\alpha = V_g \left(y_{1,\alpha} - z_\alpha \right) Y_\alpha. \qquad (E.2.7)$$

Substituting Eq. (E.2.7) in Eq. (E.1.2), we obtain

$$\lambda_\alpha = -\left(A^* y_{1,\alpha} + V_g z_\alpha\right) Y_\alpha. \tag{E.2.8}$$

Now, we will examine the terms of Eq. (E.1.22) individually and prove that their angular part is reduced to the integral

$$Z_{\alpha\beta\gamma} = \iint Y_\alpha^* Y_\beta Y_\gamma \sin\theta d\theta d\phi. \tag{E.2.9}$$

This is a known integral (e.g., see Sakurai and Napolitano 2011, Sect. 3.8) and is equal to

$$Z_{\alpha\beta\gamma} = \sqrt{\frac{(2l_\beta + 1)(2l_\gamma + 1)}{4\pi(2l_\alpha + 1)}} \langle l_\beta l_\gamma 00 | l_\beta l_\gamma l_\alpha 0 \rangle \langle l_\beta l_\gamma m_\beta m_\gamma | l_\beta l_\gamma l_\alpha m_\alpha \rangle, \tag{E.2.10}$$

where

$$\langle l_\beta l_\gamma m_\beta m_\gamma | l_\beta l_\gamma l_\alpha m_\alpha \rangle = \frac{\delta_{m_\alpha, m_\beta + m_\gamma} \sqrt{2l_\alpha + 1}}{\sqrt{(l_\alpha + l_\beta + l_\gamma + 1)!}} \prod_k{}' \sqrt{(-l_k + l_{k'} + l_{k''})!} \sqrt{(l_k + m_k)!} \sqrt{(l_k - m_k)!}$$

$$\sum_{j=j_-}^{j_+} \frac{(-1)^j}{j!(-j - l_\alpha + l_\beta + l_\gamma)!(-j + l_\beta - m_\beta)!(-j + l_\gamma + m_\gamma)!(j + l_\alpha - l_\gamma + m_\beta)!(j + l_\alpha - l_\beta - m_\gamma)!}$$

are the *Clebsch-Gordan coefficients*, with

$$j_- = \max(-l_\alpha + l_\gamma - m_\beta, -l_\alpha + l_\beta + m_\gamma, 0) \quad \text{and} \quad j_+ = \min(l_\beta - m_\beta, l_\gamma + m_\gamma).$$

The index k in the product successively takes one of the values α, β, γ, whereas the indices k' and k'' take the values that come next and after next respectively, namely

$$(k, k', k'') = \begin{cases} (\alpha, \beta, \gamma) \\ (\beta, \gamma, \alpha) \\ (\gamma, \alpha, \beta). \end{cases} \tag{E.2.11}$$

$K_{\alpha\beta\gamma}$

Using Eq. (E.2.8), Eq. (E.1.9) becomes

$$K_{\alpha\beta\gamma} = -\int \left(A^* y_{1,\alpha} + V_g z_\alpha\right) \left[y_{1,\beta} y_{1,\gamma} Z_{\alpha\beta\gamma} + y_{2,\beta} y_{2,\gamma} \frac{X_{\alpha\beta\gamma}}{\left(c_1 \tilde{\omega}_\beta^{(0)} \tilde{\omega}_\gamma^{(0)}\right)^2} \right] \rho r^4 dr,$$

where

$$X_{\alpha\beta\gamma} = \iint Y_\alpha^* \nabla_\perp Y_\beta \cdot \nabla_\perp Y_\gamma r^2 \sin\theta d\theta d\phi.$$

Integrating by parts and making use of Eq. (2.3.13), we get

$$X_{\alpha\beta\gamma} + X_{\beta\alpha\gamma} = \Lambda_\gamma Z_{\alpha\beta\gamma},$$

where

$$\Lambda = l(l+1). \tag{E.2.12}$$

In a similar manner, we find

$$X_{\alpha\beta\gamma} + X_{\gamma\alpha\beta} = \Lambda_\beta Z_{\alpha\beta\gamma}$$

and

$$X_{\beta\alpha\gamma} + X_{\gamma\alpha\beta} = \Lambda_\alpha Z_{\alpha\beta\gamma}.$$

Thus,

$$X_{\alpha\beta\gamma} = \frac{-\Lambda_\alpha + \Lambda_\beta + \Lambda_\gamma}{2} Z_{\alpha\beta\gamma}. \tag{E.2.13}$$

So, $K_{\alpha\beta\gamma}$ is written as

$$K_{\alpha\beta\gamma} = -Z_{\alpha\beta\gamma} \int \left(A^* y_{1,\alpha} + V_g z_\alpha\right) \left[y_{1,\beta} y_{1,\gamma} + y_{2,\beta} y_{2,\gamma} \frac{\Lambda_\beta + \Lambda_\gamma - \Lambda_\alpha}{2 \left(c_1 \tilde{\omega}_\beta^{(0)} \tilde{\omega}_\gamma^{(0)} \right)^2} \right] \rho r^4 dr. \tag{E.2.14}$$

$N_{\alpha\beta\gamma}$

Taking the curl of Eq. (E.1.5), we get

$$\omega_\alpha^{(0)2} \nabla \times \boldsymbol{\xi}_\alpha = \nabla \left(\frac{1}{\rho}\right) \times \nabla \mu_\alpha + \nabla \lambda_\alpha \times \left(\frac{\nabla p}{\rho}\right),$$

or, using Eqs. (E.2.5) and (E.2.8),

$$\omega_\alpha^{(0)2} \nabla \times \boldsymbol{\xi}_\alpha = gA^* \left(z_\alpha - y_{1,\alpha}\right) \boldsymbol{e}_r \times \nabla_\perp Y_\alpha.$$

So, making use of the vectorial identity

$$\boldsymbol{A} \times (\boldsymbol{B} \times \boldsymbol{C}) = (\boldsymbol{A} \cdot \boldsymbol{C}) \boldsymbol{B} - (\boldsymbol{A} \cdot \boldsymbol{B}) \boldsymbol{C},$$

we obtain

$$\omega_\gamma^{(0)2} \boldsymbol{\xi}_\beta \times (\nabla \times \boldsymbol{\xi}_\gamma) = rgA^* \left(z_\gamma - y_{1,\gamma}\right) \left[(\nabla_\perp Y_\beta) \cdot (\nabla_\perp Y_\gamma) \frac{r y_{2,\beta}}{c_1 \tilde{\omega}_\beta^{(0)2}} \boldsymbol{e}_r - y_{1,\beta} Y_\beta \nabla_\perp Y_\gamma \right].$$

Finally, with the help of Eq. (E.2.13), Eq. (E.1.11) is written as

$$N_{\alpha\beta\gamma} = \frac{Z_{\alpha\beta\gamma}}{\omega_\gamma^{(0)2}} \int (z_\gamma - y_{1,\gamma}) \left[y_{1,\alpha} y_{2,\beta} \frac{\Lambda_\beta + \Lambda_\gamma - \Lambda_\alpha}{2c_1 \tilde{\omega}_\beta^{(0)2}} - y_{1,\beta} y_{2,\alpha} \frac{\Lambda_\gamma + \Lambda_\alpha - \Lambda_\beta}{2c_1 \tilde{\omega}_\alpha^{(0)2}} \right] gA^* \rho r^3 dr.$$

$$(E.2.15)$$

$M_{\alpha\beta\gamma}$

Substituting Eq. (E.2.7) in Eq. (E.1.19), we get

$$M_{\alpha\beta\gamma} = -Z_{\alpha\beta\gamma} \int \prod_k (y_{1,k} - z_k) \, \chi V_g^2 g \rho r^3 dr, \qquad (E.2.16)$$

with k successively taking the values α, β, γ.

$J_{\alpha\beta\gamma}$

Equations (E.2.5) and (E.2.8) may be combined to give

$$\lambda_\alpha = -A^* y_{1,\alpha} Y_\alpha - \frac{\mu_\alpha}{p\Gamma_1}. \qquad (E.2.17)$$

Using Eq. (E.2.17), Eq. (E.1.13) becomes

$$J_{\alpha\beta\gamma} = -\int \boldsymbol{\xi}_\alpha^* \cdot \left[A^* \left(y_{1,\beta} Y_\beta \nabla \mu_\gamma + y_{1,\gamma} Y_\gamma \nabla \mu_\beta \right) - \lambda_\beta \lambda_\gamma \nabla p + \frac{\nabla \left(\mu_\beta \mu_\gamma \right)}{p\Gamma_1} \right] d^3 \boldsymbol{r}.$$

Applying Gauss's theorem to the last term, we get

$$-\int \frac{\nabla \left(\mu_\beta \mu_\gamma \right)}{p\Gamma_1} \cdot \boldsymbol{\xi}_\alpha^* d^3 \boldsymbol{r} = -\oint \frac{\mu_\beta \mu_\gamma}{p\Gamma_1} \boldsymbol{\xi}_\alpha^* \cdot d\boldsymbol{S} + \int \frac{\mu_\beta \mu_\gamma}{p\Gamma_1} \left[\nabla \cdot \boldsymbol{\xi}_\alpha^* - \boldsymbol{\xi}_\alpha^* \cdot \nabla \ln(p\Gamma_1) \right] d^3 \boldsymbol{r}$$

$$= Z_{\alpha\beta\gamma} \int z_\beta z_\gamma \left[-z_\alpha V_g + y_{1,\alpha} \left(V + V_g - \frac{d \ln \Gamma_1}{d \ln r} \right) \right] V_g g \rho r^3 dr,$$

where we used Eqs. (E.2.5) and (E.2.7), and the surface integral was neglected. Next, we calculate the radial component of $\nabla \mu_\alpha$ via Eq. (E.1.5), as

$$\frac{\partial \mu_\alpha}{\partial r} = \rho \left[\omega_\alpha^{(0)2} r y_{1,\alpha} Y_\alpha + g \left(\lambda_\alpha - y_{4,\alpha} Y_\alpha \right) \right]$$

$$= g\rho Y_\alpha \left[y_{1,\alpha} \left(c_1 \tilde{\omega}_\alpha^{(0)2} - A^* \right) - V_g z_\alpha - y_{4,\alpha} \right], \qquad (E.2.18)$$

where Eq. (E.2.8) was used. With the help of the relations above, and using Eqs. (E.2.8) and (E.2.13), we obtain the final expression for $J_{\alpha\beta\gamma}$, which is

$$J_{\alpha\beta\gamma} = Z_{\alpha\beta\gamma} \int \left\{ A^* y_{2,\alpha} \left(\frac{\Lambda_\beta - \Lambda_\gamma - \Lambda_\alpha}{2c_1 \tilde{\omega}_\alpha^{(0)2}} y_{1,\beta} z_\gamma + \frac{\Lambda_\gamma - \Lambda_\beta - \Lambda_\alpha}{2c_1 \tilde{\omega}_\alpha^{(0)2}} y_{1,\gamma} z_\beta \right) \right.$$

$$+ A^* \left[y_{1,\alpha} y_{1,\beta} y_{1,\gamma} \left(A^* - c_1 \tilde{\omega}_\beta^{(0)2} - c_1 \tilde{\omega}_\gamma^{(0)2} \right) + y_{1,\alpha} \left(y_{1,\beta} y_{4,\gamma} + y_{1,\gamma} y_{4,\beta} \right) \right]$$

$$\left. + V_g \left[y_{1,\alpha} z_\beta z_\gamma \left(V - \frac{d \ln \Gamma_1}{d \ln r} \right) - V_g z_\alpha z_\beta z_\gamma \right] \right\} g\rho r^3 dr. \qquad (E.2.19)$$

$L_{\beta\alpha\gamma}$

Applying Gauss's theorem to the first and fourth terms of Eq. (E.1.21), while neglecting the surface integrals, we get

$$L_{\beta\alpha\gamma} = \int \{ \boldsymbol{\xi}_\beta \cdot [\lambda_\gamma \nabla \left(-p\Gamma_1 \nabla \cdot \boldsymbol{\xi}_\alpha^* \right) - \mu_\gamma \nabla \left(\nabla \cdot \boldsymbol{\xi}_\alpha^* \right)] - \mu_\gamma \left(\nabla \cdot \boldsymbol{\xi}_\alpha^* \right) \left(\nabla \cdot \boldsymbol{\xi}_\beta \right)$$
$$+ \nabla p \cdot \left(\boldsymbol{\xi}_\alpha^* \lambda_\gamma \nabla \cdot \boldsymbol{\xi}_\beta + \lambda_\alpha^* \lambda_\gamma \boldsymbol{\xi}_\beta \right) - \nabla (p\Gamma_1) \cdot \boldsymbol{\xi}_\gamma \left(\nabla \cdot \boldsymbol{\xi}_\alpha^* \right) \left(\nabla \cdot \boldsymbol{\xi}_\beta \right) \} d^3 r,$$

where we used Eq. (E.2.5). Furthermore, with the help of Eq. (E.2.17), we obtain

$$L_{\beta\alpha\gamma} = \int \Big\{ p\Gamma_1 A^* y_{1,\gamma} Y_\gamma \boldsymbol{\xi}_\beta \cdot \nabla \left(\nabla \cdot \boldsymbol{\xi}_\alpha^* \right) + \nabla p \cdot \left(\boldsymbol{\xi}_\alpha^* \lambda_\gamma \nabla \cdot \boldsymbol{\xi}_\beta + \lambda_\alpha^* \lambda_\gamma \boldsymbol{\xi}_\beta \right)$$
$$- \mu_\gamma \left(\nabla \cdot \boldsymbol{\xi}_\alpha^* \right) \left(\nabla \cdot \boldsymbol{\xi}_\beta \right) - \nabla (p\Gamma_1) \cdot \left[\lambda_\gamma \left(\nabla \cdot \boldsymbol{\xi}_\alpha^* \right) \boldsymbol{\xi}_\beta + \left(\nabla \cdot \boldsymbol{\xi}_\alpha^* \right) \left(\nabla \cdot \boldsymbol{\xi}_\beta \right) \boldsymbol{\xi}_\gamma \right] \Big\} d^3 r.$$

Next, we calculate the radial component of $\nabla \left(\nabla \cdot \boldsymbol{\xi}_\alpha \right)$. First, we take the derivative of Eq. (E.1.1) with respect to r and solve for $\partial \left(\nabla \cdot \boldsymbol{\xi}_\alpha \right) / \partial r$, i.e.,

$$\frac{\partial}{\partial r} \left(\nabla \cdot \boldsymbol{\xi}_\alpha \right) = \frac{1}{p\Gamma_1} \left[-\frac{\partial \mu_\alpha}{\partial r} - \frac{d}{dr} \left(\boldsymbol{\xi}_\alpha \cdot \nabla p \right) - \frac{d(p\Gamma_1)}{dr} \nabla \cdot \boldsymbol{\xi}_\alpha \right].$$

We evaluate the first and third terms of the relation above by making use of Eqs. (E.2.18) and (E.2.7) respectively. From the second term we get

$$-\frac{d}{dr} \left(\boldsymbol{\xi}_\alpha \cdot \nabla p \right) = \rho g Y_\alpha \frac{d}{dr} \left(r y_{1,\alpha} \right) + y_{1,\alpha} Y_\alpha r \frac{d(\rho g)}{dr}.$$

Taking the divergence of $\boldsymbol{\xi}_\alpha$ via Eq. (E.2.1), we obtain an expression for $d \left(r y_{1,\alpha} \right) / dr$. So, finally,

$$\frac{\partial}{\partial r} \left(\nabla \cdot \boldsymbol{\xi}_\alpha \right) = \frac{V_g}{r} Y_\alpha \bigg[y_{1,\alpha} \left(U - 4 - c_1 \tilde{\omega}_\alpha^{(0)2} \right) + \left(y_{1,\alpha} - z_\alpha \right) \left(V - \frac{d \ln \Gamma_1}{d \ln r} \right)$$
$$+ \frac{\Lambda_\alpha}{c_1 \tilde{\omega}_\alpha^{(0)2}} y_{2,\alpha} + y_{4,\alpha} \bigg].$$

Using the relation above, together with Eqs. (E.2.7) and (E.2.8), and appropriately rearranging the terms, the final expression we get for $L_{\beta\alpha\gamma}$ is

$$L_{\beta\alpha\gamma} = Z_{\alpha\beta\gamma} \int \Big\{ - y_{1,\alpha} y_{1,\beta} y_{1,\gamma} \left[\left(A^* + V_g \right)^2 + A^* \left(4 + c_1 \tilde{\omega}_\alpha^{(0)2} - U \right) \right] + A^* y_{4,\alpha} y_{1,\beta} y_{1,\gamma}$$
$$+ V_g \sum_k y_{1,k} \left[\left(A^* + V_g \right) y_{1,k'} \left(y_{1,k''} - z_{k''} \right) - V_g \left(y_{1,k'} - z_{k'} \right) \left(y_{1,k''} - z_{k''} \right) \right]$$

$$+ V_g \left(V - \frac{d \ln \Gamma_1}{d \ln r} \right) (y_{1,\alpha} - z_\alpha) \left[y_{1,\beta} \left(y_{1,\gamma} - z_\gamma \right) + y_{1,\gamma} \left(y_{1,\beta} - z_\beta \right) - y_{1,\beta} y_{1,\gamma} \right]$$

$$+ A^* y_{1,\gamma} \left[\frac{\Lambda_1 + \Lambda_2 - \Lambda_3}{2 c_1 \tilde{\omega}_\beta^{(0)2}} y_{2,\beta} \left(y_{1,\alpha} - z_\alpha \right) + \frac{\Lambda_\alpha}{c_1 \tilde{\omega}_\alpha^{(0)2}} y_{2,\alpha} y_{1,\beta} \right]$$

$$+ V_g^2 \prod_k \left(y_{1,k} - z_k \right) \Big\} g \rho r^3 dr, \tag{E.2.20}$$

with the indices k, k', and k'' behaving as in Eq. (E.2.11).

E.3 Radial Part

We will now proceed with the evaluation of $F_{\alpha\beta\gamma}$ from Eq. (E.1.22), based on the formulae derived in the previous section for the various parameters. Once again, we will study each term separately. For reasons explained later, we will divide $F_{\alpha\beta\gamma}$ by GM/R^3.

$K_{\alpha\beta\gamma}$

Using Eq. (E.2.14), the combination of the K terms in Eq. (E.1.22) gives

$$\frac{R^3}{GM} \left[K_{\alpha\beta\gamma} \omega_\beta \omega_\gamma - K_{\gamma\alpha\beta} \frac{\omega_\alpha^{(0)2} \omega_\beta}{\omega_\alpha} - K_{\beta\alpha\gamma} \frac{\omega_\alpha^{(0)2} \omega_\gamma}{\omega_\alpha} \right] = \tilde{\mathcal{H}}_1,$$

where

$$\tilde{\mathcal{H}}_1 = Z_{\alpha\beta\gamma} \int \Bigg\{ - \sum_k \left(A^* y_{1,k} + V_g z_k \right) \Psi_k$$

$$\times \left(\varpi_{k'} \varpi_{k''} y_{1,k'} y_{1,k''} + \frac{QC_k}{c_1^2} \psi_{k'} \psi_{k''} y_{2,k'} y_{2,k''} \right) \Bigg\} \rho r^4 dr, \tag{E.3.1}$$

with

$$\varpi_k = \begin{cases} \tilde{\omega}_k & \\ -\tilde{\omega}_k & \end{cases} \text{for} \quad \begin{matrix} k = \alpha \\ k = \beta, \gamma \end{matrix} \tag{E.3.2}$$

and

$$QC_k = \frac{-\Lambda_k + \Lambda_{k'} + \Lambda_{k''}}{2 \varpi_{k'} \varpi_{k''}} \tag{E.3.3}$$

[Λ has been defined in Eq. (E.2.12)], whereas

$$\psi_k = \left(\frac{\tilde{\omega}_k}{\tilde{\omega}_k^{(0)}} \right)^2 \tag{E.3.4}$$

and

$$\Psi_k = \begin{cases} 1 & \\ 1/\psi_\alpha & \end{cases} \text{ for } \begin{matrix} k = \alpha \\ k = \beta, \gamma. \end{matrix} \tag{E.3.5}$$

The behaviour of the indices k, k', and k'' is explained in Eq. (E.2.11).

$N_{\alpha\beta\gamma}$

From Eqs. (E.1.22) and (E.2.15) we obtain

$$\frac{R^3}{GM} N_{\alpha\beta\gamma} \omega_\beta \omega_\gamma = Z_{\alpha\beta\gamma} \frac{\tilde{\omega}_\beta}{\tilde{\omega}_\gamma} \psi_\gamma \int \frac{A^*}{c_1^2} \left(z_\gamma - y_{1,\gamma} \right)$$

$$\times \left(y_{1,\alpha} y_{2,\beta} \psi_\beta \frac{\Lambda_\beta + \Lambda_\gamma - \Lambda_\alpha}{2\tilde{\omega}_\beta^2} - y_{1,\beta} y_{2,\alpha} \psi_\alpha \frac{\Lambda_\gamma + \Lambda_\alpha - \Lambda_\beta}{2\tilde{\omega}_\alpha^2} \right) \rho r^4 dr.$$

The corresponding expression for $N_{\alpha\gamma\beta}\omega_\beta\omega_\gamma/(GM/R^3)$ is the same, with the indices β and γ interchanged. For later convenience, we will write the correction factor ψ_γ (but not ψ_α and ψ_β) as

$$\psi_\gamma = 1 + \Xi_\gamma,$$

where

$$\Xi_\gamma = \frac{\tilde{\omega}_\gamma^2 - \tilde{\omega}_\gamma^{(0)2}}{\tilde{\omega}_\gamma^{(0)2}}. \tag{E.3.6}$$

Doing the same for the equivalent factor ψ_β in $N_{\alpha\gamma\beta}\omega_\beta\omega_\gamma/(GM/R^3)$, we get for the N terms

$$\frac{R^3}{GM} \left(N_{\alpha\beta\gamma} + N_{\alpha\gamma\beta} \right) \omega_\beta \omega_\gamma = Z_{\alpha\beta\gamma} \frac{\tilde{\omega}_\beta}{\tilde{\omega}_\gamma} \left(1 + \Xi_\gamma \right) \int \frac{A^*}{c_1^2} \left(z_\gamma - y_{1,\gamma} \right)$$

$$\times \left(y_{1,\alpha} y_{2,\beta} \psi_\beta \frac{\Lambda_\beta + \Lambda_\gamma - \Lambda_\alpha}{2\tilde{\omega}_\beta^2} - y_{1,\beta} y_{2,\alpha} \psi_\alpha \frac{\Lambda_\gamma + \Lambda_\alpha - \Lambda_\beta}{2\tilde{\omega}_\alpha^2} \right) \rho r^4 dr$$

$$+ Z_{\alpha\beta\gamma} \frac{\tilde{\omega}_\gamma}{\tilde{\omega}_\beta} \left(1 + \Xi_\beta \right) \int \frac{A^*}{c_1^2} \left(z_\beta - y_{1,\beta} \right)$$

$$\times \left(y_{1,\alpha} y_{2,\gamma} \psi_\gamma \frac{\Lambda_\gamma + \Lambda_\beta - \Lambda_\alpha}{2\tilde{\omega}_\gamma^2} - y_{1,\gamma} y_{2,\alpha} \psi_\alpha \frac{\Lambda_\beta + \Lambda_\alpha - \Lambda_\gamma}{2\tilde{\omega}_\alpha^2} \right) \rho r^4 dr.$$

$M_{\alpha\beta\gamma}$

Equations (E.1.22) and (E.2.16) give

$$\frac{R^3}{GM} M_{\alpha\beta\gamma} = -Z_{\alpha\beta\gamma} \int \frac{1}{c_1} \prod_k \left(y_{1,k} - z_k \right) \chi V_g^2 \rho r^4 dr.$$

$J_{\alpha\beta\gamma}$

From Eqs. (E.1.22) and (E.2.19) we get

$$\frac{R^3}{GM} J_{\alpha\beta\gamma} = Z_{\alpha\beta\gamma} \int \Bigg\{ \underbrace{\frac{A^*}{c_1^2} y_{2,\alpha}\psi_\alpha \left(\frac{\Lambda_\beta - \Lambda_\gamma - \Lambda_\alpha}{2\tilde{\omega}_\alpha^2} y_{1,\beta} z_\gamma + \frac{\Lambda_\gamma - \Lambda_\beta - \Lambda_\alpha}{2\tilde{\omega}_\alpha^2} y_{1,\gamma} z_\beta \right)}_{J_\Lambda}$$

$$+ \frac{A^*}{c_1} \Bigg[\underbrace{y_{1,\alpha} y_{1,\beta} y_{1,\gamma} \left(A^* - c_1 \frac{\tilde{\omega}_\beta^2}{\psi_\beta} - c_1 \frac{\tilde{\omega}_\gamma^2}{\psi_\gamma} \right)}_{J_1} + \underbrace{y_{1,\alpha} \left(y_{1,\beta} y_{4,\gamma} + y_{1,\gamma} y_{4,\beta} \right)}_{J_2} \Bigg]$$

$$+ \frac{V_g}{c_1} \Bigg[\underbrace{y_{1,\alpha} z_\beta z_\gamma \left(V - \frac{d \ln \Gamma_1}{d \ln r} \right)}_{J_3} \underbrace{- V_g z_\alpha z_\beta z_\gamma}_{J_4} \Bigg] \Bigg\} \rho r^4 dr.$$

$L_{\beta\alpha\gamma}$

With the help of Eq. (E.2.20), the L terms in Eq. (E.1.22) yield

$$\frac{R^3}{GM} \left[\frac{\omega_\beta}{\omega_\alpha} L_{\beta\alpha\gamma} + \frac{\omega_\gamma}{\omega_\alpha} L_{\gamma\alpha\beta} \right] = Z_{\alpha\beta\gamma} \int \frac{1}{c_1} \Bigg\{ \underbrace{v_g^2 \prod_k \left(y_{1,k} - z_k \right)}_{L_1}$$

$$\underbrace{+ V_g \sum_k y_{1,k} \left[\left(A^* + V_g \right) y_{1,k'} \left(y_{1,k''} - z_{k''} \right) - V_g \left(y_{1,k'} - z_{k'} \right) \left(y_{1,k''} - z_{k''} \right) \right]}_{L_2}$$

$$\underbrace{+ V_g \left(V - \frac{d \ln \Gamma_1}{d \ln r} \right) \left(y_{1,\alpha} - z_\alpha \right) \left[y_{1,\beta} \left(y_{1,\gamma} - z_\gamma \right) + y_{1,\gamma} \left(y_{1,\beta} - z_\beta \right) - y_{1,\beta} y_{1,\gamma} \right]}_{L_3}$$

$$\underbrace{- y_{1,\alpha} y_{1,\beta} y_{1,\gamma} \left[\left(A^* + V_g \right)^2 + A^* \left(4 + c_1 \frac{\tilde{\omega}_\alpha^2}{\psi_\alpha} - U \right) \right]}_{L_4} \underbrace{+ A^* y_{4,\alpha} y_{1,\beta} y_{1,\gamma}}_{L_5}$$

$$+ A^* y_{1,\gamma} \left[\frac{\Lambda_\alpha + \Lambda_\beta - \Lambda_\gamma}{2 c_1 \tilde{\omega}_\alpha \tilde{\omega}_\beta} \psi_\beta y_{2,\beta} \left(y_{1,\alpha} - z_\alpha \right) + \frac{\Lambda_\alpha}{c_1 \tilde{\omega}_\alpha^2} \psi_\alpha y_{2,\alpha} y_{1,\beta} \right]$$

$$\left. + A^* y_{1,\beta} \frac{\Lambda_\gamma + \Lambda_\alpha - \Lambda_\beta}{2 c_1 \tilde{\omega}_\alpha \tilde{\omega}_\gamma} \psi_\gamma y_{2,\gamma} \left(y_{1,\alpha} - z_\alpha \right) \right\} g \rho r^3 dr. \Bigg\}_{L_\Lambda}$$

Adding the N terms to J_Λ and L_Λ, we get

$$\tilde{\mathcal{H}}_4 = Z_{\alpha\beta\gamma} \int \frac{A^*}{c_1^2} \sum_k \psi_k y_{2,k} \left(GC_k y_{1,k'} y_{1,k''} + QC_{k'} y_{1,k'} z_{k''} + QC_{k''} y_{1,k''} z_{k'} \right) \rho r^4 dr + N_{cor},$$

$$\text{(E.3.7)}$$

where

$$GC_k = \frac{\Lambda_k \varpi_k + \left(\Lambda_{k'} - \Lambda_{k''} \right) \left(\varpi_{k'} - \varpi_{k''} \right)}{2 \varpi_k \varpi_{k'} \varpi_{k''}} \qquad \text{(E.3.8)}$$

and

$$N_{\text{cor}} = Z_{\alpha\beta\gamma}\,\Xi_\beta \int \frac{A^*}{c_1^2}\left(z_\beta - y_{1,\beta}\right)\left(Q C_\alpha y_{1,\alpha} y_{2,\gamma}\psi_\gamma - \frac{\varpi_\gamma}{\varpi_\alpha} Q C_\gamma y_{1,\gamma} y_{2,\alpha}\psi_\alpha\right)\rho r^4 dr$$

$$+ Z_{\alpha\beta\gamma}\,\Xi_\gamma \int \frac{A^*}{c_1^2}\left(z_\gamma - y_{1,\gamma}\right)\left(Q C_\alpha y_{1,\alpha} y_{2,\beta}\psi_\beta - \frac{\varpi_\beta}{\varpi_\alpha} Q C_\beta y_{1,\beta} y_{2,\alpha}\psi_\alpha\right)\rho r^4 dr.$$

$$\text{(E.3.9)}$$

Furthermore, $J_2 + L_5$ gives

$$\tilde{\mathcal{H}}_{3.3} = Z_{\alpha\beta\gamma} \int \frac{A^*}{c_1} \sum_k y_{4,k} y_{1,k'} y_{1,k''} \rho r^4 dr. \tag{E.3.10}$$

Also, from $J_1 + L_4$, we obtain

$$\tilde{\mathcal{H}}_{3.1} \underbrace{- Z_{\alpha\beta\gamma} \int \frac{V_g^2}{c_1} \prod_k y_{1,k} \rho r^4 dr}_{R_1} \underbrace{- Z_{\alpha\beta\gamma} \int \frac{3 V_g A^*}{c_1} \prod_k y_{1,k} \rho r^4 dr}_{R_2},$$

where

$$\tilde{\mathcal{H}}_{3.1} = Z_{\alpha\beta\gamma} \int \frac{A^*}{c_1}\left(V_g + U - 4 - c_1 \sum_k \frac{\varpi_k^2}{\psi_k}\right) \prod_k y_{1,k} \rho r^4 dr. \tag{E.3.11}$$

Adding the terms J_3, M, L_1, and L_3, we get

$$\tilde{\mathcal{H}}_{2.2} \underbrace{+ Z_{\alpha\beta\gamma} \int \frac{V_g}{c_1}\left[V_g \prod_k (y_{1,k} - z_k) + \left(V - \frac{d\ln\Gamma_1}{d\ln r}\right)\prod_k z_k\right]\rho r^4 r}_{R_3},$$

where

$$\tilde{\mathcal{H}}_{2.2} = Z_{\alpha\beta\gamma} \int \frac{V_g}{c_1} A_g \prod_k (y_{1,k} - z_k)\,\rho r^4 dr, \tag{E.3.12}$$

with

$$A_g = -\frac{d\ln\Gamma_1}{d\ln r} - V_g \left(\frac{\partial\ln\Gamma_1}{\partial\ln\rho}\right)_\mu. \tag{E.3.13}$$

Moreover, from $J_4 + R_1 + R_3$, we obtain

$$\tilde{\mathcal{H}}_{2.1} + \underbrace{Z_{\alpha\beta\gamma} \int \frac{V_g^2}{c_1} \sum_k \left(y_{1,k} z_{k'} z_{k''} - y_{1,k} y_{1,k'} z_{k''} \right) \rho r^4 dr,}_{R_4}$$

where

$$\tilde{\mathcal{H}}_{2.1} = Z_{\alpha\beta\gamma} \int \frac{V_g}{c_1} \left(V - 2V_g - \frac{d \ln \Gamma_1}{d \ln r} \right) \prod_k z_k \rho r^4 dr. \tag{E.3.14}$$

Adding up the terms L_2, R_2, and R_4 also gives

$$\tilde{\mathcal{H}}_{3.2} = -Z_{\alpha\beta\gamma} \int \frac{A^*}{c_1} V_g \sum_k z_k y_{1,k'} y_{1,k''} \rho r^4 dr. \tag{E.3.15}$$

Finally (!), collecting all the $\tilde{\mathcal{H}}$ terms, as defined by Eqs. (E.3.1), (E.3.7), (E.3.10)–(E.3.12), (E.3.14) and (E.3.15), we obtain the desired expression for the zeroth-order component of the coupling coefficient (corrected due to rotation only through the eigenfrequencies), for polar mode coupling, as

$$
\begin{aligned}
\tilde{\mathcal{H}} = Z_{\alpha\beta\gamma} \int \Bigg\{ & -\sum_k \left(A^* y_{1,k} + V_g z_k \right) \Psi_k \left(\varpi_{k'} \varpi_{k''} y_{1,k'} y_{1,k''} + \frac{QC_k}{c_1^2} \psi_{k'} \psi_{k''} y_{2,k'} y_{2,k''} \right) \\
& + \frac{V_g}{c_1} \left[\left(V - 2V_g - \frac{d \ln \Gamma_1}{d \ln r} \right) \prod_k z_k + A_g \prod_k \left(y_{1,k} - z_k \right) \right] \\
& + \frac{A^*}{c_1} \left[\left(V_g + U - 4 - c_1 \sum_k \frac{\varpi_k^2}{\psi_k} \right) \prod_k y_{1,k} - V_g \sum_k z_k y_{1,k'} y_{1,k''} + \sum_k y_{4,k} y_{1,k'} y_{1,k''} \right] \\
& + \frac{A^*}{c_1^2} \sum_k \psi_k y_{2,k} \left(GC_k y_{1,k'} y_{1,k''} + QC_{k'} y_{1,k'} z_{k''} + QC_{k''} y_{1,k''} z_{k'} \right) \Bigg\} \rho r^4 dr + N_{\text{cor}}.
\end{aligned}
\tag{E.3.16}
$$

For the reader who skipped the derivation of Eq. (E.3.16), we have used the dimensionless formulation of Sect. 2.3.3. The auxiliary parameters z_k, ϖ_k, QC_k, GC_k, and A_g are given by Eqs. (E.2.6), (E.3.2), (E.3.3), (E.3.8) and (E.3.13), respectively, and the indices k, k', and k'' behave according to Eq. (E.2.11). The angular part of the coupling coefficient has been denoted by $Z_{\alpha\beta\gamma}$, which is given by Eq. (E.2.9). Equation (E.3.16) is identical with Eq. (3.12) in Dziembowski (1982), with the exception of some variables which parametrise rotational corrections to the eigenfrequencies, namely, ψ_k, Ψ_k, and N_{cor}, given by Eqs. (E.3.4), (E.3.5) and (E.3.9), respectively. If eigenfrequency corrections are not considered (in which case we obtain the actual zeroth-order component of the coupling coefficient), then $\psi_k \to 1$, $\Psi_k \to 1$, and $N_{\text{cor}} \to 0$.

Since, as mentioned in the beginning of the section, we have divided the coupling coefficient $F_{\alpha\beta\gamma}$ by GM/R^3 for this derivation, $\tilde{\mathcal{H}}$ is related to $F_{\alpha\beta\gamma}$ as

$$F_{\alpha\beta\gamma} = \frac{GM}{R^3}\tilde{\mathcal{H}} \equiv \mathcal{H}.$$

Also, it can be easily seen that Eq. (E.3.16) is invariant to the transformations

$$Y_\alpha \rightleftarrows Y_\beta, \quad y_{i,\alpha} \rightleftarrows y_{i,\beta}, \quad Y_\gamma \to Y_\gamma^*, \quad \tilde{\omega}_\gamma \to -\tilde{\omega}_\gamma,$$

and

$$Y_\alpha \rightleftarrows Y_\gamma, \quad y_{i,\alpha} \rightleftarrows y_{i,\gamma}, \quad Y_\beta \to Y_\beta^*, \quad \tilde{\omega}_\beta \to -\tilde{\omega}_\beta,$$

which proves that

$$F_{\alpha\beta\gamma} = F_{\beta\bar{\gamma}\alpha} = F_{\gamma\alpha\bar{\beta}} \equiv \mathcal{H}.$$

$\mathcal{H}$ has units of energy; the normalisation in Eq. (E.3.16) is useful when all quantities in the coupled triplet equations of motion (4.2.3) are normalised accordingly. Defining a dimensionless time $\tau = t\sqrt{GM/R^3}$, the equations of motion are written as

$$Q'_\alpha = \tilde{\gamma}_\alpha Q_\alpha + \frac{i\tilde{\mathcal{H}}}{\tilde{b}_\alpha} Q_\beta Q_\gamma e^{-i\Delta\tilde{\omega}\tau}, \tag{E.3.17a}$$

$$Q'_\beta = \tilde{\gamma}_\beta Q_\beta + \frac{i\tilde{\mathcal{H}}}{\tilde{b}_\beta} Q_\gamma^* Q_\alpha e^{i\Delta\tilde{\omega}\tau}, \tag{E.3.17b}$$

$$Q'_\gamma = \tilde{\gamma}_\gamma Q_\gamma + \frac{i\tilde{\mathcal{H}}}{\tilde{b}_\gamma} Q_\alpha Q_\beta^* e^{i\Delta\tilde{\omega}\tau}, \tag{E.3.17c}$$

where $\tilde{\gamma} = \gamma/\sqrt{GM/R^3}$, $\tilde{b} = b/\sqrt{GM/R^3}$, the dimensionless eigenfrequency $\tilde{\omega}$ is defined in Eq. (2.3.41), and the prime denotes differentiation with respect to τ. Now, Eqs. (E.3.17) coincide with Eqs. (2.25) and (2.26) in Dziembowski (1982), except they have been generalised for the case of a rotating star.[6]

References

1. Dziembowski, W. (1982). Nonlinear mode coupling in oscillating stars. I. Second order theory of the coherent mode coupling. *Acta Astronomica, 32*, 147–171. http://adsabs.harvard.edu/abs/1982AcA....32..147D.
2. Sakurai, J. J., & Napolitano, J. (2011). *Modern quantum mechanics* (2nd ed.). San Francisco: Addison-Wesley. http://adsabs.harvard.edu/abs/1985mqm..book.....S.

[6]Note that Dziembowski (1982) normalises with respect to $4\pi G\langle\rho\rangle$, where $\langle\rho\rangle$ is the mean density, instead of GM/R^3. He also uses a different convention for the spherical harmonics, which he normalises to 4π, instead of unity like us [see Eq. (2.3.14)]. Finally, he uses c_1 (in our notation) = $3(r/R)^3 M/M_r$, which is a factor of 3 larger than our c_1 [Eq. (2.3.39)].

Appendix F
Parametrically Unstable Mode Triplet

F.1 Parametric Instability Threshold

The amplitude equations of motion of the coupled triplet are given by Eqs. (4.2.12), namely

$$\dot{Q}_\alpha = \gamma_\alpha Q_\alpha + i\omega_\alpha \frac{\mathcal{H}}{E_{\text{unit}}} Q_\beta Q_\gamma e^{-i\Delta\omega t}, \qquad (\text{F.1.1a})$$

$$\dot{Q}_\beta = \gamma_\beta Q_\beta + i\omega_\beta \frac{\mathcal{H}}{E_{\text{unit}}} Q_\gamma^* Q_\alpha e^{i\Delta\omega t}, \qquad (\text{F.1.1b})$$

$$\dot{Q}_\gamma = \gamma_\gamma Q_\gamma + i\omega_\gamma \frac{\mathcal{H}}{E_{\text{unit}}} Q_\alpha Q_\beta^* e^{i\Delta\omega t}. \qquad (\text{F.1.1c})$$

In order to derive the formula for the parametric instability threshold (4.3.1), we take the amplitude equations of motion for the daughter modes (F.1.1b) and (F.1.1c) and ask what the value of the parent mode's amplitude Q_α should be, in order for the daughters' amplitudes $Q_{\beta,\gamma}$ start growing. Setting

$$Q_{\beta,\gamma} = \tilde{Q}_{\beta,\gamma} e^{i\Delta\omega t/2},$$

Eqs. (F.1.1b) and (F.1.1c) become

$$\dot{\tilde{Q}}_\beta = \left(\gamma_\beta - i\frac{\Delta\omega}{2}\right) \tilde{Q}_\beta + \frac{i\omega_\beta \mathcal{H}}{E_{\text{unit}}} Q_\alpha \tilde{Q}_\gamma^*$$

and

$$\dot{\tilde{Q}}_\gamma^* = \left(\gamma_\gamma + i\frac{\Delta\omega}{2}\right) \tilde{Q}_\gamma^* - \frac{i\omega_\gamma \mathcal{H}}{E_{\text{unit}}} Q_\alpha^* \tilde{Q}_\beta,$$

or, in matrix form,

© Springer Nature Switzerland AG 2018
P. Pnigouras, *Saturation of the f-mode Instability in Neutron Stars*,
Springer Theses, https://doi.org/10.1007/978-3-319-98258-8

$$\begin{pmatrix} \tilde{\dot{Q}}_\beta \\ \tilde{\dot{Q}}_\gamma^* \end{pmatrix} = \begin{pmatrix} \gamma_\beta - i\,\Delta\omega/2 & i\,Q_\alpha\omega_\beta\mathcal{H}/E_{\text{unit}} \\ -i\,Q_\alpha^*\omega_\gamma\mathcal{H}/E_{\text{unit}} & \gamma_\gamma + i\,\Delta\omega/2 \end{pmatrix} \begin{pmatrix} \tilde{Q}_\beta \\ \tilde{Q}_\gamma^* \end{pmatrix},$$

with Q_α treated as an unknown constant. If T is the trace and d is the determinant of the system matrix, then its eigenvalues $\lambda_{1,2}$ can be found as

$$\lambda_{1,2} = \frac{1}{2}\left(T \pm \sqrt{T^2 - 4d}\right),$$

or

$$\lambda_{1,2} = \frac{1}{2}\left[\gamma_\beta + \gamma_\gamma \pm \sqrt{\left(\gamma_\gamma - \gamma_\beta + i\,\Delta\omega\right)^2 + \frac{4\omega_\beta\omega_\gamma\mathcal{H}^2}{E_{\text{unit}}^2}|Q_\alpha|^2}\,\right]. \qquad \text{(F.1.2)}$$

For the system to admit a growing exponential solution, i.e., for the daughter modes to grow, the condition $\text{Re}(\lambda) > 0$ has to be satisfied, where Re denotes the real part, for at least one of the eigenvalues. Hence, at the onset of the parametric instability, $\text{Re}(\lambda) = 0$, or

$$\left(\gamma_\beta + \gamma_\gamma\right)^2 = \left[\text{Re}\left(\sqrt{\left(\gamma_\gamma - \gamma_\beta + i\,\Delta\omega\right)^2 + \frac{4\omega_\beta\omega_\gamma\mathcal{H}^2}{E_{\text{unit}}^2}|Q_\alpha|^2}\,\right)\right]^2. \qquad \text{(F.1.3)}$$

We set the radicand in the expression above equal to the complex number $u + iv$. Then, let $\sqrt{u + iv} = x + iy$, or $u + iv = x^2 - y^2 + i2xy$. Distinguishing the real from the imaginary part and omitting y, we get

$$4x^4 - 4ux^2 - v^2 = 0.$$

From Eq. (F.1.3), we have

$$u = \left(\gamma_\gamma - \gamma_\beta\right)^2 - \Delta\omega^2 + \frac{4\omega_\beta\omega_\gamma\mathcal{H}^2}{E_{\text{unit}}^2}|Q_\alpha|^2,$$

$$v = 2\left(\gamma_\gamma - \gamma_\beta\right)\Delta\omega,$$

and

$$x^2 = \left(\gamma_\beta + \gamma_\gamma\right)^2.$$

Thus, replacing u, v, and x in the quartic equation for x above, we obtain

$$|Q_\alpha|^2 = \frac{\gamma_\beta\gamma_\gamma}{\omega_\beta\omega_\gamma}\frac{E_{\text{unit}}^2}{\mathcal{H}^2}\left[1 + \left(\frac{\Delta\omega}{\gamma_\beta + \gamma_\gamma}\right)^2\right] \equiv |Q_{\text{PIT}}|^2. \qquad \text{(F.1.4)}$$

Note the importance of the mode eigenfrequency signs here: if $\omega_\beta \omega_\gamma < 0$, then no parametric instability can occur. This is a result of the assumed resonance (4.2.1) between the parent and the daughters. If we perform the same analysis, for example, for mode β being the parent, then $\omega_\beta \approx \omega_\alpha - \omega_\gamma$, in which case $\omega_\alpha \omega_\gamma < 0$ is a *necessary* condition for parametric instability.

F.2 Equilibrium Solution

The amplitude equations of motion (4.2.12) [or (F.1.1)] admit an easy-to-obtain equilibrium solution. Expressing the complex amplitudes Q in terms of real amplitude and phase variables, we can introduce the variable transformation

$$Q_\alpha = \frac{1}{\sqrt{\omega_\beta \omega_\gamma}} \frac{E_{\text{unit}}}{\mathcal{H}} \varepsilon_\alpha e^{i\vartheta_\alpha}, \tag{F.2.1a}$$

$$Q_\beta = \frac{1}{\sqrt{\omega_\gamma \omega_\alpha}} \frac{E_{\text{unit}}}{\mathcal{H}} \varepsilon_\beta e^{i\vartheta_\beta}, \tag{F.2.1b}$$

$$Q_\gamma = \frac{1}{\sqrt{\omega_\alpha \omega_\beta}} \frac{E_{\text{unit}}}{\mathcal{H}} \varepsilon_\gamma e^{i\vartheta_\gamma}. \tag{F.2.1c}$$

Then, Eqs. (F.1.1) are written as

$$\dot{\varepsilon}_\alpha = \gamma_\alpha \varepsilon_\alpha + \varepsilon_\beta \varepsilon_\gamma \sin \varphi, \tag{F.2.2a}$$

$$\dot{\varepsilon}_\beta = \gamma_\beta \varepsilon_\beta - \varepsilon_\gamma \varepsilon_\alpha \sin \varphi, \tag{F.2.2b}$$

$$\dot{\varepsilon}_\gamma = \gamma_\gamma \varepsilon_\gamma - \varepsilon_\alpha \varepsilon_\beta \sin \varphi, \tag{F.2.2c}$$

and

$$\dot{\varphi} = \cos \varphi \left(\frac{\varepsilon_\beta \varepsilon_\gamma}{\varepsilon_\alpha} - \frac{\varepsilon_\gamma \varepsilon_\alpha}{\varepsilon_\beta} - \frac{\varepsilon_\alpha \varepsilon_\beta}{\varepsilon_\gamma} \right) + \Delta\omega, \tag{F.2.2d}$$

or, equivalently,

$$\dot{\varphi} = \cot \varphi \left(\frac{\dot{\varepsilon}_\alpha}{\varepsilon_\alpha} + \frac{\dot{\varepsilon}_\beta}{\varepsilon_\beta} + \frac{\dot{\varepsilon}_\gamma}{\varepsilon_\gamma} - \gamma \right) + \Delta\omega, \tag{F.2.2d$'$}$$

where

$$\varphi = \vartheta_\alpha - \vartheta_\beta - \vartheta_\gamma + \Delta\omega t \tag{F.2.3}$$

and

$$\gamma = \gamma_\alpha + \gamma_\beta + \gamma_\gamma. \tag{F.2.4}$$

Setting the time derivatives in Eqs. (F.2.2) to zero, we get

$$-\gamma_\alpha \varepsilon_\alpha = \varepsilon_\beta \varepsilon_\gamma \sin\varphi, \tag{F.2.5a}$$

$$\gamma_\beta \varepsilon_\beta = \varepsilon_\gamma \varepsilon_\alpha \sin\varphi, \tag{F.2.5b}$$

$$\gamma_\gamma \varepsilon_\gamma = \varepsilon_\alpha \varepsilon_\beta \sin\varphi, \tag{F.2.5c}$$

and

$$\cot\varphi = \kappa, \tag{F.2.5d}$$

where

$$\kappa = \frac{\Delta\omega}{\gamma}. \tag{F.2.6}$$

Then, combining Eqs. (F.2.5a)–(F.2.5c) in pairs and using the trigonometric identity

$$\frac{1}{\sin^2\varphi} = 1 + \cot^2\varphi = 1 + \kappa^2,$$

we obtain the equilibrium solution

$$\varepsilon_\alpha^2 = \gamma_\beta \gamma_\gamma \left(1 + \kappa^2\right), \tag{F.2.7a}$$

$$\varepsilon_\beta^2 = -\gamma_\gamma \gamma_\alpha \left(1 + \kappa^2\right), \tag{F.2.7b}$$

$$\varepsilon_\gamma^2 = -\gamma_\alpha \gamma_\beta \left(1 + \kappa^2\right), \tag{F.2.7c}$$

which, in terms of the original variables Q, admits the form (4.3.3).

Combining Eqs. (F.2.2b) and (F.2.2c) with Eq. (F.2.2a), we can further show that

$$\frac{1}{2}\frac{d}{dt}\left(\varepsilon_\alpha^2 + \varepsilon_\beta^2\right) = \gamma_\alpha \varepsilon_\alpha^2 + \gamma_\beta \varepsilon_\beta^2$$

and

$$\frac{1}{2}\frac{d}{dt}\left(\varepsilon_\alpha^2 + \varepsilon_\gamma^2\right) = \gamma_\alpha \varepsilon_\alpha^2 + \gamma_\gamma \varepsilon_\gamma^2,$$

or, restoring the original variables Q with the help of Eqs. (F.2.1),

$$\frac{1}{2}\frac{d}{dt}\left(\omega_\beta |Q_\alpha|^2 + \omega_\alpha |Q_\beta|^2\right) = \gamma_\alpha \omega_\beta |Q_\alpha|^2 + \gamma_\beta \omega_\alpha |Q_\beta|^2$$

and

$$\frac{1}{2}\frac{d}{dt}\left(\omega_\gamma |Q_\alpha|^2 + \omega_\alpha |Q_\gamma|^2\right) = \gamma_\alpha \omega_\gamma |Q_\alpha|^2 + \gamma_\gamma \omega_\alpha |Q_\gamma|^2.$$

Adding the equations above and using $\omega_\beta + \omega_\gamma \approx \omega_\alpha$, we obtain

$$\frac{1}{2}\frac{d}{dt}\left(|Q_\alpha|^2 + |Q_\beta|^2 + |Q_\gamma|^2\right) = \gamma_\alpha|Q_\alpha|^2 + \gamma_\beta|Q_\beta|^2 + \gamma_\gamma|Q_\gamma|^2, \qquad \text{(F.2.8)}$$

which, multiplied by E_{unit}, gives the rate of change of the triplet energy (4.4.11).

Finally, we can incorporate the phases ϑ, as defined in Eqs. (F.2.1), in the harmonic time dependence of the modes, as

$$\xi_k \propto e^{i(\omega_k t + \vartheta_k)},$$

where $k = \alpha, \beta, \gamma$. This implies that the eigenfrequency of the mode is shifted to

$$\omega'_k = \omega_k + \dot{\vartheta}_k. \qquad \text{(F.2.9)}$$

From Eq. (F.2.2d) we already know that

$$\dot{\vartheta}_\alpha = \frac{\varepsilon_\beta \varepsilon_\gamma}{\varepsilon_\alpha} \cos\varphi,$$

$$\dot{\vartheta}_\beta = \frac{\varepsilon_\gamma \varepsilon_\alpha}{\varepsilon_\beta} \cos\varphi,$$

$$\dot{\vartheta}_\gamma = \frac{\varepsilon_\alpha \varepsilon_\beta}{\varepsilon_\gamma} \cos\varphi,$$

or, in compact form,

$$\dot{\vartheta}_k = \frac{\varepsilon_\alpha \varepsilon_\beta \varepsilon_\gamma}{\varepsilon_k^2} \cos\varphi. \qquad \text{(F.2.10)}$$

In terms of the original variables Q, Eq. (F.2.10) is written as Eq. (4.4.13). Replacing the equilibrium values (F.2.7) and (F.2.5d) in Eq. (F.2.10), we get

$$\dot{\vartheta}_k = |\gamma_k|\sqrt{1 + \kappa^2} \cos\varphi_0,$$

where

$$\cot\varphi_0 = \kappa.$$

We use the trigonometric identity

$$\cos^2\varphi = \frac{\cot^2\varphi}{1 + \cot^2\varphi}$$

to obtain

$$\cos\varphi_0 = \pm\frac{|\kappa|}{\sqrt{1 + \kappa^2}}.$$

We notice from Eq. (F.2.5a) that, in equilibrium, $\sin \varphi_0 < 0$, which means that, if $\kappa = \cot \varphi_0 > 0$, then $\cos \varphi_0 < 0$, and vice versa. In other words, $\mathrm{sgn}(\cos \varphi_0) = -\mathrm{sgn}(\kappa)$, where sgn is the sign function. So, finally, we obtain for the eigenfrequency shift in equilibrium

$$\dot{\vartheta}_k = -|\gamma_k|\kappa. \tag{F.2.11}$$

F.3 Linear Stability Analysis

We will linearise Eqs. (F.2.2) by imposing small perturbations about their equilibrium solutions (F.2.5). Denoting these perturbations by δ (not to be confused with a Eulerian perturbation), we get

$$\frac{\mathrm{d}\delta\varepsilon_\alpha}{\mathrm{d}t} - \gamma_\alpha \delta\varepsilon_\alpha = \delta\varepsilon_\beta\varepsilon_\gamma \sin \varphi + \varepsilon_\beta \delta\varepsilon_\gamma \sin \varphi + \varepsilon_\beta\varepsilon_\gamma \cos \varphi \delta\varphi.$$

Dividing by ε_α and using Eq. (F.2.5a), we obtain

$$\frac{\mathrm{d}}{\mathrm{d}t}\left(\frac{\delta\varepsilon_\alpha}{\varepsilon_\alpha}\right) = -\gamma_\alpha \left(-\frac{\delta\varepsilon_\alpha}{\varepsilon_\alpha} + \frac{\delta\varepsilon_\beta}{\varepsilon_\beta} + \frac{\delta\varepsilon_\gamma}{\varepsilon_\gamma} + \kappa\delta\varphi\right). \tag{F.3.1a}$$

Note that ε_α above corresponds to the equilibrium solution and, hence, is a constant. In a similar manner, we get

$$\frac{\mathrm{d}}{\mathrm{d}t}\left(\frac{\delta\varepsilon_\beta}{\varepsilon_\beta}\right) = -\gamma_\beta \left(\frac{\delta\varepsilon_\alpha}{\varepsilon_\alpha} - \frac{\delta\varepsilon_\beta}{\varepsilon_\beta} + \frac{\delta\varepsilon_\gamma}{\varepsilon_\gamma} + \kappa\delta\varphi\right), \tag{F.3.1b}$$

$$\frac{\mathrm{d}}{\mathrm{d}t}\left(\frac{\delta\varepsilon_\gamma}{\varepsilon_\gamma}\right) = -\gamma_\gamma \left(\frac{\delta\varepsilon_\alpha}{\varepsilon_\alpha} + \frac{\delta\varepsilon_\beta}{\varepsilon_\beta} - \frac{\delta\varepsilon_\gamma}{\varepsilon_\gamma} + \kappa\delta\varphi\right). \tag{F.3.1c}$$

The linearisation of Eq. (F.2.2d') about its equilibrium (F.2.5d) yields

$$\frac{\mathrm{d}\delta\varphi}{\mathrm{d}t} - \kappa\frac{\mathrm{d}}{\mathrm{d}t}\left(\frac{\delta\varepsilon_\alpha}{\varepsilon_\alpha} + \frac{\delta\varepsilon_\beta}{\varepsilon_\beta} + \frac{\delta\varepsilon_\gamma}{\varepsilon_\gamma}\right) = \gamma\left(1 + \kappa^2\right)\delta\varphi,$$

where we used the fact that

$$\cot(\varphi + \delta\varphi) = \kappa - \left(1 + \kappa^2\right)\delta\varphi + \mathcal{O}\left(\delta\varphi^2\right).$$

With the help of Eqs. (F.3.1a)–(F.3.1c), the equation above becomes

$$\frac{\mathrm{d}\delta\varphi}{\mathrm{d}t} = \kappa \sum_k \Gamma_k \frac{\delta\varepsilon_k}{\varepsilon_k} + \gamma\delta\varphi, \tag{F.3.1d}$$

where

$$\Gamma_k = 2\gamma_k - \gamma, \tag{F.3.2}$$

with the index k successively taking the values α, β, γ.

The matrix of the linear system (F.3.1) is

$$
A = \begin{pmatrix}
\gamma_\alpha & -\gamma_\alpha & -\gamma_\alpha & -\kappa\gamma_\alpha \\
-\gamma_\beta & \gamma_\beta & -\gamma_\beta & -\kappa\gamma_\beta \\
-\gamma_\gamma & -\gamma_\gamma & \gamma_\gamma & -\kappa\gamma_\gamma \\
\kappa\Gamma_\alpha & \kappa\Gamma_\beta & \kappa\Gamma_\gamma & \gamma
\end{pmatrix},
$$

with the help of which we can find the system's characteristic polynomial, via the relation $|A - \lambda I| = 0$, where λ are the eigenvalues of A and I is the identity matrix. The polynomial has the form

$$\lambda^4 + a_1\lambda^3 + a_2\lambda^2 + a_3\lambda + a_4 = 0,$$

where

$$a_1 = -2\gamma,$$
$$a_2 = \gamma^2\left(1 + \kappa^2\right) - 4\kappa^2 \sum_k \gamma_k\gamma_{k'},$$
$$a_3 = 4\left(1 + 3\kappa^2\right)\prod_k \gamma_k,$$

and

$$a_4 = -4\left(1 + \kappa^2\right)\gamma\prod_k \gamma_k,$$

with the index k' taking the value that comes after k's value, as explained in Eq. (E.2.11).

Now, we can use the Routh-Hurwitz stability criteria (see, for instance, Horn and Johnson 1991, Sect. 2.3), in order to determine the behaviour of the system. First, we construct the Routh-Hurwitz matrix, using the polynomial coefficients, as

$$
M = \begin{pmatrix}
a_1 & 1 & 0 & 0 \\
a_3 & a_2 & a_1 & 1 \\
0 & a_4 & a_3 & a_2 \\
0 & 0 & 0 & a_4
\end{pmatrix}.
$$

Then, the stability criteria are given by

$$W_1 \equiv a_1 > 0, \tag{F.3.3}$$

$$W_2 \equiv \begin{vmatrix} a_1 & 1 \\ a_3 & a_2 \end{vmatrix} = a_1 a_2 - a_3 > 0, \tag{F.3.4}$$

$$W_3 \equiv \begin{vmatrix} a_1 & 1 & 0 \\ a_3 & a_2 & a_1 \\ 0 & a_4 & a_3 \end{vmatrix} = a_3 W_2 - a_1^2 a_4 > 0, \tag{F.3.5}$$

and

$$W_4 \equiv |M| = a_4 W_3 > 0. \tag{F.3.6}$$

Since $\gamma_{\beta,\gamma} < 0$, it can be easily shown that the second and fourth criteria are redundant and follow from the other ones. Indeed, if $W_1 > 0$ then a_4 is also positive, which, combined with $W_3 > 0$, makes the fourth criterion true. Also, $W_3 > 0$ yields $W_2 > a_1^2 a_4 / a_3$, but since $a_3 > 0$, the second criterion is also true. So, finally, from Eqs. (F.3.3) and (F.3.5), we obtain the stability conditions (4.4.1) and (4.4.2).

Reference

1. Horn, R. A., & Johnson, C. R. (1991). *Topics in matrix analysis*. Cambridge, England: Cambridge University Press. https://doi.org/10.1017/CBO9780511840371.

Appendix G
Coupling Spectrum

In the following table we present the *coupling spectrum* of the octupole ($l = m = 3$) f-mode, in a typical neutron star with $M \approx 1.4\, M_\odot$ and $R \approx 10\,\mathrm{km}$, described by a polytropic equation of state with a polytropic exponent $\Gamma = 3$ and an adiabatic exponent $\Gamma_1 = 3.1$ (see Table 5.1). This is the coupling spectrum used to generate Figs. 5.2 (right) and 5.4. The following data is presented in the table, by column:

1. Angular velocity Ω, normalised to the Kepler limit Ω_K.
2. Temperature (decimal) logarithm $\log(T/1\,\mathrm{K})$.
3–4. Daughter pair; the notation $^m_l f$ and $^m_l g_n$ is used for f- and g-modes respectively.
5. Triplet detuning $\Delta\tilde{\omega} = \Delta\omega/\sqrt{GM/R^3}$.
6. Coupling coefficient $\mathcal{H}$, normalised to $E_\mathrm{unit} = Mc^2$.
7–9. Parent (α) and daughter (β, γ) growth/damping rates $\tilde{\gamma}_k = \gamma_k/\sqrt{GM/R^3}$.
10. Lowest stable parametric instability threshold $|Q_\mathrm{PIT}|$.

For this model, the normalisation factor $\sqrt{GM/R^3}$ evaluates to $13\,416\,\mathrm{rad\,s^{-1}} \approx 2\,\mathrm{kHz}$.

For all models used in Chap. 5, the instability window in the ($\log T$, Ω/Ω_K) plane was divided into blocks with dimensions $(0.1, 0.002)$, forming a grid. For typical neutron star models, the maximum allowed detuning (see Sect. 5.1.3) was set to $\Delta\tilde{\omega}_\mathrm{max} = 0.1$, whereas for supramassive neutron star models we chose $\Delta\tilde{\omega}_\mathrm{max} = 0.2$.

© Springer Nature Switzerland AG 2018

P. Pnigouras, *Saturation of the f-mode Instability in Neutron Stars*,
Springer Theses, https://doi.org/10.1007/978-3-319-98258-8

Table G.1 Coupling spectrum of the octupole f-mode in a typical neutron star (see text for model details)

Ω/Ω_K (%)	$\log T$ (K)	Daughters		$\Delta\tilde{\omega}$	$\mathcal{H}/E_{unit}$	$\tilde{\gamma}_\alpha$	$\tilde{\gamma}_\beta$	$\tilde{\gamma}_\gamma$	$|Q_{PIT}|$
93.4	9.4	$^{-3}_{3}f$	$^{6}_{6}g_{10}$	-2.1×10^{-2}	3.3	6.1×10^{-14}	-3.4×10^{-4}	-2.5×10^{-13}	2.3×10^{-7}
93.4	9.5	$^{-3}_{3}f$	$^{6}_{6}g_{10}$	-2.1×10^{-2}	3.3	7.5×10^{-14}	-3.4×10^{-4}	-8.9×10^{-13}	4.4×10^{-7}
93.6	9.3	$^{-3}_{3}f$	$^{6}_{6}g_{10}$	-2.4×10^{-2}	3.3	8.1×10^{-14}	-3.4×10^{-4}	-9.6×10^{-14}	1.6×10^{-7}
93.6	9.4	$^{-3}_{3}f$	$^{6}_{6}g_{10}$	-2.4×10^{-2}	3.3	1.7×10^{-13}	-3.4×10^{-4}	-2.4×10^{-13}	2.6×10^{-7}
93.6	9.5	$^{-3}_{3}f$	$^{6}_{6}g_{10}$	-2.4×10^{-2}	3.3	1.8×10^{-13}	-3.4×10^{-4}	-8.9×10^{-13}	5.0×10^{-7}
93.6	9.6	$^{-3}_{3}f$	$^{6}_{6}g_{10}$	-2.4×10^{-2}	3.3	3.3×10^{-14}	-3.4×10^{-4}	-3.5×10^{-12}	9.8×10^{-7}
93.8	9.2	$^{-3}_{3}f$	$^{6}_{6}g_{10}$	-2.8×10^{-2}	3.4	7.5×10^{-14}	-3.4×10^{-4}	-7.8×10^{-14}	1.7×10^{-7}
93.8	9.3	$^{-3}_{3}f$	$^{6}_{6}g_{10}$	-2.8×10^{-2}	3.4	2.3×10^{-13}	-3.4×10^{-4}	-9.5×10^{-14}	1.8×10^{-7}
93.8	9.4	$^{-3}_{3}f$	$^{6}_{6}g_{10}$	-2.8×10^{-2}	3.4	3.1×10^{-13}	-3.4×10^{-4}	-2.4×10^{-13}	2.9×10^{-7}
93.8	9.5	$^{-3}_{3}f$	$^{6}_{6}g_{10}$	-2.8×10^{-2}	3.4	3.3×10^{-13}	-3.4×10^{-4}	-8.8×10^{-13}	5.5×10^{-7}
93.8	9.6	$^{-3}_{3}f$	$^{6}_{6}g_{10}$	-2.8×10^{-2}	3.4	1.8×10^{-13}	-3.4×10^{-4}	-3.5×10^{-12}	1.1×10^{-6}
94.0	9.1	$^{-3}_{3}f$	$^{6}_{6}g_{10}$	-3.1×10^{-2}	3.4	3.1×10^{-14}	-3.4×10^{-4}	-1.0×10^{-13}	2.1×10^{-7}
94.0	9.2	$^{-3}_{3}f$	$^{6}_{6}g_{10}$	-3.1×10^{-2}	3.4	2.8×10^{-13}	-3.4×10^{-4}	-7.7×10^{-14}	1.8×10^{-7}
94.0	9.3	$^{-3}_{3}f$	$^{6}_{6}g_{10}$	-3.1×10^{-2}	3.4	4.3×10^{-13}	-3.4×10^{-4}	-9.4×10^{-14}	2.0×10^{-7}
94.0	9.4	$^{-3}_{3}f$	$^{6}_{6}g_{10}$	-3.1×10^{-2}	3.4	5.2×10^{-13}	-3.4×10^{-4}	-2.4×10^{-13}	3.2×10^{-7}
94.0	9.5	$^{-3}_{3}f$	$^{6}_{6}g_{10}$	-3.1×10^{-2}	3.4	5.3×10^{-13}	-3.4×10^{-4}	-8.7×10^{-13}	6.1×10^{-7}
94.0	9.6	$^{-3}_{3}f$	$^{6}_{6}g_{10}$	-3.1×10^{-2}	3.4	3.8×10^{-13}	-3.4×10^{-4}	-3.4×10^{-12}	1.2×10^{-6}
94.2	9.1	$^{-4}_{4}f$	$^{7}_{7}g_{7}$	4.8×10^{-3}	-2.9	3.1×10^{-13}	-4.9×10^{-5}	-3.9×10^{-13}	2.0×10^{-7}
94.2	9.2	$^{-4}_{4}f$	$^{7}_{7}g_{7}$	4.8×10^{-3}	-2.9	5.6×10^{-13}	-4.9×10^{-5}	-3.4×10^{-13}	1.9×10^{-7}
94.2	9.3	$^{-3}_{3}f$	$^{6}_{6}g_{10}$	-3.5×10^{-2}	3.4	7.1×10^{-13}	-3.4×10^{-4}	-9.3×10^{-14}	2.2×10^{-7}
94.2	9.4	$^{-3}_{3}f$	$^{6}_{6}g_{10}$	-3.5×10^{-2}	3.4	8.0×10^{-13}	-3.4×10^{-4}	-2.4×10^{-13}	3.5×10^{-7}
94.2	9.5	$^{-3}_{3}f$	$^{6}_{6}g_{10}$	-3.5×10^{-2}	3.4	8.1×10^{-13}	-3.4×10^{-4}	-8.7×10^{-13}	6.7×10^{-7}

(continued)

Table G.1 (continued)

Ω/Ω_K (%)	$\log T$ (K)	Daughters		$\Delta\tilde{\omega}$	$\mathcal{H}/E_{unit}$	$\tilde{\gamma}_\alpha$	$\tilde{\gamma}_\beta$	$\tilde{\gamma}_\gamma$	$|\mathcal{Q}_{PIT}|$
94.2	9.6	$^{-3}_{3}f$	$^{6}_{6}g_{10}$	-3.5×10^{-2}	3.4	6.6×10^{-4}	-3.4×10^{-13}	-3.4×10^{-12}	1.3×10^{-6}
94.4	9.0	$^{-4}_{4}f$	$^{7}_{7}g_{7}$	2.7×10^{-3}	-2.9	2.9×10^{-13}	-5.0×10^{-5}	-5.8×10^{-13}	1.4×10^{-7}
94.4	9.1	$^{-4}_{4}f$	$^{7}_{7}g_{7}$	2.7×10^{-3}	-2.9	6.9×10^{-13}	-5.0×10^{-5}	-3.9×10^{-13}	1.1×10^{-7}
94.4	9.2	$^{-2}_{2}f$	$^{5}_{5}g_{10}$	9.7×10^{-2}	-6.7	9.4×10^{-13}	-2.6×10^{-3}	-3.6×10^{-14}	1.0×10^{-7}
94.4	9.3	$^{-2}_{2}f$	$^{5}_{5}g_{10}$	9.7×10^{-2}	-6.7	1.1×10^{-12}	-2.6×10^{-3}	-4.6×10^{-14}	1.2×10^{-7}
94.4	9.4	$^{-2}_{2}f$	$^{5}_{5}g_{10}$	9.7×10^{-2}	-6.7	1.2×10^{-12}	-2.6×10^{-3}	-1.2×10^{-13}	1.9×10^{-7}
94.4	9.5	$^{-2}_{2}f$	$^{5}_{5}g_{10}$	9.7×10^{-2}	-6.7	1.2×10^{-12}	-2.6×10^{-3}	-4.5×10^{-13}	3.6×10^{-7}
94.4	9.6	$^{-2}_{2}f$	$^{5}_{5}g_{10}$	9.7×10^{-2}	-6.7	1.0×10^{-12}	-2.6×10^{-3}	-1.8×10^{-12}	7.1×10^{-7}
94.4	9.7	$^{-2}_{2}f$	$^{5}_{5}g_{10}$	9.7×10^{-2}	-6.7	3.1×10^{-13}	-2.6×10^{-3}	-7.0×10^{-12}	1.4×10^{-6}
94.6	8.9	$^{-4}_{4}f$	$^{7}_{7}g_{7}$	4.9×10^{-4}	-3.0	1.7×10^{-13}	-5.0×10^{-5}	-9.1×10^{-13}	3.1×10^{-8}
94.6	9.0	$^{-4}_{4}f$	$^{7}_{7}g_{7}$	4.9×10^{-4}	-3.0	8.0×10^{-13}	-5.0×10^{-5}	-5.8×10^{-13}	2.5×10^{-8}
94.6	9.1	$^{-4}_{4}f$	$^{7}_{7}g_{7}$	4.9×10^{-4}	-3.0	1.2×10^{-12}	-5.0×10^{-5}	-3.9×10^{-13}	2.0×10^{-8}
94.6	9.2	$^{-4}_{4}f$	$^{7}_{7}g_{7}$	4.9×10^{-4}	-3.0	1.5×10^{-12}	-5.0×10^{-5}	-3.3×10^{-13}	1.9×10^{-8}
94.6	9.3	$^{-4}_{4}f$	$^{7}_{7}g_{7}$	4.9×10^{-4}	-3.0	1.6×10^{-12}	-5.0×10^{-5}	-5.6×10^{-13}	2.4×10^{-8}
94.6	9.4	$^{-4}_{4}f$	$^{7}_{7}g_{7}$	4.9×10^{-4}	-3.0	1.7×10^{-12}	-5.0×10^{-5}	-1.8×10^{-12}	4.3×10^{-8}
94.6	9.5	$^{-4}_{4}f$	$^{7}_{7}g_{7}$	4.9×10^{-4}	-3.0	1.7×10^{-12}	-5.0×10^{-5}	-6.7×10^{-12}	8.4×10^{-8}
94.6	9.6	$^{-4}_{4}f$	$^{7}_{7}g_{7}$	4.9×10^{-4}	-3.0	1.6×10^{-12}	-5.0×10^{-5}	-2.7×10^{-11}	1.7×10^{-7}
94.6	9.7	$^{-4}_{4}f$	$^{7}_{7}g_{7}$	4.9×10^{-4}	-3.0	8.2×10^{-13}	-5.0×10^{-5}	-1.1×10^{-10}	3.3×10^{-7}
94.8	8.9	$^{-4}_{4}f$	$^{7}_{7}g_{7}$	-1.9×10^{-3}	-3.0	8.6×10^{-13}	-5.0×10^{-5}	-9.0×10^{-13}	1.2×10^{-7}
94.8	9.0	$^{-4}_{4}f$	$^{7}_{7}g_{7}$	-1.9×10^{-3}	-3.0	1.5×10^{-12}	-5.0×10^{-5}	-5.7×10^{-13}	9.4×10^{-8}
94.8	9.1	$^{-4}_{4}f$	$^{7}_{7}g_{7}$	-1.9×10^{-3}	-3.0	1.9×10^{-12}	-5.0×10^{-5}	-3.8×10^{-13}	7.7×10^{-8}
94.8	9.2	$^{-4}_{4}f$	$^{7}_{7}g_{7}$	-1.9×10^{-3}	-3.0	2.2×10^{-12}	-5.0×10^{-5}	-3.3×10^{-13}	7.1×10^{-8}

(continued)

Table G.1 (continued)

Ω/Ω_K (%)	$\log T$ (K)	Daughters		$\Delta\tilde{\omega}$	$\mathcal{H}/E_{\text{unit}}$	$\tilde{\gamma}_\alpha$	$\tilde{\gamma}_\beta$	$\tilde{\gamma}_\gamma$	$\lvert Q_{\text{PIT}}\rvert$
94.8	9.3	$^{-4}_{4}f$	$^{7}_{7}g7$	-1.9×10^{-3}	-3.0	2.3×10^{-12}	-5.0×10^{-5}	-5.6×10^{-13}	9.3×10^{-8}
94.8	9.4	$^{-2}_{2}f$	$^{5}_{5}g10$	8.7×10^{-2}	-6.9	2.4×10^{-12}	-2.6×10^{-3}	-1.2×10^{-13}	1.6×10^{-7}
94.8	9.5	$^{-2}_{2}f$	$^{5}_{5}g10$	8.7×10^{-2}	-6.9	2.4×10^{-12}	-2.6×10^{-3}	-4.4×10^{-13}	3.1×10^{-7}
94.8	9.6	$^{-2}_{2}f$	$^{5}_{5}g10$	8.7×10^{-2}	-6.9	2.3×10^{-12}	-2.6×10^{-3}	-1.7×10^{-12}	6.2×10^{-7}
94.8	9.7	$^{-2}_{2}f$	$^{5}_{5}g10$	8.7×10^{-2}	-6.9	1.5×10^{-12}	-2.6×10^{-3}	-6.8×10^{-12}	1.2×10^{-6}
95.0	8.8	$^{-5}_{5}f$	$^{8}_{8}g3$	2.4×10^{-4}	-4.9	7.9×10^{-13}	-7.5×10^{-6}	-4.2×10^{-12}	5.7×10^{-8}
95.0	8.9	$^{-5}_{5}f$	$^{8}_{8}g3$	2.4×10^{-4}	-4.9	1.8×10^{-12}	-7.5×10^{-6}	-2.7×10^{-12}	4.6×10^{-8}
95.0	9.0	$^{-5}_{5}f$	$^{8}_{8}g3$	2.4×10^{-4}	-4.9	2.4×10^{-12}	-7.5×10^{-6}	-1.7×10^{-12}	3.7×10^{-8}
95.0	9.1	$^{-5}_{5}f$	$^{8}_{8}g3$	2.4×10^{-4}	-4.9	2.9×10^{-12}	-7.5×10^{-6}	-1.3×10^{-12}	3.2×10^{-8}
95.0	9.2	$^{-5}_{5}f$	$^{8}_{8}g3$	2.4×10^{-4}	-4.9	3.1×10^{-12}	-7.5×10^{-6}	-1.7×10^{-12}	3.6×10^{-8}
95.0	9.3	$^{-5}_{5}f$	$^{8}_{8}g3$	2.4×10^{-4}	-4.9	3.3×10^{-12}	-7.5×10^{-6}	-4.4×10^{-12}	5.8×10^{-8}
95.0	9.4	$^{-5}_{5}f$	$^{8}_{8}g3$	2.4×10^{-4}	-4.9	3.4×10^{-12}	-7.5×10^{-6}	-1.6×10^{-11}	1.1×10^{-7}
95.0	9.5	$^{-5}_{5}f$	$^{8}_{8}g3$	2.4×10^{-4}	-4.9	3.4×10^{-12}	-7.5×10^{-6}	-6.3×10^{-11}	2.2×10^{-7}
95.0	9.6	$^{-5}_{5}f$	$^{8}_{8}g3$	2.4×10^{-4}	-4.9	3.2×10^{-12}	-7.5×10^{-6}	-2.5×10^{-10}	4.4×10^{-7}
95.0	9.7	$^{-5}_{5}f$	$^{8}_{8}g3$	2.4×10^{-4}	-4.9	2.5×10^{-12}	-7.5×10^{-6}	-9.9×10^{-10}	8.8×10^{-7}
95.2	8.7	$^{-2}_{2}f$	$^{5}_{5}g10$	7.5×10^{-2}	-7.0	4.2×10^{-13}	-2.6×10^{-3}	-2.8×10^{-13}	2.1×10^{-7}
95.2	8.8	$^{-2}_{2}f$	$^{5}_{5}g10$	7.5×10^{-2}	-7.0	2.0×10^{-12}	-2.6×10^{-3}	-1.8×10^{-13}	1.7×10^{-7}
95.2	8.9	$^{-2}_{2}f$	$^{5}_{5}g10$	7.5×10^{-2}	-7.0	3.1×10^{-12}	-2.6×10^{-3}	-1.1×10^{-13}	1.3×10^{-7}
95.2	9.0	$^{-2}_{2}f$	$^{5}_{5}g10$	7.5×10^{-2}	-7.0	3.7×10^{-12}	-2.6×10^{-3}	-7.1×10^{-14}	1.1×10^{-7}
95.2	9.1	$^{-2}_{2}f$	$^{5}_{5}g10$	7.5×10^{-2}	-7.0	4.1×10^{-12}	-2.6×10^{-3}	-4.6×10^{-14}	8.6×10^{-8}
95.2	9.2	$^{-2}_{2}f$	$^{5}_{5}g10$	7.5×10^{-2}	-7.0	4.4×10^{-12}	-2.6×10^{-3}	-3.5×10^{-14}	7.5×10^{-8}
95.2	9.3	$^{-2}_{2}f$	$^{5}_{5}g10$	7.5×10^{-2}	-7.0	4.5×10^{-12}	-2.6×10^{-3}	-4.4×10^{-14}	8.5×10^{-8}

(continued)

Table G.1 (continued)

Ω/Ω_K (%)	$\log T$ (K)	Daughters		$\Delta\tilde{\omega}$	$\mathcal{H}/E_{\text{unit}}$	$\tilde{\gamma}_\alpha$	$\tilde{\gamma}_\beta$	$\tilde{\gamma}_\gamma$	$\lvert Q_{\text{PIT}}\rvert$
95.2	9.4	${}^{-2}_{2}f$	${}^{5}_{5}g_{10}$	7.5×10^{-2}	-7.0	4.6×10^{-12}	-2.6×10^{-3}	-1.2×10^{-13}	1.4×10^{-7}
95.2	9.5	${}^{-2}_{2}f$	${}^{5}_{5}g_{10}$	7.5×10^{-2}	-7.0	4.7×10^{-12}	-2.6×10^{-3}	-4.3×10^{-13}	2.6×10^{-7}
95.2	9.6	${}^{-2}_{2}f$	${}^{5}_{5}g_{10}$	7.5×10^{-2}	-7.0	4.5×10^{-12}	-2.6×10^{-3}	-1.7×10^{-12}	5.2×10^{-7}
95.2	9.7	${}^{-2}_{2}f$	${}^{5}_{5}g_{10}$	7.5×10^{-2}	-7.0	3.7×10^{-12}	-2.6×10^{-3}	-6.7×10^{-12}	1.0×10^{-6}
95.2	9.8	${}^{-2}_{2}f$	${}^{5}_{5}g_{10}$	7.5×10^{-2}	-7.0	6.3×10^{-13}	-2.6×10^{-3}	-2.7×10^{-11}	2.1×10^{-6}
95.4	8.7	${}^{-2}_{2}f$	${}^{5}_{5}g_{10}$	6.9×10^{-2}	-7.1	2.1×10^{-12}	-2.6×10^{-3}	-2.8×10^{-13}	1.9×10^{-7}
95.4	8.8	${}^{-2}_{2}f$	${}^{5}_{5}g_{10}$	6.9×10^{-2}	-7.1	3.7×10^{-12}	-2.6×10^{-3}	-1.8×10^{-13}	1.5×10^{-7}
95.4	8.9	${}^{-2}_{2}f$	${}^{5}_{5}g_{10}$	6.9×10^{-2}	-7.1	4.8×10^{-12}	-2.6×10^{-3}	-1.1×10^{-13}	1.2×10^{-7}
95.4	9.0	${}^{-2}_{2}f$	${}^{5}_{5}g_{10}$	6.9×10^{-2}	-7.1	5.4×10^{-12}	-2.6×10^{-3}	-7.0×10^{-14}	9.7×10^{-8}
95.4	9.1	${}^{-2}_{2}f$	${}^{5}_{5}g_{10}$	6.9×10^{-2}	-7.1	5.8×10^{-12}	-2.6×10^{-3}	-4.6×10^{-14}	7.8×10^{-8}
95.4	9.2	${}^{-2}_{2}f$	${}^{5}_{5}g_{10}$	6.9×10^{-2}	-7.1	6.1×10^{-12}	-2.6×10^{-3}	-3.4×10^{-14}	6.8×10^{-8}
95.4	9.3	${}^{-2}_{2}f$	${}^{5}_{5}g_{10}$	6.9×10^{-2}	-7.1	6.3×10^{-12}	-2.6×10^{-3}	-4.4×10^{-14}	7.7×10^{-8}
95.4	9.4	${}^{-2}_{2}f$	${}^{5}_{5}g_{10}$	6.9×10^{-2}	-7.1	6.4×10^{-12}	-2.6×10^{-3}	-1.2×10^{-13}	1.2×10^{-7}
95.4	9.5	${}^{-2}_{2}f$	${}^{5}_{5}g_{10}$	6.9×10^{-2}	-7.1	6.4×10^{-12}	-2.6×10^{-3}	-4.3×10^{-13}	2.4×10^{-7}
95.4	9.6	${}^{-2}_{2}f$	${}^{5}_{5}g_{10}$	6.9×10^{-2}	-7.1	6.2×10^{-12}	-2.6×10^{-3}	-1.7×10^{-12}	4.7×10^{-7}
95.4	9.7	${}^{-2}_{2}f$	${}^{5}_{5}g_{10}$	6.9×10^{-2}	-7.1	5.4×10^{-12}	-2.6×10^{-3}	-6.7×10^{-12}	9.4×10^{-7}
95.4	9.8	${}^{-2}_{2}f$	${}^{5}_{5}g_{10}$	6.9×10^{-2}	-7.1	2.3×10^{-12}	-2.6×10^{-3}	-2.7×10^{-11}	1.9×10^{-6}
95.6	8.6	${}^{-4}_{4}g_7$	${}^{7}_{7}g_4$	1.0×10^{-6}	4.0	1.8×10^{-12}	-1.3×10^{-12}	-6.0×10^{-12}	1.5×10^{-7}
95.6	8.7	${}^{-2}_{2}f$	${}^{5}_{5}g_{10}$	6.3×10^{-2}	-7.2	4.4×10^{-12}	-2.6×10^{-3}	-2.8×10^{-13}	1.7×10^{-7}
95.6	8.8	${}^{-2}_{2}f$	${}^{5}_{5}g_{10}$	6.3×10^{-2}	-7.2	6.0×10^{-12}	-2.6×10^{-3}	-1.7×10^{-13}	1.4×10^{-7}
95.6	8.9	${}^{-2}_{2}f$	${}^{5}_{5}g_{10}$	6.3×10^{-2}	-7.2	7.1×10^{-12}	-2.6×10^{-3}	-1.1×10^{-13}	1.1×10^{-7}
95.6	9.0	${}^{-2}_{2}f$	${}^{5}_{5}g_{10}$	6.3×10^{-2}	-7.2	7.7×10^{-12}	-2.6×10^{-3}	-7.0×10^{-14}	8.7×10^{-8}

(continued)

Table G.1 (continued)

$\Omega/\Omega_{\rm K}$ (%)	$\log T$ (K)	Daughters		$\Delta\tilde{\omega}$	$\mathcal{H}/E_{\rm unit}$	$\tilde{\gamma}_\alpha$	$\tilde{\gamma}_\beta$	$\tilde{\gamma}_\gamma$	$\lvert Q_{\rm PIT}\rvert$
95.6	9.1	$^{-2}_2 f$	$^5_5 g_{10}$	6.3×10^{-2}	-7.2	8.1×10^{-12}	-2.6×10^{-3}	-4.5×10^{-14}	7.0×10^{-8}
95.6	9.2	$^{-2}_2 f$	$^5_5 g_{10}$	6.3×10^{-2}	-7.2	8.4×10^{-12}	-2.6×10^{-3}	-3.4×10^{-14}	6.1×10^{-8}
95.6	9.3	$^{-2}_2 f$	$^5_5 g_{10}$	6.3×10^{-2}	-7.2	8.6×10^{-12}	-2.6×10^{-3}	-4.4×10^{-14}	6.9×10^{-8}
95.6	9.4	$^{-2}_2 f$	$^5_5 g_{10}$	6.3×10^{-2}	-7.2	8.7×10^{-12}	-2.6×10^{-3}	-1.2×10^{-13}	1.1×10^{-7}
95.6	9.5	$^{-4}_4 g_7$	$^7_7 g_4$	1.0×10^{-6}	4.0	8.7×10^{-12}	-1.9×10^{-12}	-2.7×10^{-11}	9.3×10^{-8}
95.6	9.6	$^{-4}_4 g_7$	$^7_7 g_4$	1.0×10^{-6}	4.0	8.5×10^{-12}	-4.5×10^{-12}	-1.1×10^{-10}	7.5×10^{-8}
95.6	9.7	$^{-4}_4 g_7$	$^7_7 g_4$	1.0×10^{-6}	4.0	7.7×10^{-12}	-1.5×10^{-11}	-4.2×10^{-10}	6.9×10^{-8}
95.6	9.8	$^{-4}_4 g_7$	$^7_7 g_4$	1.0×10^{-6}	4.0	4.6×10^{-12}	-5.7×10^{-11}	-1.7×10^{-9}	6.7×10^{-8}
95.8	8.5	$^{-2}_2 f$	$^5_5 g_{10}$	5.6×10^{-2}	-7.3	5.9×10^{-13}	-2.6×10^{-3}	-6.9×10^{-13}	2.4×10^{-7}
95.8	8.6	$^{-2}_2 f$	$^5_5 g_{10}$	5.6×10^{-2}	-7.3	4.8×10^{-12}	-2.6×10^{-3}	-4.3×10^{-13}	1.9×10^{-7}
95.8	8.7	$^{-2}_2 f$	$^5_5 g_{10}$	5.6×10^{-2}	-7.3	7.4×10^{-12}	-2.6×10^{-3}	-2.7×10^{-13}	1.5×10^{-7}
95.8	8.8	$^{-2}_2 f$	$^5_5 g_{10}$	5.6×10^{-2}	-7.3	9.1×10^{-12}	-2.6×10^{-3}	-1.7×10^{-13}	1.2×10^{-7}
95.8	8.9	$^{-2}_2 f$	$^5_5 g_{10}$	5.6×10^{-2}	-7.3	1.0×10^{-11}	-2.6×10^{-3}	-1.1×10^{-13}	9.6×10^{-8}
95.8	9.0	$^{-2}_2 f$	$^5_5 g_{10}$	5.6×10^{-2}	-7.3	1.1×10^{-11}	-2.6×10^{-3}	-6.9×10^{-14}	7.7×10^{-8}
95.8	9.1	$^{-2}_2 f$	$^5_5 g_{10}$	5.6×10^{-2}	-7.3	1.1×10^{-11}	-2.6×10^{-3}	-4.5×10^{-14}	6.2×10^{-8}
95.8	9.2	$^{-2}_2 f$	$^5_5 g_{10}$	5.6×10^{-2}	-7.3	1.1×10^{-11}	-2.6×10^{-3}	-3.4×10^{-14}	5.4×10^{-8}
95.8	9.3	$^{-2}_2 f$	$^5_5 g_{10}$	5.6×10^{-2}	-7.3	1.2×10^{-11}	-2.6×10^{-3}	-4.3×10^{-14}	6.1×10^{-8}
95.8	9.4	$^{-2}_2 f$	$^5_5 g_{10}$	5.6×10^{-2}	-7.3	1.2×10^{-11}	-2.6×10^{-3}	-1.2×10^{-13}	9.9×10^{-8}
95.8	9.5	$^{-2}_2 f$	$^5_5 g_{10}$	5.6×10^{-2}	-7.3	1.2×10^{-11}	-2.6×10^{-3}	-4.2×10^{-13}	1.9×10^{-7}
95.8	9.6	$^{-2}_2 f$	$^5_5 g_{10}$	5.6×10^{-2}	-7.3	1.2×10^{-11}	-2.6×10^{-3}	-1.7×10^{-12}	3.7×10^{-7}
95.8	9.7	$^{-2}_2 f$	$^5_5 g_{10}$	5.6×10^{-2}	-7.3	1.1×10^{-11}	-2.6×10^{-3}	-6.6×10^{-12}	7.5×10^{-7}
95.8	9.8	$^{-2}_2 f$	$^5_5 g_{10}$	5.6×10^{-2}	-7.3	7.6×10^{-12}	-2.6×10^{-3}	-2.6×10^{-11}	1.5×10^{-6}

(continued)

Table G.1 (continued)

Ω/Ω_K (%)	$\log T$ (K)	Daughters		$\Delta\tilde{\omega}$	$\mathcal{H}/E_{\text{unit}}$	$\tilde{\gamma}_\alpha$	$\tilde{\gamma}_\beta$	$\tilde{\gamma}_\gamma$	$|Q_{\text{PIT}}|$
96.0	8.5	$_{-2}^{-2}f$	$_5^5 g_{10}$	4.9×10^{-2}	-7.4	4.6×10^{-12}	-2.6×10^{-3}	-6.8×10^{-13}	2.1×10^{-7}
96.0	8.6	$_{-2}^{-2}f$	$_5^5 g_{10}$	4.9×10^{-2}	-7.4	8.8×10^{-12}	-2.6×10^{-3}	-4.3×10^{-13}	1.7×10^{-7}
96.0	8.7	$_{-2}^{-2}f$	$_5^5 g_{10}$	4.9×10^{-2}	-7.4	1.2×10^{-11}	-2.6×10^{-3}	-2.7×10^{-13}	1.3×10^{-7}
96.0	8.8	$_{-2}^{-2}f$	$_5^5 g_{10}$	4.9×10^{-2}	-7.4	1.3×10^{-11}	-2.6×10^{-3}	-1.7×10^{-13}	1.0×10^{-7}
96.0	8.9	$_{-2}^{-2}f$	$_5^5 g_{10}$	4.9×10^{-2}	-7.4	1.4×10^{-11}	-2.6×10^{-3}	-1.1×10^{-13}	8.3×10^{-8}
96.0	9.0	$_{-2}^{-2}f$	$_5^5 g_{10}$	4.9×10^{-2}	-7.4	1.5×10^{-11}	-2.6×10^{-3}	-6.9×10^{-14}	6.6×10^{-8}
96.0	9.1	$_{-2}^{-2}f$	$_5^5 g_{10}$	4.9×10^{-2}	-7.4	1.5×10^{-11}	-2.6×10^{-3}	-4.5×10^{-14}	5.3×10^{-8}
96.0	9.2	$_{-2}^{-2}f$	$_5^5 g_{10}$	4.9×10^{-2}	-7.4	1.6×10^{-11}	-2.6×10^{-3}	-3.4×10^{-14}	4.6×10^{-8}
96.0	9.3	$_{-2}^{-2}f$	$_5^5 g_{10}$	4.9×10^{-2}	-7.4	1.6×10^{-11}	-2.6×10^{-3}	-4.3×10^{-14}	5.3×10^{-8}
96.0	9.4	$_{-2}^{-2}f$	$_5^5 g_{10}$	4.9×10^{-2}	-7.4	1.6×10^{-11}	-2.6×10^{-3}	-1.1×10^{-13}	8.5×10^{-8}
96.0	9.5	$_{-2}^{-2}f$	$_5^5 g_{10}$	4.9×10^{-2}	-7.4	1.6×10^{-11}	-2.6×10^{-3}	-4.2×10^{-13}	1.6×10^{-7}
96.0	9.6	$_{-2}^{-2}f$	$_5^5 g_{10}$	4.9×10^{-2}	-7.4	1.6×10^{-11}	-2.6×10^{-3}	-1.6×10^{-12}	3.2×10^{-7}
96.0	9.7	$_{-2}^{-2}f$	$_5^5 g_{10}$	4.9×10^{-2}	-7.4	1.5×10^{-11}	-2.6×10^{-3}	-6.5×10^{-12}	6.5×10^{-7}
96.0	9.8	$_{-2}^{-2}f$	$_5^5 g_{10}$	4.9×10^{-2}	-7.4	1.2×10^{-11}	-2.6×10^{-3}	-2.6×10^{-11}	1.3×10^{-6}
96.2	8.4	$_{-2}^{-2}f$	$_5^5 g_{10}$	4.2×10^{-2}	-7.5	3.2×10^{-12}	-2.6×10^{-3}	-1.1×10^{-12}	2.2×10^{-7}
96.2	8.5	$_{-2}^{-2}f$	$_5^5 g_{10}$	4.2×10^{-2}	-7.5	1.0×10^{-11}	-2.6×10^{-3}	-6.8×10^{-13}	1.8×10^{-7}
96.2	8.6	$_{-2}^{-2}f$	$_5^5 g_{10}$	4.2×10^{-2}	-7.5	1.4×10^{-11}	-2.6×10^{-3}	-4.3×10^{-13}	1.4×10^{-7}
96.2	8.7	$_{-2}^{-2}f$	$_5^5 g_{10}$	4.2×10^{-2}	-7.5	1.7×10^{-11}	-2.6×10^{-3}	-2.7×10^{-13}	1.1×10^{-7}
96.2	8.8	$_{-2}^{-2}f$	$_5^5 g_{10}$	4.2×10^{-2}	-7.5	1.9×10^{-11}	-2.6×10^{-3}	-1.7×10^{-13}	8.8×10^{-8}
96.2	8.9	$_{-2}^{-2}f$	$_5^5 g_{10}$	4.2×10^{-2}	-7.5	2.0×10^{-11}	-2.6×10^{-3}	-1.1×10^{-13}	7.0×10^{-8}
96.2	9.0	$_{-2}^{-2}f$	$_5^5 g_{10}$	4.2×10^{-2}	-7.5	2.0×10^{-11}	-2.6×10^{-3}	-6.8×10^{-14}	5.6×10^{-8}
96.2	9.1	$_{-2}^{-2}f$	$_5^5 g_{10}$	4.2×10^{-2}	-7.5	2.1×10^{-11}	-2.6×10^{-3}	-4.4×10^{-14}	4.5×10^{-8}

(continued)

Table G.1 (continued)

Ω/Ω_K (%)	$\log T$ (K)	Daughters		$\Delta\tilde{\omega}$	$\mathcal{H}/E_{unit}$	$\tilde{\gamma}_\alpha$	$\tilde{\gamma}_\beta$	$\tilde{\gamma}_\gamma$	$|\mathcal{Q}_{PIT}|$
96.2	9.2	$^{-2}_{2}f$	$^{5}_{5}g_{10}$	4.2×10^{-2}	-7.5	2.1×10^{-11}	-2.6×10^{-3}	-3.3×10^{-14}	3.9×10^{-8}
96.2	9.3	$^{-2}_{2}f$	$^{5}_{5}g_{10}$	4.2×10^{-2}	-7.5	2.1×10^{-11}	-2.6×10^{-3}	-4.3×10^{-14}	4.4×10^{-8}
96.2	9.4	$^{-2}_{2}f$	$^{5}_{5}g_{10}$	4.2×10^{-2}	-7.5	2.1×10^{-11}	-2.6×10^{-3}	-1.1×10^{-13}	7.2×10^{-8}
96.2	9.5	$^{-2}_{2}f$	$^{5}_{5}g_{10}$	4.2×10^{-2}	-7.5	2.1×10^{-11}	-2.6×10^{-3}	-4.2×10^{-13}	1.4×10^{-7}
96.2	9.6	$^{-2}_{2}f$	$^{5}_{5}g_{10}$	4.2×10^{-2}	-7.5	2.1×10^{-11}	-2.6×10^{-3}	-1.6×10^{-12}	2.7×10^{-7}
96.2	9.7	$^{-2}_{2}f$	$^{5}_{5}g_{10}$	4.2×10^{-2}	-7.5	2.0×10^{-11}	-2.6×10^{-3}	-6.5×10^{-12}	5.5×10^{-7}
96.2	9.8	$^{-2}_{2}f$	$^{5}_{5}g_{10}$	4.2×10^{-2}	-7.5	1.7×10^{-11}	-2.6×10^{-3}	-2.6×10^{-11}	1.1×10^{-6}
96.2	9.9	$^{-2}_{2}f$	$^{5}_{5}g_{10}$	4.2×10^{-2}	-7.5	4.2×10^{-12}	-2.6×10^{-3}	-1.0×10^{-10}	2.2×10^{-6}
96.4	8.4	$^{-2}_{2}f$	$^{5}_{5}g_{10}$	3.5×10^{-2}	-7.7	1.0×10^{-11}	-2.6×10^{-3}	-1.1×10^{-12}	1.8×10^{-7}
96.4	8.5	$^{-2}_{2}f$	$^{5}_{5}g_{10}$	3.5×10^{-2}	-7.7	1.7×10^{-11}	-2.6×10^{-3}	-6.7×10^{-13}	1.4×10^{-7}
96.4	8.6	$^{-2}_{2}f$	$^{5}_{5}g_{10}$	3.5×10^{-2}	-7.7	2.2×10^{-11}	-2.6×10^{-3}	-4.2×10^{-13}	1.1×10^{-7}
96.4	8.7	$^{-2}_{2}f$	$^{5}_{5}g_{10}$	3.5×10^{-2}	-7.7	2.4×10^{-11}	-2.6×10^{-3}	-2.7×10^{-13}	9.1×10^{-8}
96.4	8.8	$^{-2}_{2}f$	$^{5}_{5}g_{10}$	3.5×10^{-2}	-7.7	2.6×10^{-11}	-2.6×10^{-3}	-1.7×10^{-13}	7.2×10^{-8}
96.4	8.9	$^{-2}_{2}f$	$^{5}_{5}g_{10}$	3.5×10^{-2}	-7.7	2.7×10^{-11}	-2.6×10^{-3}	-1.1×10^{-13}	5.7×10^{-8}
96.4	9.0	$^{-2}_{2}f$	$^{5}_{5}g_{10}$	3.5×10^{-2}	-7.7	2.8×10^{-11}	-2.6×10^{-3}	-6.8×10^{-14}	4.5×10^{-8}
96.4	9.1	$^{-2}_{2}f$	$^{5}_{5}g_{10}$	3.5×10^{-2}	-7.7	2.8×10^{-11}	-2.6×10^{-3}	-4.4×10^{-14}	3.7×10^{-8}
96.4	9.2	$^{-2}_{2}f$	$^{5}_{5}g_{10}$	3.5×10^{-2}	-7.7	2.8×10^{-11}	-2.6×10^{-3}	-3.3×10^{-14}	3.2×10^{-8}
96.4	9.3	$^{-2}_{2}f$	$^{5}_{5}g_{10}$	3.5×10^{-2}	-7.7	2.9×10^{-11}	-2.6×10^{-3}	-4.2×10^{-14}	3.6×10^{-8}
96.4	9.4	$^{-2}_{2}f$	$^{5}_{5}g_{10}$	3.5×10^{-2}	-7.7	2.9×10^{-11}	-2.6×10^{-3}	-1.1×10^{-13}	5.9×10^{-8}
96.4	9.5	$^{-2}_{2}f$	$^{5}_{5}g_{10}$	3.5×10^{-2}	-7.7	2.9×10^{-11}	-2.6×10^{-3}	-4.1×10^{-13}	1.1×10^{-7}
96.4	9.6	$^{-2}_{2}f$	$^{5}_{5}g_{10}$	3.5×10^{-2}	-7.7	2.9×10^{-11}	-2.6×10^{-3}	-1.6×10^{-12}	2.2×10^{-7}
96.4	9.7	$^{-2}_{2}f$	$^{5}_{5}g_{10}$	3.5×10^{-2}	-7.7	2.8×10^{-11}	-2.6×10^{-3}	-6.4×10^{-12}	4.4×10^{-7}

(continued)

Table G.1 (continued)

Ω/Ω_K (%)	$\log T$ (K)	Daughters		$\Delta\tilde{\omega}$	$\mathcal{H}/E_{unit}$	$\tilde{\gamma}_\alpha$	$\tilde{\gamma}_\beta$	$\tilde{\gamma}_\gamma$	$|Q_{PIT}|$
96.4	9.8	$-\frac{2}{2}f$	$\frac{5}{5}g_{10}$	3.5×10^{-2}	-7.7	2.4×10^{-11}	-2.6×10^{-3}	-2.6×10^{-11}	8.9×10^{-7}
96.4	9.9	$-\frac{2}{2}f$	$\frac{5}{5}g_{10}$	3.5×10^{-2}	-7.7	1.1×10^{-11}	-2.6×10^{-3}	-1.0×10^{-10}	1.8×10^{-6}
96.6	8.3	$-\frac{2}{2}f$	$\frac{5}{5}g_{10}$	2.7×10^{-2}	-7.8	9.1×10^{-12}	-2.6×10^{-3}	-1.7×10^{-12}	1.7×10^{-7}
96.6	8.4	$-\frac{2}{2}f$	$\frac{5}{5}g_{10}$	2.7×10^{-2}	-7.8	2.0×10^{-11}	-2.6×10^{-3}	-1.1×10^{-12}	1.4×10^{-7}
96.6	8.5	$-\frac{2}{2}f$	$\frac{5}{5}g_{10}$	2.7×10^{-2}	-7.8	2.7×10^{-11}	-2.6×10^{-3}	-6.7×10^{-13}	1.1×10^{-7}
96.6	8.6	$-\frac{2}{2}f$	$\frac{5}{5}g_{10}$	2.7×10^{-2}	-7.8	3.1×10^{-11}	-2.6×10^{-3}	-4.2×10^{-13}	8.8×10^{-8}
96.6	8.7	$-\frac{2}{2}f$	$\frac{5}{5}g_{10}$	2.7×10^{-2}	-7.8	3.4×10^{-11}	-2.6×10^{-3}	-2.7×10^{-13}	7.0×10^{-8}
96.6	8.8	$-\frac{2}{2}f$	$\frac{5}{5}g_{10}$	2.7×10^{-2}	-7.8	3.6×10^{-11}	-2.6×10^{-3}	-1.7×10^{-13}	5.5×10^{-8}
96.6	8.9	$-\frac{2}{2}f$	$\frac{5}{5}g_{10}$	2.7×10^{-2}	-7.8	3.7×10^{-11}	-2.6×10^{-3}	-1.1×10^{-13}	4.4×10^{-8}
96.6	9.0	$-\frac{2}{2}f$	$\frac{5}{5}g_{10}$	2.7×10^{-2}	-7.8	3.8×10^{-11}	-2.6×10^{-3}	-6.7×10^{-14}	3.5×10^{-8}
96.6	9.1	$-\frac{2}{2}f$	$\frac{5}{5}g_{10}$	2.7×10^{-2}	-7.8	3.8×10^{-11}	-2.6×10^{-3}	-4.4×10^{-14}	2.8×10^{-8}
96.6	9.2	$-\frac{2}{2}f$	$\frac{5}{5}g_{10}$	2.7×10^{-2}	-7.8	3.8×10^{-11}	-2.6×10^{-3}	-3.3×10^{-14}	2.5×10^{-8}
96.6	9.3	$-\frac{2}{2}f$	$\frac{5}{5}g_{10}$	2.7×10^{-2}	7.0	3.8×10^{-11}	-2.6×10^{-3}	-4.2×10^{-14}	2.8×10^{-8}
96.6	9.4	$-\frac{2}{2}f$	$\frac{5}{5}g_{10}$	2.7×10^{-2}	-7.8	3.9×10^{-11}	-2.6×10^{-3}	-1.1×10^{-13}	4.5×10^{-8}
96.6	9.5	$-\frac{2}{2}f$	$\frac{5}{5}g_{10}$	2.7×10^{-2}	-7.8	3.9×10^{-11}	-2.6×10^{-3}	-4.1×10^{-13}	8.6×10^{-8}
96.6	9.6	$-\frac{2}{2}f$	$\frac{5}{5}g_{10}$	2.7×10^{-2}	-7.8	3.8×10^{-11}	-2.6×10^{-3}	-1.6×10^{-12}	1.7×10^{-7}
96.6	9.7	$-\frac{2}{2}f$	$\frac{5}{5}g_{10}$	2.7×10^{-2}	-7.8	3.8×10^{-11}	-2.6×10^{-3}	-6.4×10^{-12}	3.4×10^{-7}
96.6	9.8	$-\frac{2}{2}f$	$\frac{5}{5}g_{10}$	2.7×10^{-2}	-7.8	3.4×10^{-11}	-2.6×10^{-3}	-2.5×10^{-11}	6.8×10^{-7}
96.6	9.9	$-\frac{2}{2}f$	$\frac{5}{5}g_{10}$	2.7×10^{-2}	-7.8	2.1×10^{-11}	-2.6×10^{-3}	-1.0×10^{-10}	1.4×10^{-6}
96.8	8.2	$-\frac{2}{2}f$	$\frac{5}{5}g_{10}$	1.9×10^{-2}	-7.9	4.3×10^{-12}	-2.6×10^{-3}	-2.6×10^{-12}	1.5×10^{-7}
96.8	8.3	$-\frac{2}{2}f$	$\frac{5}{5}g_{10}$	1.9×10^{-2}	-7.9	2.2×10^{-11}	-2.6×10^{-3}	-1.7×10^{-12}	1.2×10^{-7}

(continued)

Table G.1 (continued)

Ω/Ω_K (%)	$\log T$ (K)	Daughters		$\Delta\tilde{\omega}$	$\mathcal{H}/E_{\text{unit}}$	$\tilde{\gamma}_\alpha$	$\tilde{\gamma}_\beta$	$\tilde{\gamma}_\gamma$	$|Q_{\text{PIT}}|$
96.8	8.4	$^{-2}_{2}f$	$^{5}_{5}g_{10}$	1.9×10^{-2}	-7.9	3.3×10^{-11}	-2.6×10^{-3}	-1.0×10^{-12}	9.7×10^{-8}
96.8	8.5	$^{-2}_{2}f$	$^{5}_{5}g_{10}$	1.9×10^{-2}	-7.9	4.0×10^{-11}	-2.6×10^{-3}	-6.6×10^{-13}	7.7×10^{-8}
96.8	8.6	$^{-2}_{2}f$	$^{5}_{5}g_{10}$	1.9×10^{-2}	-7.9	4.4×10^{-11}	-2.6×10^{-3}	-4.2×10^{-13}	6.1×10^{-8}
96.8	8.7	$^{-2}_{2}f$	$^{5}_{5}g_{10}$	1.9×10^{-2}	-7.9	4.7×10^{-11}	-2.6×10^{-3}	-2.6×10^{-13}	4.9×10^{-8}
96.8	8.8	$^{-2}_{2}f$	$^{5}_{5}g_{10}$	1.9×10^{-2}	-7.9	4.9×10^{-11}	-2.6×10^{-3}	-1.7×10^{-13}	3.9×10^{-8}
96.8	8.9	$^{-2}_{2}f$	$^{5}_{5}g_{10}$	1.9×10^{-2}	-7.9	5.0×10^{-11}	-2.6×10^{-3}	-1.0×10^{-13}	3.1×10^{-8}
96.8	9.0	$^{-2}_{2}f$	$^{5}_{5}g_{10}$	1.9×10^{-2}	-7.9	5.1×10^{-11}	-2.6×10^{-3}	-6.7×10^{-14}	2.4×10^{-8}
96.8	9.1	$^{-2}_{2}f$	$^{5}_{5}g_{10}$	1.9×10^{-2}	-7.9	5.1×10^{-11}	-2.6×10^{-3}	-4.3×10^{-14}	2.0×10^{-8}
96.8	9.2	$^{-2}_{2}f$	$^{5}_{5}g_{10}$	1.9×10^{-2}	-7.9	5.1×10^{-11}	-2.6×10^{-3}	-3.3×10^{-14}	1.7×10^{-8}
96.8	9.3	$^{-2}_{2}f$	$^{5}_{5}g_{10}$	1.9×10^{-2}	-7.9	5.2×10^{-11}	-2.6×10^{-3}	-4.2×10^{-14}	1.9×10^{-8}
96.8	9.4	$^{-2}_{2}f$	$^{5}_{5}g_{10}$	1.9×10^{-2}	-7.9	5.2×10^{-11}	-2.6×10^{-3}	-1.1×10^{-13}	3.2×10^{-8}
96.8	9.5	$^{-2}_{2}f$	$^{5}_{5}g_{10}$	1.9×10^{-2}	-7.9	5.2×10^{-11}	-2.6×10^{-3}	-4.1×10^{-13}	6.0×10^{-8}
96.8	9.6	$^{-2}_{2}f$	$^{5}_{5}g_{10}$	1.9×10^{-2}	-7.9	5.1×10^{-11}	-2.6×10^{-3}	-1.6×10^{-12}	1.2×10^{-7}
96.8	9.7	$^{-2}_{2}f$	$^{5}_{5}g_{10}$	1.9×10^{-2}	-7.9	5.1×10^{-11}	-2.6×10^{-3}	-6.3×10^{-12}	2.4×10^{-7}
96.8	9.8	$^{-2}_{2}f$	$^{5}_{5}g_{10}$	1.9×10^{-2}	-7.9	4.7×10^{-11}	-2.6×10^{-3}	-2.5×10^{-11}	4.8×10^{-7}
96.8	9.9	$^{-2}_{2}f$	$^{5}_{5}g_{10}$	1.9×10^{-2}	-7.9	3.4×10^{-11}	-2.6×10^{-3}	-1.0×10^{-10}	9.5×10^{-7}
97.0	8.2	$^{-3}_{3}f$	$^{6}_{6}g_{9}$	-1.2×10^{-3}	4.3	2.1×10^{-11}	-3.6×10^{-4}	-7.8×10^{-12}	6.1×10^{-8}
97.0	8.3	$^{-3}_{3}f$	$^{6}_{6}g_{9}$	-1.2×10^{-3}	4.3	3.9×10^{-11}	-3.6×10^{-4}	-4.9×10^{-12}	4.9×10^{-8}
97.0	8.4	$^{-3}_{3}f$	$^{6}_{6}g_{9}$	-1.2×10^{-3}	4.3	5.0×10^{-11}	-3.6×10^{-4}	-3.1×10^{-12}	3.9×10^{-8}
97.0	8.5	$^{-3}_{3}f$	$^{6}_{6}g_{9}$	-1.2×10^{-3}	4.3	5.7×10^{-11}	-3.6×10^{-4}	-2.0×10^{-12}	3.1×10^{-8}
97.0	8.6	$^{-3}_{3}f$	$^{6}_{6}g_{9}$	-1.2×10^{-3}	4.3	6.2×10^{-11}	-3.6×10^{-4}	-1.2×10^{-12}	2.4×10^{-8}

(continued)

Table G.1 (continued)

Ω/Ω_K (%)	$\log T$ (K)	Daughters		$\Delta\tilde{\omega}$	$\mathcal{H}/E_{\mathrm{unit}}$	$\tilde{\gamma}_\alpha$	$\tilde{\gamma}_\beta$	$\tilde{\gamma}_\gamma$	$\lvert Q_{\mathrm{PIT}}\rvert$
97.0	8.7	$^{-3}_{3}f$	$^{6}_{6}g_9$	-1.2×10^{-3}	4.3	6.4×10^{-11}	-3.6×10^{-4}	-7.8×10^{-13}	1.9×10^{-8}
97.0	8.8	$^{-3}_{3}f$	$^{6}_{6}g_9$	-1.2×10^{-3}	4.3	6.6×10^{-11}	-3.6×10^{-4}	-4.9×10^{-13}	1.5×10^{-8}
97.0	8.9	$^{-3}_{3}f$	$^{6}_{6}g_9$	-1.2×10^{-3}	4.3	6.7×10^{-11}	-3.6×10^{-4}	-3.1×10^{-13}	1.2×10^{-8}
97.0	9.0	$^{-3}_{3}f$	$^{6}_{6}g_9$	-1.2×10^{-3}	4.3	6.8×10^{-11}	-3.6×10^{-4}	-2.0×10^{-13}	9.7×10^{-9}
97.0	9.1	$^{-3}_{3}f$	$^{6}_{6}g_9$	-1.2×10^{-3}	4.3	6.9×10^{-11}	-3.6×10^{-4}	-1.3×10^{-13}	7.9×10^{-9}
97.0	9.2	$^{-3}_{3}f$	$^{6}_{6}g_9$	-1.2×10^{-3}	4.3	6.9×10^{-11}	-3.6×10^{-4}	-1.0×10^{-13}	6.9×10^{-9}
97.0	9.3	$^{-3}_{3}f$	$^{6}_{6}g_9$	-1.2×10^{-3}	4.3	6.9×10^{-11}	-3.6×10^{-4}	-1.4×10^{-13}	8.2×10^{-9}
97.0	9.4	$^{-3}_{3}f$	$^{6}_{6}g_9$	-1.2×10^{-3}	4.3	6.9×10^{-11}	-3.6×10^{-4}	-3.9×10^{-13}	1.4×10^{-8}
97.0	9.5	$^{-3}_{3}f$	$^{6}_{6}g_9$	-1.2×10^{-3}	4.3	6.9×10^{-11}	-3.6×10^{-4}	-1.4×10^{-12}	2.6×10^{-8}
97.0	9.6	$^{-3}_{3}f$	$^{6}_{6}g_9$	-1.2×10^{-3}	4.3	6.9×10^{-11}	-3.6×10^{-4}	-5.7×10^{-12}	5.2×10^{-8}
97.0	9.7	$^{-3}_{3}f$	$^{6}_{6}g_9$	-1.2×10^{-3}	4.3	6.8×10^{-11}	-3.6×10^{-4}	-2.3×10^{-11}	1.0×10^{-7}
97.0	9.8	$^{-3}_{3}f$	$^{6}_{6}g_9$	-1.2×10^{-3}	v4.3	6.5×10^{-11}	-3.6×10^{-4}	-9.0×10^{-11}	2.1×10^{-7}
97.0	9.9	$^{-3}_{3}f$	$^{6}_{6}g_9$	-1.2×10^{-3}	4.3	5.1×10^{-11}	-3.6×10^{-4}	-3.6×10^{-10}	4.1×10^{-7}
97.2	8.1	$^{-2}_{2}f$	$^{5}_{5}g_{10}$	2.3×10^{-3}	−8.2	1.5×10^{-11}	-2.6×10^{-3}	-4.1×10^{-12}	3.3×10^{-8}
97.2	8.2	$^{-2}_{2}f$	$^{5}_{5}g_{10}$	2.3×10^{-3}	−8.2	4.4×10^{-11}	-2.6×10^{-3}	-2.6×10^{-12}	2.7×10^{-8}
97.2	8.3	$^{-2}_{2}f$	$^{5}_{5}g_{10}$	2.3×10^{-3}	−8.2	6.2×10^{-11}	-2.6×10^{-3}	-1.6×10^{-12}	2.1×10^{-8}
97.2	8.4	$^{-2}_{2}f$	$^{5}_{5}g_{10}$	2.3×10^{-3}	−8.2	7.3×10^{-11}	-2.6×10^{-3}	-1.0×10^{-12}	1.7×10^{-8}
97.2	8.5	$^{-2}_{2}f$	$^{5}_{5}g_{10}$	2.3×10^{-3}	−8.2	8.0×10^{-11}	-2.6×10^{-3}	-6.5×10^{-13}	1.3×10^{-8}
97.2	8.6	$^{-2}_{2}f$	$^{5}_{5}g_{10}$	2.3×10^{-3}	−8.2	8.5×10^{-11}	-2.6×10^{-3}	-4.1×10^{-13}	1.1×10^{-8}
97.2	8.7	$^{-2}_{2}f$	$^{5}_{5}g_{10}$	2.3×10^{-3}	−8.2	8.8×10^{-11}	-2.6×10^{-3}	-2.6×10^{-13}	8.4×10^{-9}
97.2	8.8	$^{-2}_{2}f$	$^{5}_{5}g_{10}$	2.3×10^{-3}	−8.2	8.9×10^{-11}	-2.6×10^{-3}	-1.6×10^{-13}	6.7×10^{-9}
97.2	8.9	$^{-2}_{2}f$	$^{5}_{5}g_{10}$	2.3×10^{-3}	−8.2	9.1×10^{-11}	-2.6×10^{-3}	-1.0×10^{-13}	5.3×10^{-9}

(continued)

Table G.1 (continued)

Ω/Ω_K (%)	$\log T$ (K)	Daughters		$\Delta\tilde{\omega}$	$\mathcal{H}/E_{\text{unit}}$	$\tilde{\gamma}_\alpha$	$\tilde{\gamma}_\beta$	$\tilde{\gamma}_\gamma$	$|Q_{\text{PIT}}|$
97.2	9.0	$^{-2}_{2}f$	$^{5}_{5}g_{10}$	2.3×10^{-3}	-8.2	9.1×10^{-11}	-2.6×10^{-3}	-6.6×10^{-14}	4.2×10^{-9}
97.2	9.1	$^{-2}_{2}f$	$^{5}_{5}g_{10}$	2.3×10^{-3}	-8.2	9.2×10^{-11}	-2.6×10^{-3}	-4.3×10^{-14}	3.4×10^{-9}
97.2	9.2	$^{-2}_{2}f$	$^{5}_{5}g_{10}$	2.3×10^{-3}	-8.2	9.2×10^{-11}	-2.6×10^{-3}	-3.2×10^{-14}	3.0×10^{-9}
97.2	9.3	$^{-2}_{2}f$	$^{5}_{5}g_{10}$	2.3×10^{-3}	-8.2	9.2×10^{-11}	-2.6×10^{-3}	-4.1×10^{-14}	3.3×10^{-9}
97.2	9.4	$^{-2}_{2}f$	$^{5}_{5}g_{10}$	2.3×10^{-3}	-8.2	9.2×10^{-11}	-2.6×10^{-3}	-1.1×10^{-13}	5.4×10^{-9}
97.2	9.5	$^{-2}_{2}f$	$^{5}_{5}g_{10}$	2.3×10^{-3}	-8.2	9.2×10^{-11}	-2.6×10^{-3}	-4.0×10^{-13}	1.0×10^{-8}
97.2	9.6	$^{-2}_{2}f$	$^{5}_{5}g_{10}$	2.3×10^{-3}	-8.2	9.2×10^{-11}	-2.6×10^{-3}	-1.6×10^{-12}	2.1×10^{-8}
97.2	9.7	$^{-2}_{2}f$	$^{5}_{5}g_{10}$	2.3×10^{-3}	-8.2	9.1×10^{-11}	-2.6×10^{-3}	-6.2×10^{-12}	4.1×10^{-8}
97.2	9.8	$^{-2}_{2}f$	$^{5}_{5}g_{10}$	2.3×10^{-3}	-8.2	8.8×10^{-11}	-2.6×10^{-3}	-2.5×10^{-11}	8.2×10^{-8}
97.2	9.9	$^{-2}_{2}f$	$^{5}_{5}g_{10}$	2.3×10^{-3}	-8.2	7.4×10^{-11}	-2.6×10^{-3}	-9.9×10^{-11}	1.6×10^{-7}
97.2	10.0	$^{-2}_{2}f$	$^{5}_{5}g_{10}$	2.3×10^{-3}	-8.2	1.9×10^{-11}	-2.6×10^{-3}	-3.9×10^{-10}	3.3×10^{-7}
97.4	8.1	$^{-2}_{2}f$	$^{5}_{5}g_{10}$	-6.8×10^{-3}	-8.3	4.5×10^{-11}	-2.6×10^{-3}	-4.1×10^{-12}	6.8×10^{-8}
97.4	8.2	$^{-2}_{2}f$	$^{5}_{5}g_{10}$	-6.8×10^{-3}	-8.3	7.4×10^{-11}	-2.6×10^{-3}	-2.6×10^{-12}	5.4×10^{-8}
97.4	8.3	$^{-2}_{2}f$	$^{5}_{5}g_{10}$	-6.8×10^{-3}	-8.3	9.2×10^{-11}	-2.6×10^{-3}	-1.6×10^{-12}	4.3×10^{-8}
97.4	8.4	$^{-2}_{2}f$	$^{5}_{5}g_{10}$	-6.8×10^{-3}	-8.3	1.0×10^{-10}	-2.6×10^{-3}	-1.0×10^{-12}	3.4×10^{-8}
97.4	8.5	$^{-2}_{2}f$	$^{5}_{5}g_{10}$	-6.8×10^{-3}	-8.3	1.1×10^{-10}	-2.6×10^{-3}	-6.5×10^{-13}	2.7×10^{-8}
97.4	8.6	$^{-2}_{2}f$	$^{5}_{5}g_{10}$	-6.8×10^{-3}	-8.3	1.2×10^{-10}	-2.6×10^{-3}	-4.1×10^{-13}	2.1×10^{-8}
97.4	8.7	$^{-2}_{2}f$	$^{5}_{5}g_{10}$	-6.8×10^{-3}	-8.3	1.2×10^{-10}	-2.6×10^{-3}	-2.6×10^{-13}	1.7×10^{-8}
97.4	8.8	$^{-2}_{2}f$	$^{5}_{5}g_{10}$	-6.8×10^{-3}	-8.3	1.2×10^{-10}	-2.6×10^{-3}	-1.6×10^{-13}	1.4×10^{-8}
97.4	8.9	$^{-2}_{2}f$	$^{5}_{5}g_{10}$	-6.8×10^{-3}	-8.3	1.2×10^{-10}	-2.6×10^{-3}	-1.0×10^{-13}	1.1×10^{-8}
97.4	9.0	$^{-2}_{2}f$	$^{5}_{5}g_{10}$	-6.8×10^{-3}	-8.3	1.2×10^{-10}	-2.6×10^{-3}	-6.5×10^{-14}	8.6×10^{-9}
97.4	9.1	$^{-2}_{2}f$	$^{5}_{5}g_{10}$	-6.8×10^{-3}	-8.3	1.2×10^{-10}	-2.6×10^{-3}	-4.2×10^{-14}	6.9×10^{-9}

(continued)

Table G.1 (continued)

Ω/Ω_K (%)	$\log T$ (K)	Daughters		$\Delta\tilde{\omega}$	$\mathcal{H}/E_{unit}$	$\tilde{\gamma}_\alpha$	$\tilde{\gamma}_\beta$	$\tilde{\gamma}_\gamma$	$|Q_{PIT}|$
97.4	9.2	$_{-2}^{2}f$	$_5^5 g_{10}$	-6.8×10^{-3}	-8.3	1.2×10^{-10}	-2.6×10^{-3}	-3.2×10^{-14}	6.0×10^{-9}
97.4	9.3	$_{-2}^{2}f$	$_5^5 g_{10}$	-6.8×10^{-3}	-8.3	1.2×10^{-10}	-2.6×10^{-3}	-4.1×10^{-14}	6.8×10^{-9}
97.4	9.4	$_{-2}^{2}f$	$_5^5 g_{10}$	-6.8×10^{-3}	-8.3	1.2×10^{-10}	-2.6×10^{-3}	-1.1×10^{-13}	1.1×10^{-8}
97.4	9.5	$_{-2}^{2}f$	$_5^5 g_{10}$	-6.8×10^{-3}	-8.3	1.2×10^{-10}	-2.6×10^{-3}	-4.0×10^{-13}	2.1×10^{-8}
97.4	9.6	$_{-2}^{2}f$	$_5^5 g_{10}$	-6.8×10^{-3}	-8.3	1.2×10^{-10}	-2.6×10^{-3}	-1.6×10^{-12}	4.2×10^{-8}
97.4	9.7	$_{-2}^{2}f$	$_5^5 g_{10}$	-6.8×10^{-3}	-8.3	1.2×10^{-10}	-2.6×10^{-3}	-6.2×10^{-12}	8.4×10^{-8}
97.4	9.8	$_{-2}^{2}f$	$_5^5 g_{10}$	-6.8×10^{-3}	-8.3	1.2×10^{-10}	-2.6×10^{-3}	-2.5×10^{-11}	1.7×10^{-7}
97.4	9.9	$_{-2}^{2}f$	$_5^5 g_{10}$	-6.8×10^{-3}	-8.3	1.1×10^{-10}	-2.6×10^{-3}	-9.8×10^{-11}	3.3×10^{-7}
97.4	10.0	$_{-2}^{2}f$	$_5^5 g_{10}$	-6.8×10^{-3}	-8.3	4.9×10^{-11}	-2.6×10^{-3}	-3.9×10^{-10}	6.6×10^{-7}
97.6	8.0	$_{-2}^{2}f$	$_5^5 g_{10}$	-1.6×10^{-2}	-8.5	3.9×10^{-11}	-2.6×10^{-3}	-6.4×10^{-12}	1.9×10^{-7}
97.6	8.1	$_{-2}^{2}f$	$_5^5 g_{10}$	-1.6×10^{-2}	-8.5	8.6×10^{-11}	-2.6×10^{-3}	-4.0×10^{-12}	1.5×10^{-7}
97.6	8.2	$_{-2}^{2}f$	$_5^5 g_{10}$	-1.6×10^{-2}	-8.5	1.1×10^{-10}	-2.6×10^{-3}	-2.6×10^{-12}	1.2×10^{-7}
97.6	8.3	$_{-2}^{2}f$	$_5^5 g_{10}$	-1.6×10^{-2}	-8.5	1.3×10^{-10}	-2.6×10^{-3}	-1.6×10^{-12}	9.5×10^{-8}
97.6	8.4	$_{-2}^{2}f$	$_5^5 g_{10}$	-1.6×10^{-2}	-8.5	1.5×10^{-10}	-2.6×10^{-3}	-1.0×10^{-12}	7.6×10^{-8}
97.6	8.5	$_{-2}^{2}f$	$_5^5 g_{10}$	-1.6×10^{-2}	-8.5	1.5×10^{-10}	-2.6×10^{-3}	-6.4×10^{-13}	6.0×10^{-8}
97.6	8.6	$_{-2}^{2}f$	$_5^5 g_{10}$	-1.6×10^{-2}	-8.5	1.6×10^{-10}	-2.6×10^{-3}	-4.0×10^{-13}	4.8×10^{-8}
97.6	8.7	$_{-2}^{2}f$	$_5^5 g_{10}$	-1.6×10^{-2}	-8.5	1.6×10^{-10}	-2.6×10^{-3}	-2.6×10^{-13}	3.8×10^{-8}
97.6	8.8	$_{-2}^{2}f$	$_5^5 g_{10}$	-1.6×10^{-2}	-8.5	1.6×10^{-10}	-2.6×10^{-3}	-1.6×10^{-13}	3.0×10^{-8}
97.6	8.9	$_{-2}^{2}f$	$_5^5 g_{10}$	-1.6×10^{-2}	-8.5	1.6×10^{-10}	-2.6×10^{-3}	-1.0×10^{-13}	2.4×10^{-8}
97.6	9.0	$_{-2}^{2}f$	$_5^5 g_{10}$	-1.6×10^{-2}	-8.5	1.6×10^{-10}	-2.6×10^{-3}	-6.5×10^{-14}	1.9×10^{-8}
97.6	9.1	$_{-2}^{2}f$	$_5^5 g_{10}$	-1.6×10^{-2}	-8.5	1.6×10^{-10}	-2.6×10^{-3}	-4.2×10^{-14}	1.5×10^{-8}
97.6	9.2	$_{-2}^{2}f$	$_5^5 g_{10}$	-1.6×10^{-2}	-8.5	1.6×10^{-10}	-2.6×10^{-3}	-3.2×10^{-14}	1.3×10^{-8}

(continued)

Table G.1 (continued)

Ω/Ω_K (%)	$\log T$ (K)	Daughters		$\Delta\tilde{\omega}$	$\mathcal{H}/E_{\text{unit}}$	$\tilde{\gamma}_\alpha$	$\tilde{\gamma}_\beta$	$\tilde{\gamma}_\gamma$	$|Q_{\text{PIT}}|$
97.6	9.3	$^{-2}_{2}f$	$^{5}_{5}g_{10}$	-1.6×10^{-2}	-8.5	1.6×10^{-10}	-2.6×10^{-3}	-4.1×10^{-14}	1.5×10^{-8}
97.6	9.4	$^{-2}_{2}f$	$^{5}_{5}g_{10}$	-1.6×10^{-2}	-8.5	1.6×10^{-10}	-2.6×10^{-3}	-1.1×10^{-13}	2.5×10^{-8}
97.6	9.5	$^{-2}_{2}f$	$^{5}_{5}g_{10}$	-1.6×10^{-2}	-8.5	1.6×10^{-10}	-2.6×10^{-3}	-3.9×10^{-13}	4.7×10^{-8}
97.6	9.6	$^{-2}_{2}f$	$^{5}_{5}g_{10}$	-1.6×10^{-2}	-8.5	1.6×10^{-10}	-2.6×10^{-3}	-1.5×10^{-12}	9.3×10^{-8}
97.6	9.7	$^{-2}_{2}f$	$^{5}_{5}g_{10}$	-1.6×10^{-2}	-8.5	1.6×10^{-10}	-2.6×10^{-3}	-6.1×10^{-12}	1.9×10^{-7}
97.6	9.8	$^{-2}_{2}f$	$^{5}_{5}g_{10}$	-1.6×10^{-2}	-8.5	1.6×10^{-10}	-2.6×10^{-3}	-2.4×10^{-11}	3.7×10^{-7}
97.6	9.9	$^{-2}_{2}f$	$^{5}_{5}g_{10}$	-1.6×10^{-2}	-8.5	1.5×10^{-10}	-2.6×10^{-3}	-9.7×10^{-11}	7.4×10^{-7}
97.6	10.0	$^{-2}_{2}f$	$^{5}_{5}g_{10}$	-1.6×10^{-2}	-8.5	9.0×10^{-11}	-2.6×10^{-3}	-3.9×10^{-10}	1.5×10^{-6}
97.8	7.9	$^{-3}_{3}f$	$^{6}_{6}g_{8}$	2.9×10^{-3}	4.7	1.8×10^{-11}	-3.7×10^{-4}	-3.4×10^{-11}	2.8×10^{-7}
97.8	8.0	$^{-3}_{3}f$	$^{6}_{6}g_{8}$	2.9×10^{-3}	4.7	9.2×10^{-11}	-3.7×10^{-4}	-2.2×10^{-11}	2.2×10^{-7}
97.8	8.1	$^{-3}_{3}f$	$^{6}_{6}g_{8}$	2.9×10^{-3}	4.7	1.4×10^{-10}	-3.7×10^{-4}	-1.4×10^{-11}	1.8×10^{-7}
97.8	8.2	$^{-3}_{3}f$	$^{6}_{6}g_{8}$	2.9×10^{-3}	4.7	1.7×10^{-10}	-3.7×10^{-4}	-8.6×10^{-12}	1.4×10^{-7}
97.8	8.3	$^{-3}_{3}f$	$^{6}_{6}g_{8}$	2.9×10^{-3}	4.7	1.9×10^{-10}	-3.7×10^{-4}	-5.5×10^{-12}	1.1×10^{-7}
97.8	8.4	$^{-3}_{3}f$	$^{6}_{6}g_{8}$	2.9×10^{-3}	4.7	2.0×10^{-10}	-3.7×10^{-4}	-3.4×10^{-12}	8.8×10^{-8}
97.8	8.5	$^{-3}_{3}f$	$^{6}_{6}g_{8}$	2.9×10^{-3}	4.7	2.1×10^{-10}	-3.7×10^{-4}	-2.2×10^{-12}	7.0×10^{-8}
97.8	8.6	$^{-3}_{3}f$	$^{6}_{6}g_{8}$	2.9×10^{-3}	4.7	2.1×10^{-10}	-3.7×10^{-4}	1.4×10^{-12}	5.5×10^{-8}
97.8	8.7	$^{-3}_{3}f$	$^{6}_{6}g_{8}$	2.9×10^{-3}	4.7	2.2×10^{-10}	-3.7×10^{-4}	-8.6×10^{-13}	4.4×10^{-8}
97.8	8.8	$^{-3}_{3}f$	$^{6}_{6}g_{8}$	2.9×10^{-3}	4.7	2.2×10^{-10}	-3.7×10^{-4}	-5.5×10^{-13}	3.5×10^{-8}
97.8	8.9	$^{-3}_{3}f$	$^{6}_{6}g_{8}$	2.9×10^{-3}	4.7	2.2×10^{-10}	-3.7×10^{-4}	-3.4×10^{-13}	2.8×10^{-8}
97.8	9.0	$^{-3}_{3}f$	$^{6}_{6}g_{8}$	2.9×10^{-3}	4.7	2.2×10^{-10}	-3.7×10^{-4}	-2.2×10^{-13}	2.2×10^{-8}
97.8	9.1	$^{-3}_{3}f$	$^{6}_{6}g_{8}$	2.9×10^{-3}	4.7	2.2×10^{-10}	-3.7×10^{-4}	-1.5×10^{-13}	1.8×10^{-8}
97.8	9.2	$^{-3}_{3}f$	$^{6}_{6}g_{8}$	2.9×10^{-3}	4.7	2.2×10^{-10}	-3.7×10^{-4}	-1.2×10^{-13}	1.6×10^{-8}

(continued)

Table G.1 (continued)

Ω/Ω_K (%)	$\log T$ (K)	Daughters		$\Delta\tilde{\omega}$	$\mathcal{H}/E_{\text{unit}}$	$\tilde{\gamma}_\alpha$	$\tilde{\gamma}_\beta$	$\tilde{\gamma}_\gamma$	$|Q_{\text{PIT}}|$
97.8	9.3	$^{-3}_{3}f$	$^{6}_{6}g8$	2.9×10^{-3}	4.7	2.2×10^{-10}	-3.7×10^{-4}	-1.9×10^{-13}	2.0×10^{-8}
97.8	9.4	$^{-3}_{3}f$	$^{6}_{6}g8$	2.9×10^{-3}	4.7	2.2×10^{-10}	-3.7×10^{-4}	-5.6×10^{-13}	3.5×10^{-8}
97.8	9.5	$^{-3}_{3}f$	$^{6}_{6}g8$	2.9×10^{-3}	4.7	2.2×10^{-10}	-3.7×10^{-4}	-2.1×10^{-12}	6.9×10^{-8}
97.8	9.6	$^{-3}_{3}f$	$^{6}_{6}g8$	2.9×10^{-3}	4.7	2.2×10^{-10}	-3.7×10^{-4}	-8.4×10^{-12}	1.4×10^{-7}
97.8	9.7	$^{-3}_{3}f$	$^{6}_{6}g8$	2.9×10^{-3}	4.7	2.2×10^{-10}	-3.7×10^{-4}	-3.3×10^{-11}	2.7×10^{-7}
97.8	9.8	$^{-3}_{3}f$	$^{6}_{6}g8$	2.9×10^{-3}	4.7	2.2×10^{-10}	-3.7×10^{-4}	-1.3×10^{-10}	5.5×10^{-7}
97.8	9.9	$^{-3}_{3}f$	$^{6}_{6}g8$	2.9×10^{-3}	4.7	2.0×10^{-10}	-3.7×10^{-4}	-5.3×10^{-10}	1.1×10^{-6}
97.8	10.0	$^{-3}_{3}f$	$^{6}_{6}g8$	2.9×10^{-3}	4.7	1.4×10^{-10}	-3.7×10^{-4}	-2.1×10^{-9}	2.2×10^{-6}
98.0	7.9	$^{-3}_{3}f$	$^{6}_{6}g8$	-5.4×10^{-3}	4.8	8.8×10^{-11}	-3.7×10^{-4}	-3.4×10^{-11}	5.0×10^{-7}
98.0	8.0	$^{-3}_{3}f$	$^{6}_{6}g8$	-5.4×10^{-3}	4.8	1.6×10^{-10}	-3.7×10^{-4}	-2.2×10^{-11}	3.9×10^{-7}
98.0	8.1	$^{-3}_{3}f$	$^{6}_{6}g8$	-5.4×10^{-3}	4.8	2.1×10^{-10}	-3.7×10^{-4}	-1.4×10^{-11}	3.1×10^{-7}
98.0	8.2	$^{-3}_{3}f$	$^{6}_{6}g8$	-5.4×10^{-3}	4.8	2.4×10^{-10}	-3.7×10^{-4}	-8.6×10^{-12}	2.5×10^{-7}
98.0	8.3	$^{-3}_{3}f$	$^{6}_{6}g8$	-5.4×10^{-3}	4.8	2.6×10^{-10}	-3.7×10^{-4}	-5.4×10^{-12}	2.0×10^{-7}
98.0	8.4	$^{-3}_{3}f$	$^{6}_{6}g8$	-5.4×10^{-3}	4.8	2.7×10^{-10}	-3.7×10^{-4}	-3.4×10^{-12}	1.6×10^{-7}
98.0	8.5	$^{-3}_{3}f$	$^{6}_{6}g8$	-5.4×10^{-3}	4.8	2.8×10^{-10}	-3.7×10^{-4}	-2.2×10^{-12}	1.2×10^{-7}
98.0	8.6	$^{-3}_{3}f$	$^{6}_{6}g8$	-5.4×10^{-3}	4.8	2.9×10^{-10}	-3.7×10^{-4}	-1.4×10^{-12}	9.9×10^{-8}
98.0	8.7	$^{-3}_{3}f$	$^{6}_{6}g8$	-5.4×10^{-3}	4.8	2.9×10^{-10}	-3.7×10^{-4}	-8.6×10^{-13}	7.9×10^{-8}
98.0	8.8	$^{-3}_{3}f$	$^{6}_{6}g8$	-5.4×10^{-3}	4.8	2.9×10^{-10}	-3.7×10^{-4}	-5.4×10^{-13}	6.2×10^{-8}
98.0	8.9	$^{-3}_{3}f$	$^{6}_{6}g8$	-5.4×10^{-3}	4.8	2.9×10^{-10}	-3.7×10^{-4}	-3.4×10^{-13}	5.0×10^{-8}
98.0	9.0	$^{-3}_{3}f$	$^{6}_{6}g8$	-5.4×10^{-3}	4.8	2.9×10^{-10}	-3.7×10^{-4}	-2.2×10^{-13}	4.0×10^{-8}
98.0	9.1	$^{-3}_{3}f$	$^{6}_{6}g8$	-5.4×10^{-3}	4.8	2.9×10^{-10}	-3.7×10^{-4}	-1.4×10^{-13}	3.2×10^{-8}
98.0	9.2	$^{-2}_{2}f$	$^{5}_{5}g10$	-3.6×10^{-2}	-8.8	2.9×10^{-10}	-2.6×10^{-3}	-3.1×10^{-14}	2.9×10^{-8}

(continued)

Table G.1 (continued)

Ω/Ω_K (%)	$\log T$ (K)	Daughters		$\Delta\bar{\omega}$	$\mathcal{H}/E_{\text{unit}}$	$\tilde{\gamma}_\alpha$	$\tilde{\gamma}_\beta$	$\tilde{\gamma}_\gamma$	$|\mathcal{Q}_{\text{PIT}}|$
98.0	9.3	$^{-2}_{2}f$	$^{5}_{5}g_{10}$	-3.6×10^{-2}	-8.8	2.9×10^{-10}	-2.6×10^{-3}	-4.0×10^{-14}	3.2×10^{-8}
98.0	9.4	$^{-2}_{2}f$	$^{5}_{5}g_{10}$	-3.6×10^{-2}	-8.8	2.9×10^{-10}	-2.6×10^{-3}	-1.1×10^{-13}	5.3×10^{-8}
98.0	9.5	$^{-2}_{2}f$	$^{5}_{5}g_{10}$	-3.6×10^{-2}	-8.8	2.9×10^{-10}	-2.6×10^{-3}	-3.9×10^{-13}	1.0×10^{-7}
98.0	9.6	$^{-2}_{2}f$	$^{5}_{5}g_{10}$	-3.6×10^{-2}	-8.8	2.9×10^{-10}	-2.6×10^{-3}	-1.5×10^{-12}	2.0×10^{-7}
98.0	9.7	$^{-2}_{2}f$	$^{5}_{5}g_{10}$	-3.6×10^{-2}	-8.8	2.9×10^{-10}	-2.6×10^{-3}	-6.0×10^{-12}	4.0×10^{-7}
98.0	9.8	$^{-2}_{2}f$	$^{5}_{5}g_{10}$	-3.6×10^{-2}	-8.8	2.9×10^{-10}	-2.6×10^{-3}	-2.4×10^{-11}	7.9×10^{-7}
98.0	9.9	$^{-2}_{2}f$	$^{5}_{5}g_{10}$	-3.6×10^{-2}	-8.8	2.7×10^{-10}	-2.6×10^{-3}	-9.6×10^{-11}	1.6×10^{-6}
98.0	10.0	$^{-2}_{2}f$	$^{5}_{5}g_{10}$	-3.6×10^{-2}	-8.8	2.2×10^{-10}	-2.6×10^{-3}	-3.8×10^{-10}	3.2×10^{-6}
98.2	7.8	$^{-2}_{2}f$	$^{5}_{5}g_{10}$	-4.7×10^{-2}	-9.0	6.1×10^{-11}	-2.6×10^{-3}	-1.6×10^{-11}	8.2×10^{-7}
98.2	7.9	$^{-2}_{2}f$	$^{5}_{5}g_{10}$	-4.7×10^{-2}	-9.0	1.8×10^{-10}	-2.6×10^{-3}	-9.9×10^{-12}	6.5×10^{-7}
98.2	8.0	$^{-2}_{2}f$	$^{5}_{5}g_{10}$	-4.7×10^{-2}	-9.0	2.6×10^{-10}	-2.6×10^{-3}	-6.3×10^{-12}	5.1×10^{-7}
98.2	8.1	$^{-2}_{2}f$	$^{5}_{5}g_{10}$	-4.7×10^{-2}	-9.0	3.1×10^{-10}	-2.6×10^{-3}	-4.0×10^{-12}	4.1×10^{-7}
98.2	8.2	$^{-2}_{2}f$	$^{5}_{5}g_{10}$	-4.7×10^{-2}	-9.0	3.4×10^{-10}	-2.6×10^{-3}	-2.5×10^{-12}	3.2×10^{-7}
98.2	8.3	$^{-2}_{2}f$	$^{5}_{5}g_{10}$	-4.7×10^{-2}	-9.0	3.6×10^{-10}	-2.6×10^{-3}	-1.6×10^{-12}	2.6×10^{-7}
98.2	8.4	$^{-2}_{2}f$	$^{5}_{5}g_{10}$	-4.7×10^{-2}	-9.0	3.7×10^{-10}	-2.6×10^{-3}	-9.9×10^{-13}	2.0×10^{-7}
98.2	8.5	$^{-2}_{2}f$	$^{5}_{5}g_{10}$	-4.7×10^{-2}	-9.0	3.8×10^{-10}	-2.6×10^{-3}	-6.3×10^{-13}	1.6×10^{-7}
98.2	8.6	$^{-2}_{2}f$	$^{5}_{5}g_{10}$	-4.7×10^{-2}	-9.0	3.9×10^{-10}	-2.6×10^{-3}	-4.0×10^{-13}	1.3×10^{-7}
98.2	8.7	$^{-2}_{2}f$	$^{5}_{5}g_{10}$	-4.7×10^{-2}	-9.0	3.9×10^{-10}	-2.6×10^{-3}	-2.5×10^{-13}	1.0×10^{-7}
98.2	8.8	$^{-2}_{2}f$	$^{5}_{5}g_{10}$	-4.7×10^{-2}	-9.0	3.9×10^{-10}	-2.6×10^{-3}	-1.6×10^{-13}	8.2×10^{-8}
98.2	8.9	$^{-2}_{2}f$	$^{5}_{5}g_{10}$	-4.7×10^{-2}	-9.0	3.9×10^{-10}	-2.6×10^{-3}	-9.9×10^{-14}	6.5×10^{-8}
98.2	9.0	$^{-2}_{2}f$	$^{5}_{5}g_{10}$	-4.7×10^{-2}	-9.0	3.9×10^{-10}	-2.6×10^{-3}	-6.3×10^{-14}	5.2×10^{-8}
98.2	9.1	$^{-2}_{2}f$	$^{5}_{5}g_{10}$	-4.7×10^{-2}	-9.0	3.9×10^{-10}	-2.6×10^{-3}	-4.1×10^{-14}	4.2×10^{-8}

(continued)

Table G.1 (continued)

Ω/Ω_K (%)	$\log T$ (K)	Daughters		$\Delta\tilde{\omega}$	$\mathcal{H}/E_{\text{unit}}$	$\tilde{\gamma}_\alpha$	$\tilde{\gamma}_\beta$	$\tilde{\gamma}_\gamma$	$|Q_{\text{PIT}}|$
98.2	9.2	$_{2}^{-2}f$	$_{5}^{5}g_{10}$	-4.7×10^{-2}	-9.0	3.9×10^{-10}	-2.6×10^{-3}	-3.1×10^{-14}	3.6×10^{-8}
98.2	9.3	$_{2}^{-2}f$	$_{5}^{5}g_{10}$	-4.7×10^{-2}	-9.0	3.9×10^{-10}	-2.6×10^{-3}	-4.0×10^{-14}	4.1×10^{-8}
98.2	9.4	$_{2}^{-2}f$	$_{5}^{5}g_{10}$	-4.7×10^{-2}	-9.0	3.9×10^{-10}	-2.6×10^{-3}	-1.1×10^{-13}	6.7×10^{-8}
98.2	9.5	$_{2}^{-2}f$	$_{5}^{5}g_{10}$	-4.7×10^{-2}	-9.0	3.9×10^{-10}	-2.6×10^{-3}	-3.8×10^{-13}	1.3×10^{-7}
98.2	9.6	$_{2}^{-2}f$	$_{5}^{5}g_{10}$	-4.7×10^{-2}	-9.0	3.9×10^{-10}	-2.6×10^{-3}	-1.5×10^{-12}	2.5×10^{-7}
98.2	9.7	$_{2}^{-2}f$	$_{5}^{5}g_{10}$	-4.7×10^{-2}	-9.0	3.9×10^{-10}	-2.6×10^{-3}	-6.0×10^{-12}	5.0×10^{-7}
98.2	9.8	$_{2}^{-2}f$	$_{5}^{5}g_{10}$	-4.7×10^{-2}	-9.0	3.9×10^{-10}	-2.6×10^{-3}	-2.4×10^{-11}	1.0×10^{-6}
98.2	9.9	$_{2}^{-2}f$	$_{5}^{5}g_{10}$	-4.7×10^{-2}	-9.0	3.7×10^{-10}	2.6×10^{-3}	-9.5×10^{-11}	2.0×10^{-6}
98.2	10.0	$_{2}^{-2}f$	$_{5}^{5}g_{10}$	-4.7×10^{-2}	-9.0	3.1×10^{-10}	-2.6×10^{-3}	-3.8×10^{-10}	4.0×10^{-6}
98.2	10.1	$_{2}^{-2}f$	$_{5}^{5}g_{10}$	-4.7×10^{-2}	-9.0	7.9×10^{-11}	-2.6×10^{-3}	-1.5×10^{-9}	8.0×10^{-6}
98.4	7.8	$_{2}^{-2}f$	$_{5}^{5}g_{9}$	4.7×10^{-2}	-9.5	1.9×10^{-10}	-2.6×10^{-3}	-2.0×10^{-11}	9.1×10^{-7}
98.4	7.9	$_{2}^{-2}f$	$_{5}^{5}g_{9}$	4.7×10^{-2}	-9.5	3.1×10^{-10}	-2.6×10^{-3}	-1.3×10^{-11}	7.2×10^{-7}
98.4	8.0	$_{2}^{-2}f$	$_{5}^{5}g_{9}$	4.7×10^{-2}	-9.5	3.9×10^{-10}	-2.6×10^{-3}	-8.0×10^{-12}	5.7×10^{-7}
98.4	8.1	$_{2}^{-2}f$	$_{5}^{5}g_{9}$	4.7×10^{-2}	-9.5	4.4×10^{-10}	-2.6×10^{-3}	-5.0×10^{-12}	4.5×10^{-7}
98.4	8.2	$_{2}^{-2}f$	$_{5}^{5}g_{9}$	4.7×10^{-2}	-9.5	4.7×10^{-10}	-2.6×10^{-3}	-3.2×10^{-12}	3.6×10^{-7}
98.4	8.3	$_{2}^{-2}f$	$_{5}^{5}g_{9}$	4.7×10^{-2}	-9.5	4.9×10^{-10}	-2.6×10^{-3}	-2.0×10^{-12}	2.9×10^{-7}
98.4	8.4	$_{2}^{-2}f$	$_{5}^{5}g_{9}$	4.7×10^{-2}	-9.5	5.0×10^{-10}	-2.6×10^{-3}	-1.3×10^{-12}	2.3×10^{-7}
98.4	8.5	$_{2}^{-2}f$	$_{5}^{5}g_{9}$	4.7×10^{-2}	-9.5	5.1×10^{-10}	-2.6×10^{-3}	-8.0×10^{-13}	1.8×10^{-7}
98.4	8.6	$_{2}^{-2}f$	$_{5}^{5}g_{9}$	4.7×10^{-2}	-9.5	5.2×10^{-10}	-2.6×10^{-3}	-5.0×10^{-13}	1.4×10^{-7}
98.4	8.8	$_{2}^{-2}f$	$_{5}^{5}g_{9}$	4.7×10^{-2}	-9.5	5.2×10^{-10}	-2.6×10^{-3}	-2.0×10^{-13}	9.1×10^{-8}
98.4	8.9	$_{2}^{-2}f$	$_{5}^{5}g_{9}$	4.7×10^{-2}	-9.5	5.2×10^{-10}	-2.6×10^{-3}	-1.3×10^{-13}	7.2×10^{-8}
98.4	9.0	$_{2}^{-2}f$	$_{5}^{5}g_{9}$	4.7×10^{-2}	-9.5	5.2×10^{-10}	-2.6×10^{-3}	-8.1×10^{-14}	5.7×10^{-8}
98.4	9.1	$_{2}^{-2}f$	$_{5}^{5}g_{9}$	4.7×10^{-2}	-9.5	5.2×10^{-10}	-2.6×10^{-3}	-5.3×10^{-14}	4.7×10^{-8}

(continued)

Table G.1 (continued)

Ω/Ω_K (%)	$\log T$ (K)	Daughters		$\Delta\tilde{\omega}$	$\mathcal{H}/E_{\text{unit}}$	$\tilde{\gamma}_\alpha$	$\tilde{\gamma}_\beta$	$\tilde{\gamma}_\gamma$	$|\varrho_{\text{PIT}}|$
98.4	9.2	$_2^{-2}f$	$_5^5 g_9$	4.7×10^{-2}	-9.5	5.3×10^{-10}	-2.6×10^{-3}	-4.2×10^{-14}	4.1×10^{-8}
98.4	9.3	$_2^{-2}f$	$_5^5 g_{10}$	-5.8×10^{-2}	-9.2	5.3×10^{-10}	-2.6×10^{-3}	-3.9×10^{-14}	4.9×10^{-8}
98.4	9.4	$_2^{-2}f$	$_5^5 g_{10}$	-5.8×10^{-2}	-9.2	5.3×10^{-10}	-2.6×10^{-3}	-1.0×10^{-13}	8.1×10^{-8}
98.4	9.5	$_2^{-2}f$	$_5^5 g_{10}$	-5.8×10^{-2}	-9.2	5.3×10^{-10}	-2.6×10^{-3}	-3.8×10^{-13}	1.5×10^{-7}
98.4	9.6	$_2^{-2}f$	$_5^5 g_{10}$	-5.8×10^{-2}	-9.2	5.3×10^{-10}	-2.6×10^{-3}	-1.5×10^{-12}	3.1×10^{-7}
98.4	9.7	$_2^{-2}f$	$_5^5 g_{10}$	-5.8×10^{-2}	-9.2	5.2×10^{-10}	-2.6×10^{-3}	-6.0×10^{-12}	6.1×10^{-7}
98.4	9.8	$_2^{-2}f$	$_5^5 g_{10}$	-5.8×10^{-2}	-9.2	5.2×10^{-10}	-2.6×10^{-3}	-2.4×10^{-11}	1.2×10^{-6}
98.4	9.9	$_2^{-2}f$	$_5^5 g_{10}$	-5.8×10^{-2}	-9.2	5.1×10^{-10}	-2.6×10^{-3}	-9.4×10^{-11}	2.4×10^{-6}
98.4	10.0	$_2^{-2}f$	$_5^5 g_{10}$	-5.8×10^{-2}	-9.2	4.5×10^{-10}	-2.6×10^{-3}	-3.8×10^{-10}	4.8×10^{-6}
98.4	10.1	$_2^{-2}f$	$_5^5 g_{10}$	-5.8×10^{-2}	-9.2	2.1×10^{-10}	-2.6×10^{-3}	-1.5×10^{-9}	9.7×10^{-6}
98.6	7.7	$_2^{-2}f$	$_5^5 g_9$	3.5×10^{-2}	-9.7	1.6×10^{-10}	-2.6×10^{-3}	-3.2×10^{-11}	8.4×10^{-7}
98.6	7.8	$_2^{-2}f$	$_5^5 g_9$	3.5×10^{-2}	-9.7	3.6×10^{-10}	-2.6×10^{-3}	-2.0×10^{-11}	6.7×10^{-7}
98.6	7.9	$_2^{-2}f$	$_5^5 g_9$	3.5×10^{-2}	-9.7	4.9×10^{-10}	-2.6×10^{-3}	-1.3×10^{-11}	5.3×10^{-7}
98.6	8.0	$_2^{-2}f$	$_5^5 g_9$	3.5×10^{-2}	-9.7	5.7×10^{-10}	-2.6×10^{-3}	-7.9×10^{-12}	4.2×10^{-7}
98.6	8.1	$_2^{-2}f$	$_5^5 g_9$	3.5×10^{-2}	-9.7	6.2×10^{-10}	-2.6×10^{-3}	-5.0×10^{-12}	3.4×10^{-7}
98.6	8.2	$_2^{-2}f$	$_5^5 g_9$	3.5×10^{-2}	-9.7	6.5×10^{-10}	-2.6×10^{-3}	-3.2×10^{-12}	2.7×10^{-7}
98.6	8.3	$_2^{-2}f$	$_5^5 g_9$	3.5×10^{-2}	-9.7	6.7×10^{-10}	-2.6×10^{-3}	-2.0×10^{-12}	2.1×10^{-7}
98.6	8.4	$_2^{-2}f$	$_5^5 g_9$	3.5×10^{-2}	-9.7	6.8×10^{-10}	-2.6×10^{-3}	-1.3×10^{-12}	1.7×10^{-7}
98.6	8.5	$_2^{-2}f$	$_5^5 g_9$	3.5×10^{-2}	-9.7	6.9×10^{-10}	-2.6×10^{-3}	-7.9×10^{-13}	1.3×10^{-7}
98.6	8.6	$_2^{-2}f$	$_5^5 g_9$	3.5×10^{-2}	-9.7	6.9×10^{-10}	-2.6×10^{-3}	-5.0×10^{-13}	1.1×10^{-7}
98.6	8.7	$_2^{-2}f$	$_5^5 g_9$	3.5×10^{-2}	-9.7	7.0×10^{-10}	-2.6×10^{-3}	-3.2×10^{-13}	8.4×10^{-8}
98.6	8.8	$_2^{-2}f$	$_5^5 g_9$	3.5×10^{-2}	-9.7	7.0×10^{-10}	-2.6×10^{-3}	-2.0×10^{-13}	6.7×10^{-8}
98.6	8.9	$_2^{-2}f$	$_5^5 g_9$	3.5×10^{-2}	-9.7	7.0×10^{-10}	-2.6×10^{-3}	-1.3×10^{-13}	5.3×10^{-8}

(continued)

Table G.1 (continued)

Ω/Ω_K (%)	$\log T$ (K)	Daughters		$\Delta\tilde{\omega}$	$\mathcal{H}/E_{unit}$	$\tilde{\gamma}_\alpha$	$\tilde{\gamma}_\beta$	$\tilde{\gamma}_\gamma$	$|Q_{PIT}|$
98.6	9.0	$^{-2}_{2}f$	$^{5}_{5}g_9$	3.5×10^{-2}	-9.7	7.0×10^{-10}	-2.6×10^{-3}	-8.0×10^{-14}	4.2×10^{-8}
98.6	9.1	$^{-2}_{2}f$	$^{5}_{5}g_9$	3.5×10^{-2}	-9.7	7.0×10^{-10}	-2.6×10^{-3}	-5.3×10^{-14}	3.4×10^{-8}
98.6	9.2	$^{-2}_{2}f$	$^{5}_{5}g_9$	3.5×10^{-2}	-9.7	7.0×10^{-10}	-2.6×10^{-3}	-4.2×10^{-14}	3.1×10^{-8}
98.6	9.3	$^{-2}_{2}f$	$^{5}_{5}g_9$	3.5×10^{-2}	-9.7	7.0×10^{-10}	-2.6×10^{-3}	-6.1×10^{-14}	3.7×10^{-8}
98.6	9.4	$^{-2}_{2}f$	$^{5}_{5}g_9$	3.5×10^{-2}	-9.7	7.0×10^{-10}	-2.6×10^{-3}	-1.7×10^{-13}	6.3×10^{-8}
98.6	9.5	$^{-2}_{2}f$	$^{5}_{5}g_9$	3.5×10^{-2}	-9.7	7.0×10^{-10}	-2.6×10^{-3}	-6.5×10^{-13}	1.2×10^{-7}
98.6	9.6	$^{-2}_{2}f$	$^{5}_{5}g_9$	3.5×10^{-2}	-9.7	7.0×10^{-10}	-2.6×10^{-3}	-2.6×10^{-12}	2.4×10^{-7}
98.6	9.7	$^{-2}_{2}f$	$^{5}_{5}g_9$	3.5×10^{-2}	-9.7	7.0×10^{-10}	-2.6×10^{-3}	-1.0×10^{-11}	4.8×10^{-7}
98.6	9.8	$^{-2}_{2}f$	$^{5}_{5}g_9$	3.5×10^{-2}	-9.7	7.0×10^{-10}	-2.6×10^{-3}	-4.1×10^{-11}	9.6×10^{-7}
98.6	9.9	$^{-2}_{2}f$	$^{5}_{5}g_9$	3.5×10^{-2}	-9.7	6.8×10^{-10}	-2.6×10^{-3}	-1.6×10^{-10}	1.9×10^{-6}
98.6	10.0	$^{-6}_{6}g_{10}$	$^{9}_{9}g_{10}$	-2.3×10^{-4}	-42.4	6.2×10^{-10}	-8.6×10^{-10}	-3.5×10^{-9}	3.2×10^{-6}
98.6	10.1	$^{-6}_{6}g_{10}$	$^{9}_{9}g_{10}$	-2.3×10^{-4}	-42.4	3.8×10^{-10}	-3.4×10^{-9}	-1.4×10^{-8}	3.2×10^{-6}
98.8	7.6	$^{-2}_{2}f$	$^{5}_{5}g_9$	2.4×10^{-2}	-9.9	5.4×10^{-11}	-2.6×10^{-3}	-5.0×10^{-11}	6.9×10^{-7}
98.8	7.7	$^{-2}_{2}f$	$^{5}_{5}g_9$	2.4×10^{-2}	-9.9	3.8×10^{-10}	-2.6×10^{-3}	-3.1×10^{-11}	5.5×10^{-7}
98.8	7.8	$^{-2}_{2}f$	$^{5}_{5}g_9$	2.4×10^{-2}	-9.9	5.9×10^{-10}	-2.6×10^{-3}	-2.0×10^{-11}	4.4×10^{-7}
98.8	7.9	$^{-2}_{2}f$	$^{5}_{5}g_9$	2.4×10^{-2}	-9.9	7.2×10^{-10}	-2.6×10^{-3}	-1.2×10^{-11}	3.5×10^{-7}
98.8	8.0	$^{-2}_{2}f$	$^{5}_{5}g_9$	2.4×10^{-2}	-9.9	8.0×10^{-10}	-2.6×10^{-3}	-7.9×10^{-12}	2.8×10^{-7}
98.8	8.1	$^{-2}_{2}f$	$^{5}_{5}g_9$	2.4×10^{-2}	-9.9	8.5×10^{-10}	-2.6×10^{-3}	-5.0×10^{-12}	2.2×10^{-7}
98.8	8.2	$^{-2}_{2}f$	$^{5}_{5}g_9$	2.4×10^{-2}	-9.9	8.8×10^{-10}	-2.6×10^{-3}	-3.1×10^{-12}	1.7×10^{-7}
98.8	8.3	$^{-2}_{2}f$	$^{5}_{5}g_9$	2.4×10^{-2}	-9.9	9.1×10^{-10}	-2.6×10^{-3}	-2.0×10^{-12}	1.4×10^{-7}
98.8	8.4	$^{-2}_{2}f$	$^{5}_{5}g_9$	2.4×10^{-2}	-9.9	9.2×10^{-10}	-2.6×10^{-3}	-1.2×10^{-12}	1.1×10^{-7}
98.8	8.5	$^{-2}_{2}f$	$^{5}_{5}g_9$	2.4×10^{-2}	-9.9	9.3×10^{-10}	-2.6×10^{-3}	-7.9×10^{-13}	8.7×10^{-8}
98.8	8.6	$^{-2}_{2}f$	$^{5}_{5}g_9$	2.4×10^{-2}	-9.9	9.3×10^{-10}	-2.6×10^{-3}	-5.0×10^{-13}	6.9×10^{-8}
98.8	8.7	$^{-2}_{2}f$	$^{5}_{5}g_9$	2.4×10^{-2}	-9.9	9.4×10^{-10}	-2.6×10^{-3}	-3.1×10^{-13}	5.5×10^{-8}
98.8	8.8	$^{-2}_{2}f$	$^{5}_{5}g_9$	2.4×10^{-2}	-9.9	9.4×10^{-10}	-2.6×10^{-3}	-2.0×10^{-13}	4.4×10^{-8}

(continued)

Table G.1 (continued)

$\Omega/\Omega_{\mathrm{K}}$ (%)	$\log T$ (K)	Daughters		$\Delta\tilde{\omega}$	$\mathcal{H}/E_{\mathrm{unit}}$	$\tilde{\gamma}_\alpha$	$\tilde{\gamma}_\beta$	$\tilde{\gamma}_\gamma$	$|Q_{\mathrm{PIT}}|$
98.8	8.9	${}^{-2}_{2}f$	${}^{5}_{5}g_9$	2.4×10^{-2}	-9.9	9.4×10^{-10}	-2.6×10^{-3}	-1.2×10^{-13}	3.5×10^{-8}
98.8	9.0	${}^{-2}_{2}f$	${}^{5}_{5}g_9$	2.4×10^{-2}	-9.9	9.4×10^{-10}	-2.6×10^{-3}	-7.9×10^{-14}	2.8×10^{-8}
98.8	9.1	${}^{-2}_{2}f$	${}^{5}_{5}g_9$	2.4×10^{-2}	-9.9	9.4×10^{-10}	-2.6×10^{-3}	-5.2×10^{-14}	2.2×10^{-8}
98.8	9.2	${}^{-2}_{2}f$	${}^{5}_{5}g_9$	2.4×10^{-2}	-9.9	9.4×10^{-10}	-2.6×10^{-3}	-4.1×10^{-14}	2.0×10^{-8}
98.8	9.3	${}^{-2}_{2}f$	${}^{5}_{5}g_9$	2.4×10^{-2}	-9.9	9.4×10^{-10}	-2.6×10^{-3}	-6.0×10^{-14}	2.4×10^{-8}
98.8	9.4	${}^{-2}_{2}f$	${}^{5}_{5}g_9$	2.4×10^{-2}	-9.9	9.4×10^{-10}	-2.6×10^{-3}	-1.7×10^{-13}	4.1×10^{-8}
98.8	9.5	${}^{-2}_{2}f$	${}^{5}_{5}g_9$	2.4×10^{-2}	-9.9	9.4×10^{-10}	-2.6×10^{-3}	-6.5×10^{-13}	7.9×10^{-8}
98.8	9.6	${}^{-2}_{2}f$	${}^{5}_{5}g_9$	2.4×10^{-2}	-9.9	9.4×10^{-10}	-2.6×10^{-3}	-2.5×10^{-12}	1.6×10^{-7}
98.8	9.7	${}^{-2}_{2}f$	${}^{5}_{5}g_9$	2.4×10^{-2}	-9.9	9.4×10^{-10}	-2.6×10^{-3}	-1.0×10^{-11}	3.1×10^{-7}
98.8	9.8	${}^{-2}_{2}f$	${}^{5}_{5}g_9$	2.4×10^{-2}	-9.9	9.4×10^{-10}	-2.6×10^{-3}	-4.0×10^{-11}	6.2×10^{-7}
98.8	9.9	${}^{-2}_{2}f$	${}^{5}_{5}g_9$	2.4×10^{-2}	-9.9	9.2×10^{-10}	-2.6×10^{-3}	-1.6×10^{-10}	1.2×10^{-6}
98.8	10.0	${}^{-2}_{2}f$	${}^{5}_{5}g_9$	2.4×10^{-2}	-9.9	8.6×10^{-10}	-2.6×10^{-3}	-6.4×10^{-10}	2.5×10^{-6}
98.8	10.1	${}^{-2}_{2}f$	${}^{5}_{5}g_9$	2.4×10^{-2}	-9.9	6.1×10^{-10}	-2.6×10^{-3}	-2.5×10^{-9}	5.0×10^{-6}
99.0	7.6	${}^{-2}_{2}f$	${}^{5}_{5}g_9$	1.1×10^{-2}	-10.2	3.5×10^{-10}	-2.7×10^{-3}	-4.9×10^{-11}	3.3×10^{-7}
99.0	7.7	${}^{-2}_{2}f$	${}^{5}_{5}g_9$	1.1×10^{-2}	-10.2	6.9×10^{-10}	-2.7×10^{-3}	-3.1×10^{-11}	2.6×10^{-7}
99.0	7.8	${}^{-2}_{2}f$	${}^{5}_{5}g_9$	1.1×10^{-2}	-10.2	9.0×10^{-10}	-2.7×10^{-3}	-2.0×10^{-11}	2.1×10^{-7}
99.0	7.9	${}^{-2}_{2}f$	${}^{5}_{5}g_9$	1.1×10^{-2}	-10.2	1.0×10^{-9}	-2.7×10^{-3}	-1.2×10^{-11}	1.6×10^{-7}
99.0	8.0	${}^{-2}_{2}f$	${}^{5}_{5}g_9$	1.1×10^{-2}	-10.2	1.1×10^{-9}	-2.7×10^{-3}	-7.8×10^{-12}	1.3×10^{-7}
99.0	8.1	${}^{-2}_{2}f$	${}^{5}_{5}g_9$	1.1×10^{-2}	-10.2	1.2×10^{-9}	-2.7×10^{-3}	-4.9×10^{-12}	1.0×10^{-7}
99.0	8.2	${}^{-2}_{2}f$	${}^{5}_{5}g_9$	1.1×10^{-2}	-10.2	1.2×10^{-9}	-2.7×10^{-3}	-3.1×10^{-12}	8.2×10^{-8}
99.0	8.3	${}^{-2}_{2}f$	${}^{5}_{5}g_9$	1.1×10^{-2}	-10.2	1.2×10^{-9}	-2.7×10^{-3}	-2.0×10^{-12}	6.5×10^{-8}
99.0	8.4	${}^{-2}_{2}f$	${}^{5}_{5}g_9$	1.1×10^{-2}	-10.2	1.2×10^{-9}	-2.7×10^{-3}	-1.2×10^{-12}	5.2×10^{-8}
99.0	8.5	${}^{-2}_{2}f$	${}^{5}_{5}g_9$	1.1×10^{-2}	-10.2	1.2×10^{-9}	-2.7×10^{-3}	-7.8×10^{-13}	4.1×10^{-8}

(continued)

Table G.1 (continued)

Ω/Ω_K (%)	$\log T$ (K)	Daughters		$\Delta\tilde{\omega}$	$\mathcal{H}/E_{\text{unit}}$	$\tilde{\gamma}_\alpha$	$\tilde{\gamma}_\beta$	$\tilde{\gamma}_\gamma$	$|Q_{\text{PIT}}|$
99.0	8.6	$^{-2}_{2}f$	$^{5}_{5}g9$	1.1×10^{-2}	-10.2	1.3×10^{-9}	-2.7×10^{-3}	-4.9×10^{-13}	3.3×10^{-8}
99.0	8.7	$^{-2}_{2}f$	$^{5}_{5}g9$	1.1×10^{-2}	-10.2	1.3×10^{-9}	-2.7×10^{-3}	-3.1×10^{-13}	2.6×10^{-8}
99.0	8.8	$^{-2}_{2}f$	$^{5}_{5}g9$	1.1×10^{-2}	-10.2	1.3×10^{-9}	-2.7×10^{-3}	-2.0×10^{-13}	2.1×10^{-8}
99.0	8.9	$^{-2}_{2}f$	$^{5}_{5}g9$	1.1×10^{-2}	-10.2	1.3×10^{-9}	-2.7×10^{-3}	-1.2×10^{-13}	1.6×10^{-8}
99.0	9.0	$^{-2}_{2}f$	$^{5}_{5}g9$	1.1×10^{-2}	-10.2	1.3×10^{-9}	-2.7×10^{-3}	-7.9×10^{-14}	1.3×10^{-8}
99.0	9.1	$^{-2}_{2}f$	$^{5}_{5}g9$	1.1×10^{-2}	-10.2	1.3×10^{-9}	-2.7×10^{-3}	-5.2×10^{-14}	1.1×10^{-8}
99.0	9.2	$^{-2}_{2}f$	$^{5}_{5}g9$	1.1×10^{-2}	-10.2	1.3×10^{-9}	-2.7×10^{-3}	-4.1×10^{-14}	9.5×10^{-9}
99.0	9.3	$^{-2}_{2}f$	$^{5}_{5}g9$	1.1×10^{-2}	-10.2	1.3×10^{-9}	2.7×10^{-3}	-6.0×10^{-14}	1.1×10^{-8}
99.0	9.4	$^{-2}_{2}f$	$^{5}_{5}g9$	1.1×10^{-2}	-10.2	1.3×10^{-9}	-2.7×10^{-3}	-1.7×10^{-13}	1.9×10^{-8}
99.0	9.5	$^{-2}_{2}f$	$^{5}_{5}g9$	1.1×10^{-2}	-10.2	1.3×10^{-9}	-2.7×10^{-3}	-6.4×10^{-13}	3.7×10^{-8}
99.0	9.6	$^{-2}_{2}f$	$^{5}_{5}g9$	1.1×10^{-2}	-10.2	1.3×10^{-9}	-2.7×10^{-3}	-2.5×10^{-12}	7.4×10^{-8}
99.0	9.7	$^{-2}_{2}f$	$^{5}_{5}g9$	1.1×10^{-2}	-10.2	1.3×10^{-9}	-2.7×10^{-3}	-1.0×10^{-11}	1.5×10^{-7}
99.0	9.8	$^{-2}_{2}f$	$^{5}_{5}g9$	1.1×10^{-2}	-10.2	1.3×10^{-9}	-2.7×10^{-3}	-4.0×10^{-11}	3.0×10^{-7}
99.0	9.9	$^{-2}_{2}f$	$^{5}_{5}g9$	1.1×10^{-2}	-10.2	1.2×10^{-9}	-2.7×10^{-3}	-1.6×10^{-10}	5.9×10^{-7}
99.0	10.0	$^{-2}_{2}f$	$^{5}_{5}g9$	1.1×10^{-2}	-10.2	1.2×10^{-9}	-2.7×10^{-3}	-6.3×10^{-10}	1.2×10^{-6}
99.0	10.1	$^{-2}_{2}f$	$^{5}_{5}g9$	1.1×10^{-2}	-10.2	9.2×10^{-10}	-2.7×10^{-3}	-2.5×10^{-9}	2.3×10^{-6}
99.2	7.5	$^{-2}_{2}f$	$^{5}_{5}g9$	-1.7×10^{-3}	-10.4	2.2×10^{-10}	-2.7×10^{-3}	-7.8×10^{-11}	1.1×10^{-7}
99.2	7.6	$^{-2}_{2}f$	$^{5}_{5}g9$	-1.7×10^{-3}	-10.4	7.6×10^{-10}	-2.7×10^{-3}	-4.9×10^{-11}	8.8×10^{-8}
99.2	7.7	$^{-2}_{2}f$	$^{5}_{5}g9$	-1.7×10^{-3}	-10.4	1.1×10^{-9}	-2.7×10^{-3}	-3.1×10^{-11}	7.0×10^{-8}
99.2	7.8	$^{-2}_{2}f$	$^{5}_{5}g9$	-1.7×10^{-3}	-10.4	1.3×10^{-9}	-2.7×10^{-3}	-1.9×10^{-11}	5.6×10^{-8}
99.2	7.9	$^{-2}_{2}f$	$^{5}_{5}g9$	-1.7×10^{-3}	-10.4	1.5×10^{-9}	-2.7×10^{-3}	-1.2×10^{-11}	4.4×10^{-8}
99.2	8.0	$^{-2}_{2}f$	$^{5}_{5}g9$	-1.7×10^{-3}	-10.4	1.5×10^{-9}	-2.7×10^{-3}	-7.8×10^{-12}	3.5×10^{-8}

(continued)

Table G.1 (continued)

Ω/Ω_K (%)	$\log T$ (K)	Daughters		$\Delta\tilde{\omega}$	$\mathcal{H}/E_{unit}$	$\tilde{\gamma}_\alpha$	$\tilde{\gamma}_\beta$	$\tilde{\gamma}_\gamma$	$|Q_{PIT}|$
99.2	8.1	$^{-2}_{2}f$	$^{5}_{5}g9$	-1.7×10^{-3}	-10.4	1.6×10^{-9}	-2.7×10^{-3}	-4.9×10^{-12}	2.8×10^{-8}
99.2	8.2	$^{-2}_{2}f$	$^{5}_{5}g9$	-1.7×10^{-3}	-10.4	1.6×10^{-9}	-2.7×10^{-3}	-3.1×10^{-12}	2.2×10^{-8}
99.2	8.3	$^{-2}_{2}f$	$^{5}_{5}g9$	-1.7×10^{-3}	-10.4	1.7×10^{-9}	-2.7×10^{-3}	-1.9×10^{-12}	1.8×10^{-8}
99.2	8.4	$^{-2}_{2}f$	$^{5}_{5}g9$	-1.7×10^{-3}	-10.4	1.7×10^{-9}	-2.7×10^{-3}	-1.2×10^{-12}	1.4×10^{-8}
99.2	8.5	$^{-2}_{2}f$	$^{5}_{5}g9$	-1.7×10^{-3}	-10.4	1.7×10^{-9}	-2.7×10^{-3}	-7.8×10^{-13}	1.1×10^{-8}
99.2	8.6	$^{-2}_{2}f$	$^{5}_{5}g9$	-1.7×10^{-3}	-10.4	1.7×10^{-9}	-2.7×10^{-3}	-4.9×10^{-13}	8.8×10^{-9}
99.2	8.7	$^{-2}_{2}f$	$^{5}_{5}g9$	-1.7×10^{-3}	-10.4	1.7×10^{-9}	-2.7×10^{-3}	-3.1×10^{-13}	7.0×10^{-9}
99.2	8.8	$^{-2}_{2}f$	$^{5}_{5}g9$	-1.7×10^{-3}	-10.4	1.7×10^{-9}	-2.7×10^{-3}	-1.9×10^{-13}	5.6×10^{-9}
99.2	8.9	$^{-2}_{2}f$	$^{5}_{5}g9$	-1.7×10^{-3}	-10.4	1.7×10^{-9}	-2.7×10^{-3}	-1.2×10^{-13}	4.4×10^{-9}
99.2	9.0	$^{-2}_{2}f$	$^{5}_{5}g9$	-1.7×10^{-3}	-10.4	1.7×10^{-9}	-2.7×10^{-3}	-7.8×10^{-14}	3.5×10^{-9}
99.2	9.1	$^{-2}_{2}f$	$^{5}_{5}g9$	-1.7×10^{-3}	-10.4	1.7×10^{-9}	-2.7×10^{-3}	-5.1×10^{-14}	2.9×10^{-9}
99.2	9.2	$^{-2}_{2}f$	$^{5}_{5}g9$	-1.7×10^{-3}	-10.4	1.7×10^{-9}	-2.7×10^{-3}	-4.1×10^{-14}	2.6×10^{-9}
99.2	9.3	$^{-2}_{2}f$	$^{5}_{5}g9$	-1.7×10^{-3}	-10.4	1.7×10^{-9}	-2.7×10^{-3}	-5.9×10^{-14}	3.1×10^{-9}
99.2	9.4	$^{-2}_{2}f$	$^{5}_{5}g9$	-1.7×10^{-3}	-10.4	1.7×10^{-9}	-2.7×10^{-3}	-1.7×10^{-13}	5.2×10^{-9}
99.2	9.5	$^{-2}_{2}f$	$^{5}_{5}g9$	-1.7×10^{-3}	-10.4	1.7×10^{-9}	-2.7×10^{-3}	-6.4×10^{-13}	1.0×10^{-8}
99.2	9.6	$^{-2}_{2}f$	$^{5}_{5}g9$	-1.7×10^{-3}	-10.4	1.7×10^{-9}	-2.7×10^{-3}	-2.5×10^{-12}	2.0×10^{-8}
99.2	9.7	$^{-2}_{2}f$	$^{5}_{5}g9$	-1.7×10^{-3}	-10.4	1.7×10^{-9}	-2.7×10^{-3}	-1.0×10^{-11}	4.0×10^{-8}
99.2	9.8	$^{-2}_{2}f$	$^{5}_{5}g9$	-1.7×10^{-3}	-10.4	1.7×10^{-9}	-2.7×10^{-3}	-4.0×10^{-11}	8.0×10^{-8}
99.2	9.9	$^{-2}_{2}f$	$^{5}_{5}g9$	-1.7×10^{-3}	-10.4	1.7×10^{-9}	-2.7×10^{-3}	-1.6×10^{-10}	1.6×10^{-7}
99.2	10.0	$^{-2}_{2}f$	$^{5}_{5}g9$	-1.7×10^{-3}	-10.4	1.6×10^{-9}	-2.7×10^{-3}	-6.3×10^{-10}	3.2×10^{-7}
99.2	10.1	$^{-2}_{2}f$	$^{5}_{5}g9$	-1.7×10^{-3}	-10.4	1.3×10^{-9}	-2.7×10^{-3}	-2.5×10^{-9}	6.3×10^{-7}

(continued)

Table G.1 (continued)

$\Omega/\Omega_{\rm K}$ (%)	$\log T$ (K)	Daughters		$\Delta\tilde{\omega}$	$\mathcal{H}/E_{\rm unit}$	$\tilde{\gamma}_\alpha$	$\tilde{\gamma}_\beta$	$\tilde{\gamma}_\gamma$	$\lvert\varrho_{\rm PITT}\rvert$
99.2	10.2	$-\frac{2}{2}f$	$\frac{5}{5}g_9$	-1.7×10^{-3}	-10.4	3.0×10^{-10}	-2.7×10^{-3}	-1.0×10^{-8}	1.3×10^{-6}
99.4	7.5	$-\frac{2}{2}f$	$\frac{5}{5}g_9$	-1.5×10^{-2}	-10.7	7.6×10^{-10}	-2.7×10^{-3}	-7.7×10^{-11}	5.3×10^{-7}
99.4	7.6	$-\frac{2}{2}f$	$\frac{5}{5}g_9$	-1.5×10^{-2}	-10.7	1.3×10^{-9}	-2.7×10^{-3}	-4.9×10^{-11}	4.2×10^{-7}
99.4	7.7	$-\frac{2}{2}f$	$\frac{5}{5}g_9$	-1.5×10^{-2}	-10.7	1.7×10^{-9}	-2.7×10^{-3}	-3.1×10^{-11}	3.3×10^{-7}
99.4	7.8	$-\frac{2}{2}f$	$\frac{5}{5}g_9$	-1.5×10^{-2}	-10.7	1.9×10^{-9}	-2.7×10^{-3}	-1.9×10^{-11}	2.6×10^{-7}
99.4	7.9	$-\frac{2}{2}f$	$\frac{5}{5}g_9$	-1.5×10^{-2}	-10.7	2.0×10^{-9}	-2.7×10^{-3}	-1.2×10^{-11}	2.1×10^{-7}
99.4	8.0	$-\frac{2}{2}f$	$\frac{5}{5}g_9$	-1.5×10^{-2}	-10.7	2.1×10^{-9}	-2.7×10^{-3}	-7.7×10^{-12}	1.7×10^{-7}
99.4	8.1	$-\frac{2}{2}f$	$\frac{5}{5}g_9$	-1.5×10^{-2}	-10.7	2.2×10^{-9}	-2.7×10^{-3}	-4.9×10^{-12}	1.3×10^{-7}
99.4	8.2	$-\frac{2}{2}f$	$\frac{5}{5}g_9$	-1.5×10^{-2}	-10.7	2.2×10^{-9}	-2.7×10^{-3}	-3.1×10^{-12}	1.1×10^{-7}
99.4	8.3	$-\frac{2}{2}f$	$\frac{5}{5}g_9$	-1.5×10^{-2}	-10.7	2.2×10^{-9}	-2.7×10^{-3}	-1.9×10^{-12}	8.4×10^{-8}
99.4	8.4	$-\frac{2}{2}f$	$\frac{5}{5}g_9$	-1.5×10^{-2}	-10.7	2.2×10^{-9}	-2.7×10^{-3}	-1.2×10^{-12}	6.6×10^{-8}
99.4	8.5	$-\frac{2}{2}f$	$\frac{5}{5}g_9$	-1.5×10^{-2}	-10.7	2.2×10^{-9}	-2.7×10^{-3}	-7.7×10^{-13}	5.3×10^{-8}
99.4	8.6	$-\frac{2}{2}f$	$\frac{5}{5}g_9$	-1.5×10^{-2}	-10.7	2.3×10^{-9}	-2.7×10^{-3}	-4.9×10^{-13}	4.2×10^{-8}
99.4	8.7	$-\frac{2}{2}f$	$\frac{5}{5}g_9$	-1.5×10^{-2}	-10.7	2.3×10^{-9}	-2.7×10^{-3}	-3.1×10^{-13}	3.3×10^{-8}
99.4	8.8	$-\frac{2}{2}f$	$\frac{5}{5}g_9$	-1.5×10^{-2}	-10.7	2.3×10^{-9}	-2.7×10^{-3}	-1.9×10^{-13}	2.6×10^{-8}
99.4	8.9	$-\frac{2}{2}f$	$\frac{5}{5}g_9$	-1.5×10^{-2}	-10.7	2.3×10^{-9}	-2.7×10^{-3}	-1.2×10^{-13}	2.1×10^{-8}
99.4	9.0	$-\frac{2}{2}f$	$\frac{5}{5}g_9$	-1.5×10^{-2}	-10.7	2.3×10^{-9}	-2.7×10^{-3}	-7.8×10^{-14}	1.7×10^{-8}
99.4	9.1	$-\frac{2}{2}f$	$\frac{5}{5}g_9$	-1.5×10^{-2}	-10.7	2.3×10^{-9}	-2.7×10^{-3}	-5.1×10^{-14}	1.4×10^{-8}
99.4	9.2	$-\frac{2}{2}f$	$\frac{5}{5}g_9$	-1.5×10^{-2}	-10.7	2.3×10^{-9}	-2.7×10^{-3}	-4.1×10^{-14}	1.2×10^{-8}
99.4	9.3	$-\frac{2}{2}f$	$\frac{5}{5}g_9$	-1.5×10^{-2}	-10.7	2.3×10^{-9}	-2.7×10^{-3}	-5.9×10^{-14}	1.5×10^{-8}
99.4	9.4	$-\frac{2}{2}f$	$\frac{5}{5}g_9$	-1.5×10^{-2}	-10.7	2.3×10^{-9}	-2.7×10^{-3}	-1.7×10^{-13}	2.5×10^{-8}
99.4	9.5	$-\frac{2}{2}f$	$\frac{5}{5}g_9$	-1.5×10^{-2}	-10.7	2.3×10^{-9}	-2.7×10^{-3}	-6.3×10^{-13}	4.8×10^{-8}

(continued)

Table G.1 (continued)

Ω/Ω_K (%)	$\log T$ (K)	Daughters		$\Delta\tilde{\omega}$	$\mathcal{H}/E_{\text{unit}}$	$\tilde{\gamma}_\alpha$	$\tilde{\gamma}_\beta$	$\tilde{\gamma}_\gamma$	$|\mathcal{Q}_{\text{PIT}}|$
99.4	9.6	$^{-2}_2 f$	$^5_5 g_9$	-1.5×10^{-2}	-10.7	2.3×10^{-9}	-2.7×10^{-3}	-2.5×10^{-12}	9.5×10^{-8}
99.4	9.7	$^{-2}_2 f$	$^5_5 g_9$	-1.5×10^{-2}	-10.7	2.3×10^{-9}	-2.7×10^{-3}	-9.9×10^{-12}	1.9×10^{-7}
99.4	9.8	$^{-2}_2 f$	$^5_5 g_9$	-1.5×10^{-2}	-10.7	2.3×10^{-9}	-2.7×10^{-3}	-3.9×10^{-11}	3.8×10^{-7}
99.4	9.9	$^{-2}_2 f$	$^5_5 g_9$	-1.5×10^{-2}	-10.7	2.2×10^{-9}	-2.7×10^{-3}	-1.6×10^{-10}	7.5×10^{-7}
99.4	10.0	$^{-2}_2 f$	$^5_5 g_9$	-1.5×10^{-2}	-10.7	2.2×10^{-9}	-2.7×10^{-3}	-6.2×10^{-10}	1.5×10^{-6}
99.4	10.1	$^{-2}_2 f$	$^5_5 g_9$	-1.5×10^{-2}	-10.7	1.9×10^{-9}	-2.7×10^{-3}	-2.5×10^{-9}	3.0×10^{-6}
99.4	10.2	$^{-2}_2 f$	$^5_5 g_9$	-1.5×10^{-2}	-10.7	8.4×10^{-10}	-2.7×10^{-3}	-9.9×10^{-9}	6.0×10^{-6}
99.6	7.4	$^{-2}_2 f$	$^5_5 g_9$	-2.9×10^{-2}	-11.0	5.8×10^{-10}	-2.7×10^{-3}	-1.2×10^{-10}	1.2×10^{-6}
99.6	7.5	$^{-2}_2 f$	$^5_5 g_9$	-2.9×10^{-2}	-11.0	1.5×10^{-9}	-2.7×10^{-3}	-7.6×10^{-11}	9.8×10^{-7}
99.6	7.6	$^{-2}_2 f$	$^5_5 g_9$	-2.9×10^{-2}	-11.0	2.1×10^{-9}	-2.7×10^{-3}	-4.8×10^{-11}	7.8×10^{-7}
99.6	7.7	$^{-2}_2 f$	$^5_5 g_9$	-2.9×10^{-2}	-11.0	2.4×10^{-9}	-2.7×10^{-3}	-3.0×10^{-11}	6.2×10^{-7}
99.6	7.8	$^{-2}_2 f$	$^5_5 g_9$	-2.9×10^{-2}	-11.0	2.6×10^{-9}	-2.7×10^{3}	-1.9×10^{-11}	4.9×10^{-7}
99.6	7.9	$^{-2}_2 f$	$^5_5 g_9$	-2.9×10^{-2}	-11.0	2.8×10^{-9}	-2.7×10^{-3}	-1.2×10^{-11}	3.9×10^{-7}
99.6	8.0	$^{-2}_2 f$	$^5_5 g_9$	-2.9×10^{-2}	-11.0	2.9×10^{-9}	-2.7×10^{-3}	-7.6×10^{-12}	3.1×10^{-7}
99.6	8.1	$^{-2}_2 f$	$^5_5 g_9$	-2.9×10^{-2}	-11.0	2.9×10^{-9}	-2.7×10^{-3}	-4.8×10^{-12}	2.5×10^{-7}
99.6	8.2	$^{-2}_2 f$	$^5_5 g_9$	-2.9×10^{-2}	-11.0	3.0×10^{-9}	-2.7×10^{-3}	-3.0×10^{-12}	1.9×10^{-7}
99.6	8.3	$^{-2}_2 f$	$^5_5 g_9$	-2.9×10^{-2}	-11.0	3.0×10^{-9}	-2.7×10^{-3}	-1.9×10^{-12}	1.5×10^{-7}
99.6	8.4	$^{-2}_2 f$	$^5_5 g_9$	-2.9×10^{-2}	-11.0	3.0×10^{-9}	-2.7×10^{-3}	-1.2×10^{-12}	1.2×10^{-7}
99.6	8.5	$^{-2}_2 f$	$^5_5 g_9$	-2.9×10^{-2}	-11.0	3.0×10^{-9}	-2.7×10^{-3}	-7.6×10^{-13}	9.8×10^{-8}
99.6	8.6	$^{-2}_2 f$	$^5_5 g_9$	-2.9×10^{-2}	-11.0	3.0×10^{-9}	-2.7×10^{-3}	-4.8×10^{-13}	7.8×10^{-8}
99.6	8.7	$^{-2}_2 f$	$^5_5 g_9$	-2.9×10^{-2}	-11.0	3.0×10^{-9}	-2.7×10^{-3}	-3.0×10^{-13}	6.2×10^{-8}
99.6	8.8	$^{-2}_2 f$	$^5_5 g_9$	-2.9×10^{-2}	-11.0	3.0×10^{-9}	-2.7×10^{-3}	-1.9×10^{-13}	4.9×10^{-8}

(continued)

Table G.1 (continued)

Ω/Ω_K (%)	$\log T$ (K)	Daughters		$\Delta\tilde{\omega}$	$\mathcal{H}/E_{\text{unit}}$	$\tilde{\gamma}_\alpha$	$\tilde{\gamma}_\beta$	$\tilde{\gamma}_\gamma$	$\lvert\varrho_{\text{PIT}}\rvert$
99.6	8.9	$^{-2}_{2}f$	$^{5}_{5}gg$	-2.9×10^{-2}	-11.0	3.0×10^{-9}	-2.7×10^{-3}	-1.2×10^{-13}	3.9×10^{-8}
99.6	9.0	$^{-2}_{2}f$	$^{5}_{5}gg$	-2.9×10^{-2}	-11.0	3.0×10^{-9}	-2.7×10^{-3}	-7.7×10^{-14}	3.1×10^{-8}
99.6	9.1	$^{-2}_{2}f$	$^{5}_{5}gg$	-2.9×10^{-2}	-11.0	3.0×10^{-9}	-2.7×10^{-3}	-5.1×10^{-14}	2.5×10^{-8}
99.6	9.2	$^{-2}_{2}f$	$^{5}_{5}gg$	-2.9×10^{-2}	-11.0	3.0×10^{-9}	-2.7×10^{-3}	-4.0×10^{-14}	2.2×10^{-8}
99.6	9.3	$^{-2}_{2}f$	$^{5}_{5}gg$	-2.9×10^{-2}	-11.0	3.0×10^{-9}	-2.7×10^{-3}	-5.8×10^{-14}	2.7×10^{-8}
99.6	9.4	$^{-2}_{2}f$	$^{5}_{5}gg$	-2.9×10^{-2}	-11.0	3.0×10^{-9}	-2.7×10^{-3}	-1.7×10^{-13}	4.6×10^{-8}
99.6	9.5	$^{-2}_{2}f$	$^{5}_{5}gg$	-2.9×10^{-2}	-11.0	3.0×10^{-9}	-2.7×10^{-3}	-6.3×10^{-13}	8.8×10^{-8}
99.6	9.6	$^{-2}_{2}f$	$^{5}_{5}gg$	-2.9×10^{-2}	-11.0	3.0×10^{-9}	-2.7×10^{-3}	-2.5×10^{-12}	1.8×10^{-7}
99.6	9.7	$^{-2}_{2}f$	$^{5}_{5}gg$	-2.9×10^{-2}	-11.0	3.0×10^{-9}	-2.7×10^{-3}	-9.8×10^{-12}	3.5×10^{-7}
99.6	9.8	$^{-2}_{2}f$	$^{5}_{5}gg$	-2.9×10^{-2}	-11.0	3.0×10^{-9}	-2.7×10^{-3}	-3.9×10^{-11}	7.0×10^{-7}
99.6	9.9	$^{-2}_{2}f$	$^{5}_{5}gg$	-2.9×10^{-2}	-11.0	3.0×10^{-9}	-2.7×10^{-3}	-1.6×10^{-10}	1.4×10^{-6}
99.6	10.0	$^{-2}_{2}f$	$^{5}_{5}gg$	-2.9×10^{-2}	-11.0	2.9×10^{-9}	-2.7×10^{-3}	-6.2×10^{-10}	2.8×10^{-6}
99.6	10.1	$^{-2}_{2}f$	$^{5}_{5}gg$	-2.9×10^{-2}	-11.0	2.7×10^{-9}	-2.7×10^{-3}	-2.5×10^{-9}	5.5×10^{-6}
99.6	10.2	$^{-2}_{2}f$	$^{5}_{5}gg$	-2.9×10^{-2}	-11.0	1.6×10^{-9}	-2.7×10^{-3}	-9.8×10^{-9}	1.1×10^{-5}
99.8	7.3	$^{-2}_{2}f$	$^{5}_{5}gg$	-4.4×10^{-2}	-11.4	7.9×10^{-11}	-2.7×10^{-3}	-1.9×10^{-10}	2.3×10^{-6}
99.8	7.4	$^{-2}_{2}f$	$^{5}_{5}gg$	-4.4×10^{-2}	-11.4	1.6×10^{-9}	-2.7×10^{-3}	-1.2×10^{-10}	1.8×10^{-6}
99.8	7.5	$^{-2}_{2}f$	$^{5}_{5}gg$	-4.4×10^{-2}	-11.4	2.5×10^{-9}	-2.7×10^{-3}	-7.6×10^{-11}	1.4×10^{-6}
99.8	7.6	$^{-2}_{2}f$	$^{5}_{5}gg$	-4.4×10^{-2}	-11.4	3.1×10^{-9}	-2.7×10^{-3}	-4.8×10^{-11}	1.1×10^{-6}
99.8	7.7	$^{-2}_{2}f$	$^{5}_{5}gg$	-4.4×10^{-2}	-11.4	3.4×10^{-9}	-2.7×10^{-3}	-3.0×10^{-11}	9.0×10^{-7}
99.8	7.8	$^{-2}_{2}f$	$^{5}_{5}gg$	-4.4×10^{-2}	-11.4	3.7×10^{-9}	-2.7×10^{-3}	-1.9×10^{-11}	7.1×10^{-7}
99.8	7.9	$^{-2}_{2}f$	$^{5}_{5}gg$	-4.4×10^{-2}	-11.4	3.8×10^{-9}	-2.7×10^{-3}	-1.2×10^{-11}	5.7×10^{-7}
99.8	8.0	$^{-2}_{2}f$	$^{5}_{5}gg$	-4.4×10^{-2}	-11.4	3.9×10^{-9}	-2.7×10^{-3}	-7.6×10^{-12}	4.5×10^{-7}

(continued)

Table G.1 (continued)

Ω/Ω_K (%)	$\log T$ (K)	Daughters		$\Delta\tilde{\omega}$	$\mathcal{H}/E_{unit}$	$\tilde{\gamma}_\alpha$	$\tilde{\gamma}_\beta$	$\tilde{\gamma}_\gamma$	$|\mathcal{Q}_{PIT}|$
99.8	8.1	$^{-2}_{2}f$	$^{5}_{5}g9$	-4.4×10^{-2}	-11.4	4.0×10^{-9}	-2.7×10^{-3}	-4.8×10^{-12}	3.6×10^{-7}
99.8	8.2	$^{-2}_{2}f$	$^{5}_{5}g9$	-4.4×10^{-2}	-11.4	4.0×10^{-9}	-2.7×10^{-3}	-3.0×10^{-12}	2.8×10^{-7}
99.8	8.3	$^{-2}_{2}f$	$^{5}_{5}g9$	-4.4×10^{-2}	-11.4	4.0×10^{-9}	-2.7×10^{-3}	-1.9×10^{-12}	2.3×10^{-7}
99.8	8.4	$^{-2}_{2}f$	$^{5}_{5}g9$	-4.4×10^{-2}	-11.4	4.1×10^{-9}	-2.7×10^{-3}	-1.2×10^{-12}	1.8×10^{-7}
99.8	8.5	$^{-2}_{2}f$	$^{5}_{5}g9$	-4.4×10^{-2}	-11.4	4.1×10^{-9}	-2.7×10^{-3}	-7.6×10^{-13}	1.4×10^{-7}
99.8	8.6	$^{-2}_{2}f$	$^{5}_{5}g9$	-4.4×10^{-2}	-11.4	4.1×10^{-9}	-2.7×10^{-3}	-4.8×10^{-13}	1.1×10^{-7}
99.8	8.7	$^{-2}_{2}f$	$^{5}_{5}g9$	-4.4×10^{-2}	-11.4	4.1×10^{-9}	-2.7×10^{-3}	-3.0×10^{-13}	9.0×10^{-8}
99.8	8.8	$^{-2}_{2}f$	$^{5}_{5}g9$	-4.4×10^{-2}	-11.4	4.1×10^{-9}	-2.7×10^{-3}	-1.9×10^{-13}	7.1×10^{-8}
99.8	8.9	$^{-2}_{2}f$	$^{5}_{5}g9$	-4.4×10^{-2}	-11.4	4.1×10^{-9}	-2.7×10^{-3}	-1.2×10^{-13}	5.7×10^{-8}
99.8	9.0	$^{-2}_{2}f$	$^{5}_{5}g9$	-4.4×10^{-2}	-11.4	4.1×10^{-9}	-2.7×10^{-3}	-7.6×10^{-14}	4.5×10^{-8}
99.8	9.1	$^{-2}_{2}f$	$^{5}_{5}g9$	-4.4×10^{-2}	-11.4	4.1×10^{-9}	-2.7×10^{-3}	-5.0×10^{-14}	3.7×10^{-8}
99.8	9.2	$^{-2}_{2}f$	$^{5}_{5}g9$	-4.4×10^{-2}	-11.4	4.1×10^{-9}	-2.7×10^{-3}	-4.0×10^{-14}	3.3×10^{-8}
99.8	9.3	$^{-2}_{2}f$	$^{5}_{5}g9$	-4.4×10^{-2}	-11.4	4.1×10^{-9}	-2.7×10^{-3}	-5.8×10^{-14}	3.9×10^{-8}
99.8	9.4	$^{-2}_{2}f$	$^{5}_{5}g9$	-4.4×10^{-2}	-11.4	4.1×10^{-9}	-2.7×10^{-3}	-1.7×10^{-13}	6.7×10^{-8}
99.8	9.5	$^{-2}_{2}f$	$^{5}_{5}g9$	-4.4×10^{-2}	-11.4	4.1×10^{-9}	-2.7×10^{-3}	-6.2×10^{-13}	1.3×10^{-7}
99.8	9.6	$^{-2}_{2}f$	$^{5}_{5}g9$	-4.4×10^{-2}	-11.4	4.1×10^{-9}	-2.7×10^{-3}	2.5×10^{-12}	2.6×10^{-7}
99.8	9.7	$^{-2}_{2}f$	$^{5}_{5}g9$	-4.4×10^{-2}	-11.4	4.1×10^{-9}	-2.7×10^{-3}	-9.8×10^{-12}	5.1×10^{-7}
99.8	9.8	$^{-2}_{2}f$	$^{5}_{5}g9$	-4.4×10^{-2}	-11.4	4.1×10^{-9}	-2.7×10^{-3}	-3.9×10^{-11}	1.0×10^{-6}
99.8	9.9	$^{-2}_{2}f$	$^{5}_{5}g9$	-4.4×10^{-2}	-11.4	4.1×10^{-9}	-2.7×10^{-3}	-1.5×10^{-10}	2.0×10^{-6}
99.8	10.0	$^{-2}_{2}f$	$^{5}_{5}g9$	-4.4×10^{-2}	-11.4	4.0×10^{-9}	-2.7×10^{-3}	-6.2×10^{-10}	4.0×10^{-6}
99.8	10.1	$^{-2}_{2}f$	$^{5}_{5}g9$	-4.4×10^{-2}	-11.4	3.7×10^{-9}	-2.7×10^{-3}	-2.5×10^{-9}	8.1×10^{-6}
99.8	10.2	$^{-2}_{2}f$	$^{5}_{5}g9$	-4.4×10^{-2}	-11.4	2.6×10^{-9}	-2.7×10^{-3}	-9.8×10^{-9}	1.6×10^{-5}
100.0	7.3	$^{-4}_{4}f$	$^{7}_{7}g4$	-1.1×10^{-3}	6.9	1.4×10^{-9}	-5.9×10^{-5}	-2.1×10^{-9}	1.5×10^{-6}

(continued)

Table G.1 (continued)

Ω/Ω_K (%)	$\log T$ (K)	Daughters		$\Delta\tilde{\omega}$	$\mathcal{H}/E_{\text{unit}}$	$\tilde{\gamma}_\alpha$	$\tilde{\gamma}_\beta$	$\tilde{\gamma}_\gamma$	$|Q_{\text{PIT}}|$
100.0	7.4	$^{-4}_{4}f$	$^{7}_{7}94$	-1.1×10^{-3}	6.9	2.9×10^{-9}	-5.9×10^{-5}	-1.3×10^{-9}	1.2×10^{-6}
100.0	7.5	$^{-4}_{4}f$	$^{7}_{7}94$	-1.1×10^{-3}	6.9	3.8×10^{-9}	-5.9×10^{-5}	-8.4×10^{-10}	9.2×10^{-7}
100.0	7.6	$^{-4}_{4}f$	$^{7}_{7}94$	-1.1×10^{-3}	6.9	4.5×10^{-9}	-5.9×10^{-5}	-5.3×10^{-10}	7.3×10^{-7}
100.0	7.7	$^{-4}_{4}f$	$^{7}_{7}94$	-1.1×10^{-3}	6.9	4.8×10^{-9}	-5.9×10^{-5}	-3.3×10^{-10}	5.8×10^{-7}
100.0	7.8	$^{-4}_{4}f$	$^{7}_{7}94$	-1.1×10^{-3}	6.9	5.1×10^{-9}	-5.9×10^{-5}	-2.1×10^{-10}	4.6×10^{-7}
100.0	7.9	$^{-4}_{4}f$	$^{7}_{7}94$	-1.1×10^{-3}	6.9	5.2×10^{-9}	-5.9×10^{-5}	-1.3×10^{-10}	3.7×10^{-7}
100.0	8.0	$^{-4}_{4}f$	$^{7}_{7}94$	-1.1×10^{-3}	6.9	5.3×10^{-9}	-5.9×10^{-5}	-8.4×10^{-11}	2.9×10^{-7}
100.0	8.1	$^{-4}_{4}f$	$^{7}_{7}94$	-1.1×10^{-3}	6.9	5.4×10^{-9}	-5.9×10^{-5}	-5.3×10^{-11}	2.3×10^{-7}
100.0	8.2	$^{-4}_{4}f$	$^{7}_{7}94$	-1.1×10^{-3}	6.9	5.4×10^{-9}	-5.9×10^{-5}	-3.3×10^{-11}	1.8×10^{-7}
100.0	8.3	$^{-4}_{4}f$	$^{7}_{7}94$	-1.1×10^{-3}	6.9	5.4×10^{-9}	-5.9×10^{-5}	-2.1×10^{-11}	1.5×10^{-7}
100.0	8.4	$^{-4}_{4}f$	$^{7}_{7}94$	-1.1×10^{-3}	6.9	5.5×10^{-9}	-5.9×10^{-5}	-1.3×10^{-11}	1.2×10^{-7}
100.0	8.5	$^{-4}_{4}f$	$^{7}_{7}94$	-1.1×10^{-3}	6.9	5.5×10^{-9}	-5.9×10^{-5}	-8.4×10^{-12}	9.2×10^{-8}
100.0	8.6	$^{-4}_{4}f$	$^{7}_{7}94$	-1.1×10^{-3}	6.9	5.5×10^{-9}	-5.9×10^{-5}	-5.3×10^{-12}	7.3×10^{-8}
100.0	8.7	$^{-4}_{4}f$	$^{7}_{7}94$	-1.1×10^{-3}	6.9	5.5×10^{-9}	-5.9×10^{-5}	-3.3×10^{-12}	5.8×10^{-8}
100.0	8.8	$^{-4}_{4}f$	$^{7}_{7}94$	-1.1×10^{-3}	6.9	5.5×10^{-9}	-5.9×10^{-5}	-2.1×10^{-12}	4.6×10^{-8}
100.0	8.9	$^{-4}_{4}f$	$^{7}_{7}94$	-1.1×10^{-3}	6.9	5.5×10^{-9}	-5.9×10^{-5}	-1.3×10^{-12}	3.7×10^{-8}
100.0	9.0	$^{-4}_{4}f$	$^{7}_{7}94$	-1.1×10^{-3}	6.9	5.5×10^{-9}	-5.9×10^{-5}	-8.6×10^{-13}	2.9×10^{-8}
100.0	9.1	$^{-4}_{4}f$	$^{7}_{7}94$	-1.1×10^{-3}	6.9	5.5×10^{-9}	-5.9×10^{-5}	-6.2×10^{-13}	2.5×10^{-8}
100.0	9.2	$^{-4}_{4}f$	$^{7}_{7}94$	-1.1×10^{-3}	6.9	5.5×10^{-9}	-5.9×10^{-5}	-7.0×10^{-13}	2.7×10^{-8}
100.0	9.3	$^{-2}_{2}f$	$^{5}_{5}98$	3.3×10^{-2}	-12.2	5.5×10^{-9}	-2.7×10^{-3}	-8.6×10^{-14}	3.5×10^{-8}
100.0	9.4	$^{-2}_{2}f$	$^{5}_{5}98$	3.3×10^{-2}	-12.2	5.5×10^{-9}	-2.7×10^{-3}	-2.6×10^{-13}	6.1×10^{-8}
100.0	9.5	$^{-2}_{2}f$	$^{5}_{5}98$	3.3×10^{-2}	-12.2	5.5×10^{-9}	-2.7×10^{-3}	-1.0×10^{-12}	1.2×10^{-7}

(continued)

Table G.1 (continued)

Ω/Ω_K (%)	$\log T$ (K)	Daughters		$\Delta\tilde{\omega}$	$\mathcal{H}/E_{\text{unit}}$	$\tilde{\gamma}_\alpha$	$\tilde{\gamma}_\beta$	$\tilde{\gamma}_\gamma$	$|\mathcal{Q}_{\text{PIT}}|$
100.0	9.6	$^{-2}_{2}f$	$^{5}_{5}g8$	3.3×10^{-2}	-12.2	5.5×10^{-9}	-2.7×10^{-3}	-4.0×10^{-12}	2.4×10^{-7}
100.0	9.7	$^{-2}_{2}f$	$^{5}_{5}g8$	3.3×10^{-2}	-12.2	5.5×10^{-9}	-2.7×10^{-3}	-1.6×10^{-11}	4.7×10^{-7}
100.0	9.8	$^{-2}_{2}f$	$^{5}_{5}g8$	3.3×10^{-2}	-12.2	5.5×10^{-9}	-2.7×10^{-3}	-6.3×10^{-11}	9.4×10^{-7}
100.0	9.9	$^{-2}_{2}f$	$^{5}_{5}g8$	3.3×10^{-2}	-12.2	5.5×10^{-9}	-2.7×10^{-3}	-2.5×10^{-10}	1.9×10^{-6}
100.0	10.0	$^{-2}_{2}f$	$^{5}_{5}g8$	3.3×10^{-2}	-12.2	5.4×10^{-9}	-2.7×10^{-3}	-9.9×10^{-10}	3.7×10^{-6}
100.0	10.1	$^{-2}_{2}f$	$^{5}_{5}g8$	3.3×10^{-2}	-12.2	5.1×10^{-9}	-2.7×10^{-3}	-3.9×10^{-9}	7.4×10^{-6}
100.0	10.2	$^{-2}_{2}f$	$^{5}_{5}g8$	3.3×10^{-2}	-12.2	3.9×10^{-9}	-2.7×10^{-3}	-1.6×10^{-8}	1.5×10^{-5}

Printed in the United States
By Bookmasters